Advances in

Petroleum Geochemistry

Volume 1

ADVANCES IN
Petroleum Geochemistry

Volume 1

Edited by

Jim Brooks
Britoil, Glasgow, U.K.

Dietrich Welte
KFA., Jülich, West Germany

1984

ACADEMIC PRESS
(*Harcourt Brace Jovanovich, Publishers*)

London Orlando San Diego San Francisco New York
Toronto Montreal Sydney Tokyo São Paulo

ACADEMIC PRESS INC. (LONDON) LTD.
24/28 Oval Road
London NW1

United States Edition published by
ACADEMIC PRESS INC.
(Harcourt Brace Jovanovich, Inc.)
Orlando, Florida 32887

British Library Cataloguing in Publication Data
Advances in petroleum geochemistry—Vol. 1, 1984
1. Petroleum—Periodicals
2. Geochemistry—Periodicals
55.2'82 TN870.3

ISBN 0-12-032001-0
ISSN 0739-8352

Typeset and printed in Great Britain by
Page Brothers, Norwich

Contributors

J. Brooks, Exploration Division, Britoil p.l.c., 150 St. Vincent St., Glasgow G2 5LJ, U.K.

J. Connan, Direction Exploration, Elf-Aquitaine (Production), 26 Avenue des Lilas, 64018 Pau Cedex, France.

B. Horsfield, Regional Research Section, Exploration Research Division, Conoco Inc., Ponca City, Oklahoma 74601, U.S.A. *Currently with*: Geological Research, ARCO Oil and Gas Company, P.O. Box 2819, Dallas, Texas 75221, U.S.A.

F. Kokesh, Exploration Department, Phillips Petroleum Company, Bartlesville, Oklahoma 74004, U.S.A.

A. S. Mackenzie, ICH-5: Erdol und Organische Geochemie, KFA Jülich, Postfach 1913, D-5170 Jülich 1, West Germany. *Currently with*: Geochemistry Branch, Exploration and Production Division, BP Research Centre, Sunbury-on-Thames, Middlesex, TW16 7LN, U.K.

M. Schoell, Bundesanstalt für Geowissenschaften und Rohstoffe, Stilleweg 2, 3000 Hannover 51, West Germany. *Currently with*: Chevron Oil Field Research Co., P.O. Box 446, Le Habra, California 90631, U.S.A.

D. W. Waples, Mobil Research and Development Corp., Field Research Laboratory, P.O. Box 900, Dallas, Texas 75221, U.S.A. *Currently* a geochemical consultant at: 1717 Place One Lane, Garland, Texas 75042, U.S.A.

D. H. Welte, ICH-5: Erdol und Organische Geochemie, KFA Jülich, Postfach 1913, D-5170 Jülich 1, West Germany.

M. A. Yükler, Exploration Department, Phillips Petroleum Company, Bartlesville, Oklahoma 74004, U.S.A. *Currently with*: Integriete Explorations Systeme GmbH, Kartäuserstrasse 2, D-5170 Jülich, West Germany.

Preface

This new series of volumes will contain extended reviews, by experts, in recent advances in petroleum geochemistry. Previously there have been series in petroleum geology, geophysics and microbiology, but topics and advances in petroleum geochemistry have been neglected.

Petroleum exploration is an expensive and increasingly difficult, but necessary, operation. Applications of petroleum geochemistry in exploration are drastically changing from a post-mortem science to a widely accepted predictive exploration tool. Due mainly to progress in the development of new analytical techniques and interpretive concepts, application of geochemistry has become more rapid and also capable of solving more specific petroleum exploration problems. Based upon a combination of petroleum geochemistry and geology, it is now possible to make quantitative prediction of hydrocarbon potential of basins and regions during the initial stages of exploration.

During the last decade, there has been rapid progress in petroleum geochemistry. The aim of this volume, and the series, is to cover these advances in a topical, comprehensive and critical way in the more important aspects of petroleum exploration.

Petroleum geochemistry has turned out to be more than another step in the direction to quantify geology and geosciences in general. Petroleum geochemistry as it is today may very well be the triggering event that brings the other branches of geosciences like sedimentology, stratigraphy, structural geology, geophysics and others to a fruitful synthesis as evidenced by integrated basin studies. Furthermore, it may also help to open up a new era in geosciences when computer simulation of geological processes on high-speed computers is being used as an experimental tool to test and quantify geological ideas and theories.

The compilation of this first volume would not, obviously, have been possible without the contribution of the various authors. Each of the authors is very busy and active in petroleum exploration work and we sincerely thank them for their contribution and continued assistance in writing the chapters and checking proofs. Doug Waples (Mobil), Arif Yükler and K. Kokesh (Phillips), Andrew Mackenzie (BP), Martin Schoell (B für G, Hannover), Brian Horsfield (ARCO) and Jacque Connan (Elf) have all spent valuable time in reviewing recent literature and contributing their expertise and views to these chapters. The editors and publishers (and hopefully the readers) are much appreciative of all their efforts.

This series of volumes of Advances in Petroleum Geochemistry will be of interest to all those, from whatever disciplinary background, in exploration and production of petroleum and to many workers in academic institutions. We now leave the readers to explore the various chapters and trust they will appreciate some of the advances and applications in petroleum geochemistry and also some of the flavour of this interesting and rapidly advancing subject.

Jim Brooks
Britoil, Glasgow

Dietrich Welte
KFA, Jülich

February 1984

Contents

Introduction

Jim Brooks and **Dietrich H. Welte**
Britoil, Glasgow, U.K.
KFA., Jülich, West Germany

Geochemistry has been defined as "the science concerned with chemistry of the earth, as a whole, and its component parts". More specifically, geochemistry deals with the distribution and movement of chemical elements within the earth in both time and space. Geochemistry is, at one and the same time, both more restricted and also more extensive in scope than geology. It is restricted in that it is not particularly concerned with the historical or structural aspects of the earth, but it is more extensive in that it provides a much deeper and fundamental insight into geological processes and composition of geological materials. Just as chemistry can be conveniently divided into three main subjects, geochemistry can be similarly subdivided into inorganic, physical and organic geochemistry. Petroleum geochemistry is an applied, specialist section of organic geochemistry (see Fig. 1). Although applications of biochemical, microbiological, natural product chemistry and modern organic chemical analytical techniques have resulted in the rapid development of petroleum and organic geochemistry in recent years, it must not be forgotten, however, that the prefix "*geo-*" is present in geochemistry and the "chemistry" root is only a part of the geochemical investigation. The significance and success of any study is considerably enhanced, and it is essential that the study correctly evaluates and integrates the geological controls and information into work programmes and conclusions.

The science of *petroleum* geochemistry is the application of chemical principles to the study of the origin, generation, migration, accumulation and alteration of petroleum, and the use of this knowledge in exploration and recovery of oil and gas (see Fig. 2). Although the concept of petroleum originating from organic-rich sediments and migrating into sands was first observed by geologists in the late 1800's, these early theories about the controlling principles of petroleum occurrences were often limited in concept, in that they mainly addressed the question of "*where*" accumulations were located. It has become clear during the last twenty years that to be

ADVANCES IN PETROLEUM GEOCHEMISTRY Vol. 1.
ISBN 0-12-032001-0

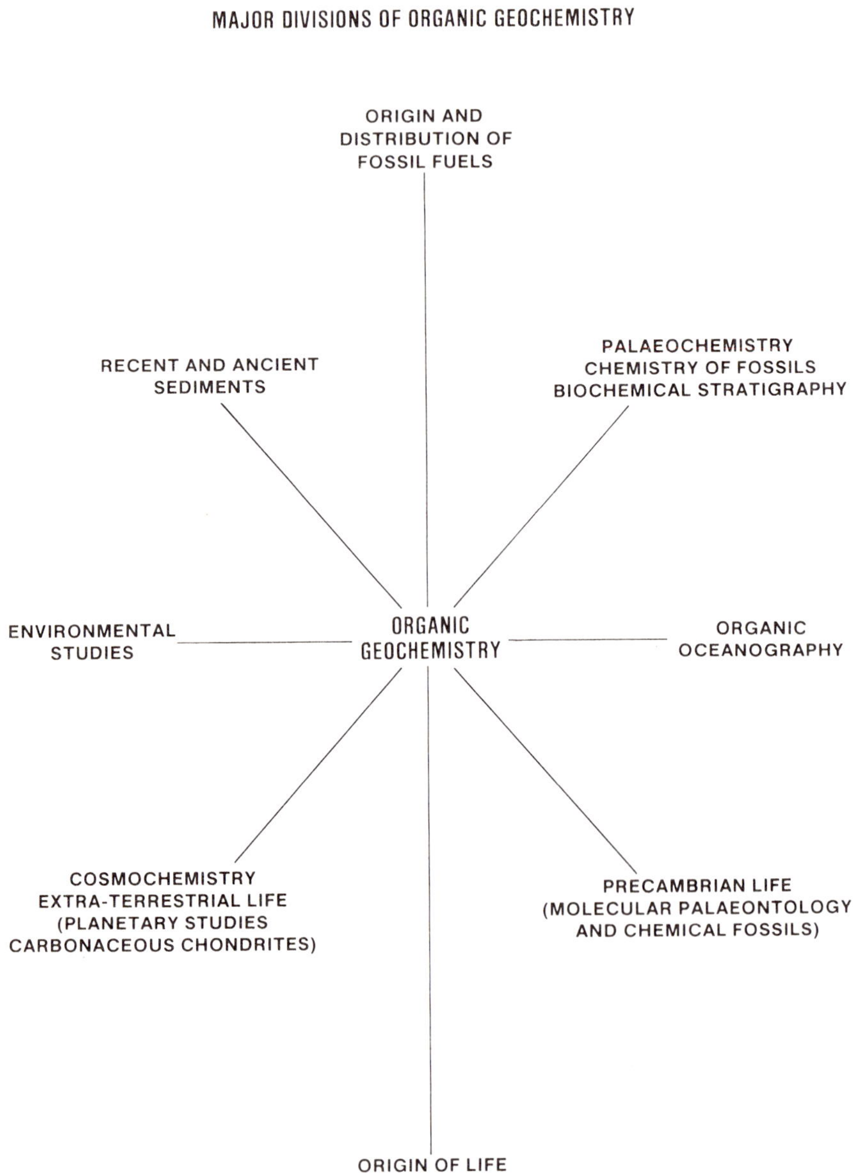

FIG. 1. Organic geochemistry.

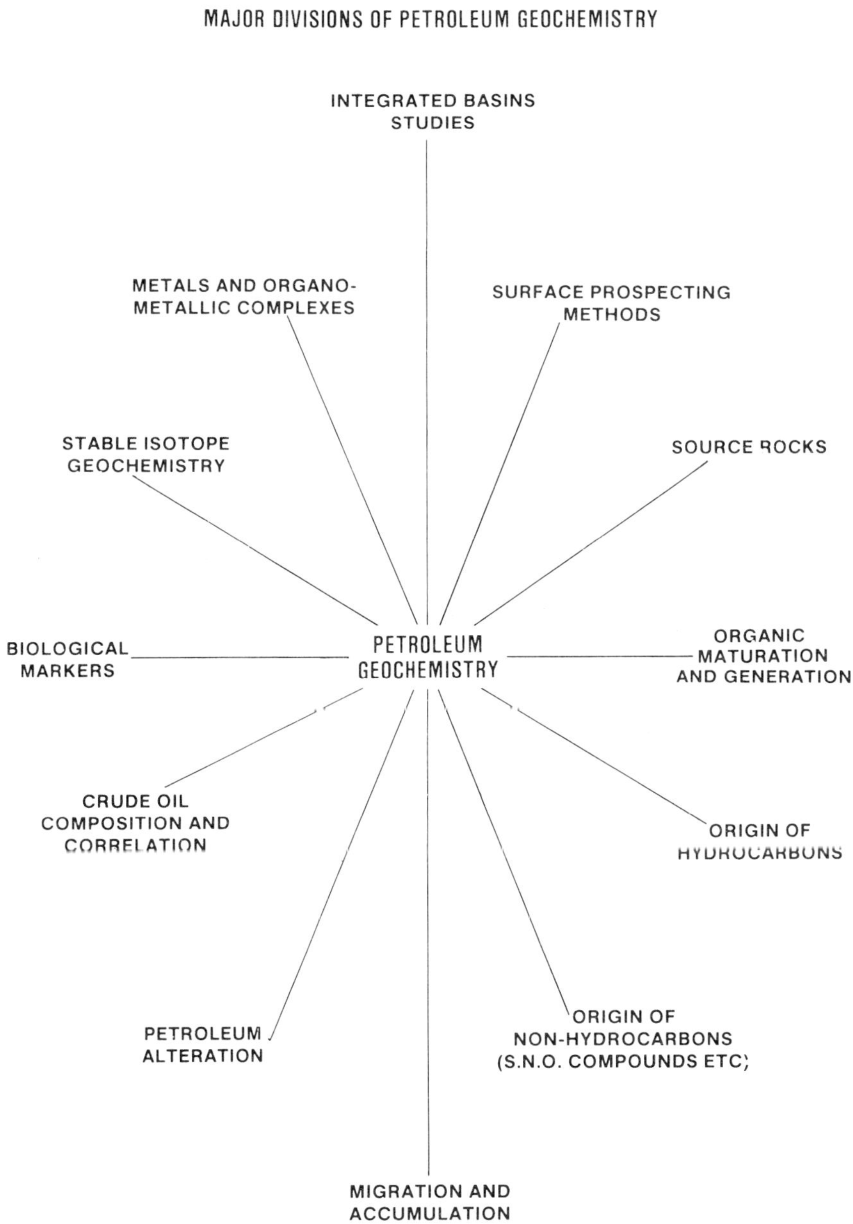

FIG. 2. Petroleum geochemistry.

able to better answer the question *"where"*, it is usually necessary to evaluate *"why, when* and *how much"* petroleum is present in a basin and to understand and establish the origin, character and distribution of hydrocarbon source rocks, and generation, migration and accumulation processes of petroleum. This understanding and application is essential if the oil industry is to improve its petroleum exploration success ratio.

Although hypotheses are periodically put forward for non-biological, earth mantle origins of petroleum, it is generally accepted that the origins are in sediments rich in biologically derived organic matter, from which there is scientifically supported evidence.

The birth of petroleum geochemistry can be considered as dating from the chemical inception of the German chemist, Alfred Treibs, who in 1934 isolated and identified biologically important organic compounds (porphyrins) from crude oils, shales and coals. These compounds, which accounted for only trace amounts of the total identified organic matter, were the first geochemicals to be directly related to known biochemicals (degradative products of chlorophylls). Treibs recognized this significance and made the following suggestions:

> Crude oils and sedimentary organic matter are of biological origin.
>
> The thermal history of the sedimentary rocks (?source rocks) and generation of oil could not have involved high temperatures (often quoted as below 200°C).

The implications of these suggestions proved far reaching, and the basic principles, methods and suggestions have provided a foundation for modern petroleum geochemistry. Much of the early impetus for the development of petroleum geochemistry came from petroleum chemists interested in the origin of petroleum and, to a lesser extent, from chemists and geologists interested in the origin of other fossil fuels (e.g. coal, shale oil). Coal chemists were probably the first group of workers to extend fundamental study and analytical techniques onto a routine and applied basis.

In the last twenty years, the subject has undergone considerable expansion, and petroleum geochemistry has become a useful and increasingly applied aid to petroleum exploration. The growth of organic and petroleum geochemistry is clearly reflected in the production of literature on the subject, particularly research and applied papers, textbooks and conference proceedings.

Although many oil explorationists are now familiar with the basic concepts of petroleum geochemistry, mainly due to the increase in the number of oil-company research laboratories, service companies and the publication of two excellent textbooks on petroleum geochemistry by Tissot and Welte

(1978, revised 1984) and Hunt (1979), the oil industry is only now becoming aware of the wide scope of modern geochemical techniques and applications to assist in petroleum exploration at many different stages, from the initial frontier basin study even to investigations of petroleum properties (e.g. heavy oil) and production evaluations.

Exploration drilling is becoming more expensive and less successful, and petroleum geochemistry has now a recognized role to play in exploration programmes. Rapidly reducing oil and gas reserves in mature basins, together with the remote locations and hostile environments of the world's remaining unexplored sedimentary basins, is placing increasing economic and technological needs on oil explorationists. These requirements, together with an increasing geochemical understanding of the origin, generation, migration and accumulation of petroleum, have resulted in the increasing use of predictive models for locating new petroleum reserves.

The topics presented in this volume will provide the reader with an up-to-date review of many of these recent advances in petroleum geochemistry. The generation of hydrocarbons within source rocks is discussed in the chapter by Doug Waples (Mobil) on *"Thermal Models for Oil Generation"*. The scope and application of predictive models is reviewed in *"Models used in Petroleum Resource Estimation and Petroleum Geochemistry"* by Arif Yükler and Fritz Kokesh (Phillips). Major geochemical advances have recently been made as a result of increasing knowledge of molecular composition (biological markers) and stable isotope (C, H, S, O and N) studies on biological systems and petroleum, coupled with fuller understanding of the origin and natural history of these compounds within the Earth. These two major topics are discussed by Andrew Mackenzie (BP) in *"Applications of Biological Markers in Petroleum Geochemistry"* and Martin Schoell (B für G) in *"Stable Isotopes in Petroleum Research"*. Pyrolysis studies are now widely applied to specific problems in petroleum exploration, and these experiments are reviewed in *"Pyrolysis Studies and Petroleum Exploration"* by Brian Horsfield (ARCO). The biodegradation of petroleum in shallow reservoirs to give heavy oil is a widespread phenomenon (and also creates production problems). Geochemical and microbiological studies of biodegradation are reviewed in *"Biodegradation of Crude Oils in Reservoirs"* by Jacque Connan (Elf-Aquitaine).

The scope of these review chapters range from qualitative and descriptive treatment of geochemical analyses and information to sophisticated applications which relate complex geology and geochemical data into comprehensive and systematic basin analyses, and crude-oil and source-rock studies which can lead to prediction of petroleum accumulation. The determination of the most favourable petroleum exploration targets depends upon the best use of our geochemical knowledge of source rocks, generation, migration, accumulation and alteration of petroleum, combined with the geology of the basin. The reviews presented in this volume should assist the expert and other interested oil explorationists to appreciate the recent advances in petroleum geochemistry, and hopefully to use and integrate these studies more meaningfully and successfully into their exploration programmes.

Furthermore, for geoscientists in general it is an important move toward synthesis and further quantification of the various disciplines in the field.

References

Hunt, J. M. (1979). "Petroleum Geochemistry and Geology". W. H. Freeman, San Francisco.

Tissot, B. P. and Welte, D. H. (1978, revised 1984). "Petroleum Formation and Occurrence". Springer Verlag, Berlin.

Treibs, A. (1934). *Anal. Chem.* **510,** 42–62.

Thermal Models For Oil Generation

Douglas W. Waples*

Mobil Research and Development Corp., Dallas, Texas, U.S.A.

I. Introduction

The vast majority of petroleum geochemists today believe that petroleum is formed by thermal transformation of organic matter preserved in fine-grained sedimentary rocks. There are, nevertheless, a few workers who believe in inorganic, cosmic, microbial, or combined organic–inorganic

* Currently a geochemical consultant at 1717 Place One Lane, Garland, Texas 75042, U.S.A.

ADVANCES IN PETROLEUM GEOCHEMISTRY Vol. 1.
ISBN 0-12-032001-0

origins for natural gas and petroleum (e.g. Gold and Soter, 1980; Porfir'ev, 1974; Hoyle, 1955; Yi-gang, 1981; Hawkes, 1972). In spite of their arguments, however, the direct and circumstantial evidence for an organic origin for all or most petroleum and natural-gas accumulations seems to me to be overwhelming. Laboratory studies of thermal transformation of organic matter to petroleum-like materials (e.g. Lijmbach, 1975), empirical studies on the depths and temperatures at which the composition of sedimentary organic matter changes (e.g. Philippi, 1965; Teichmüller, 1974; Dow, 1977) and at which petroleum and gas accumulations occur (e.g. White, 1915), the ability of time–temperature models of thermal generation of oil to account for observational fact (e.g. Lopatin, 1971; Waples, 1980), and the occurrence of highly specific organic compounds of unquestioned biological affinity (e.g. Treibs, 1936) have all played important roles in the development of the theory of the organic origin of petroleum. I shall therefore make the explicit assumption in this chapter that petroleum is of organic origin.

The problem of the origin of coal, petroleum, and oil shale has been of interest to men for millennia, although scientific approaches to answering these questions have only emerged within the last two centuries. The genetic relationships among coal, oil shale and petroleum were recognized early by a few workers (e.g. White, 1915), but until the 1960's research on each of these fossil fuels proceeded more or less independently, and oblivious of studies on the others. Coal was the subject of earliest scrutiny, both because of its long-standing economic importance, and because of its obvious affinity to plant material (see Weithofer, 1916, and Stutzer, 1940, p. 88, for reviews). Oil shale had also been known and utilized for a long time, but because of its more limited distribution it was not as well studied. The first work on oil shale, in the early twentieth century, in fact concentrated not on determining how it was formed in nature, but rather on how it could be transformed by man into a more useful material, shale oil.

Petroleum, the fossil fuel of greatest current importance, ironically was the last to be studied. An understanding of the mechanisms by which organic matter preserved in fine-grained sediments is transformed into petroleum did not begin to emerge until the middle of the twentieth century, and did not blossom until the late 1960's. The quantum leap in the last two decades in our understanding of the process of petroleum formation has been a direct result of the synthesis and application to oil generation of preceding work on coalification and oil shale pyrolysis.

Our understanding of transformations of organic matter in buried sediments and rocks has evolved very slowly. This gradual evolution produced numerous theories which ultimately were proved wrong or found to be of limited applicability, and thus became extinct. Many of these now seem naive, given our modern sophisticated perspective, but each has served as a work-point, and has focused attention on the essential requirements for coalification and petroleum-forming processes. A brief historical review of earlier theories is therefore useful.

II. Coalification

The presence in coals of macroscopic plant remains, such as leaves, twigs and logs, has made it clear for a long time that at least some portions of coal are of organic origin (e.g. Rogers, 1843). It was not clear to early workers, however, that coals of all ranks were derived from the same organic sources. In 1778 von Beroldingen (cited by Briggs, 1931) suggested that peat, lignite, bituminous coal, and anthracite all represent different stages in the transformation of organic matter buried in the earth's crust. As reasonable as this hypothesis appears now, it did not meet with universal or unconditional acceptance. Over the next century or more, various other theories emerged claiming that coals were predominantly of inorganic origin, or that coals of different ranks were not members of a single transformation sequence, but rather were products of the alteration of different organic substances under a variety of conditions. Climate, geological age, initial composition of organic matter, and microbial activity were commonly cited as factors responsible for formation of coals of different ranks. It was not until the early part of the twentieth century that the concepts of "rank" and "type" were clearly distinguished (White, 1908, 1913, 1925; Erdmann, 1924; Turner, 1925; Stadnichenko, 1934; see also Krejci-Graf, 1962).

One early theory that did survive, however, and even flourished, was the hypothesis that burial is the decisive factor in coalification. The relationship between burial and coal rank was first enunciated by Carl Hilt in 1873, who noted, without offering any explanation for the phenomenon, that fixed-carbon content in coals increased regularly with depth of burial. The utility and general validity of "Hilt's Law" led naturally to attempts to explain the dependence of coal rank on burial depth (McCreath, 1879; Reeves, 1928; Heck, 1943; Jones, 1949; Böttcher *et al.*, 1949; Teichmüller and Teichmüller, 1951; Tan, 1965). These efforts focused on three factors that also are functions of burial depth: temperature, static pressure, and time. Coalification models based on one or more of these factors have been proposed and promoted throughout the last century (see Heck, 1943, for an excellent review).

A. Temperature

Many early workers who supported the dominant influence of temperature on coalification believed that extremely high temperatures were required.

Erdmann (1924), for example, claimed that temperatures in excess of 325°C were necessary to produce bituminous coal, and Roberts (1924) proposed that formation of anthracite required temperatures of 500–550°C.

For those who believed that coalification required very high temperatures, however, there was a problem of a source for such immense amounts of heat. It was known in the late nineteenth century that geothermal gradients averaged about 2°F per 100 ft (3.6°C per 100 m) (McCreath, 1879; Burchfield, 1975, p. 35). Temperatures of 500°C could therefore only be reached by burial to immense depths, or by igneous activity.

Some early workers attempted to correlate coalification with igneous activity (e.g. Hardman, 1877). Where there were no obvious signs of intrusives, reference was often made to unseen plutons. Some of the emphasis on igneous activity was based on empirical evidence, and some on extrapolations of laboratory coalification reactions. Nevertheless, extreme burial and igneous activity were not defensible globally as the primary causes of coalification, although cases could be made for each process locally (e.g. Teichmüller and Teichmüller, 1951, 1979). Neither a cooling model for the earth, nor heat produced by exothermic chemical reactions involving organic matter or liberated by radioactive processes (see Burchfield, 1975, pp. 166–171) could account for such high temperatures.

In the first part of the twentieth century White (1913, 1925) and Bergius (1913) recognized that, given the vastness of geological time, reasonably low temperatures were adequate even for formation of anthracite. There were problems, however, with coalification schemes that relied solely on temperature as the agent of transformation. Bergius (1913), for example, noted that laboratory heating of cellulose could not produce anthracite, and therefore concluded that pressure was also required for coalification. Briggs (1931) observed that laboratory coalification reactions carried out at high temperatures tended to produce poor-quality coals, and thus deduced that the role of temperature in nature must be subordinate to that of time.

White (1908, 1913, 1925, 1935) therefore provided a very attractive alternative with his thrust-pressure hypothesis. The empirical correlation between coal rank (measured by fixed-carbon content) and orogenic folding that Rogers (1843) had first observed led him to propose that frictional heating during orogenic movements had caused the coalification. Roberts (1924) strongly supported the thrust-pressure hypothesis, but because he believed that high-rank coals were formed at very high temperatures, much greater thrust-induced heating was required by his model than by White's low-temperature model.

White's ideas were adopted by many other workers over the next few decades (e.g. Stadnichenko, 1934; Trotter, 1948; Francis, 1961; Tan, 1965; Landes, 1967), but were never universally accepted. Fixed-carbon contents of coals were found to be influenced by original chemical composition as well as by coal rank (Russell, 1925), thus complicating the application of White's ideas. Those coal workers whose favourite areas were not associated with thrusting strongly opposed the thrust-pressure hypothesis. Reeves (1928) and Gropp and Bode (1932) were among the first to offer evidence that White's model was incorrect. It was not until the middle of the twentieth century, however, that thrust pressure was finally discounted as the dominant factor in coalification (Heck, 1943; Teichmüller and Teichmüller, 1951, 1966; Huck and Karweil, 1955).

B. Static pressure

Static pressure supplied by overburden was also commonly suggested as the most important factor in coalification (e.g. Reeves, 1928), but never gained the widespread acceptance that White's dynamic-pressure hypothesis enjoyed, because of its inability to provide a source of heat. The static-pressure hypothesis has persisted with some workers almost to the present day (e.g. Ivanov, 1967), despite the numerous strong arguments against it as an important influence in the chemical transformation of coals (Bergius, 1913; Stadnichenko, 1934; White, 1935; Trotter, 1948; Jones, 1949; Huck and Karweil, 1955; Karweil, 1956; Kasatochkin, 1959; Francis, 1961; Teichmüller and Teichmüller, 1966; Berkowitz, 1967; Hanbaba and Jüntgen, 1969; Lopatin and Bostick, 1973). Most modern workers believe that although static pressure is required for the formation of true coals, it affects physical properties rather than rates of chemical reactions.

C. Time

The role of geological time has taken on a gradually increasing importance as a contributing factor in coalification since Bergius (1913) and White (1913) first recognized its power. Briggs (1931) believed that geological time was vital for natural coalification, because laboratory coalification carried out at high temperatures for short times produced coals of inferior quality.

Nevertheless, time alone has seldom been considered seriously as the dominant factor in coalification. Most coalification theories that allocate an important role to time give an even more vital one to temperature.

D. Combined time and temperature

Most early geological discussions of the effects of time in coalification processes were very qualitative; the only quantitative measurements which showed that time and temperature could to some extent be substituted for each other came from experimental work carried out in chemistry laboratories. Bergius (1913) was probably the first to suggest that time and temperature enjoy an exponential relationship in coalification by proposing that the rates of coalification reactions double with each 10°C rise in temperature. Cane (1950) showed that pyrolysis of torbanite, an algal coal, follows first-order kinetics, with a pseudo-activation energy of 48.5 kcal mol^{-1} (202.7 kJ mol^{-1}).

There were, however, some difficulties in using laboratory data to understand normal coalification processes. Although van Krevelen (1952) showed that coal pyrolysis and normal coalification are chemically distinct processes, and suggested that laboratory data may not be applicable to coals in natural settings, Hanbaba and Jüntgen (1969) and van Heek and co-workers (1972) subsequently found that such extrapolations could be made with care. Numerous workers (e.g. Fuchs and Sandhoff, 1942; Shapatina *et al.*, 1950; Stone *et al.*, 1954; Huck and Karweil, 1955; Fitzgerald, 1956; Chermin and van Krevelen, 1957; Oxley and Pitt, 1958; Kasatochkin, 1959; Jüntgen, 1964; Berkowitz, 1967; Hanbaba and Jüntgen, 1969; Mochalov and Gryaznov, 1969) showed that pyrolysis of humic coals (and thus presumably coalification) is a much more chemically complex process than had been previously appreciated, and that kinetic analyses are therefore also bound to be much more difficult to interpret. For example, activation energies measured for coal pyrolysis by the above workers varied between 4 and 59 kcal mol^{-1} (16 and 247 kJ mol^{-1}), depending upon pyrolysis conditions, the parameters measured, and the stage of reaction (Table I). The smallest of these activation energies are far too low to represent true activation energies for chemical reactions occurring during coalification. Pre-exponential (A) factors in the Arrhenius equation

$$k = A \exp(-E_a/RT) \qquad (1)$$

determined in those studies were also generally found to be far lower than those for normal unimolecular first-order decomposition reactions. These results led several workers to propose that coal-pyrolysis kinetics are controlled by the rates of physical processes, such as diffusion or removal of steric hindrance, rather than by rates of chemical reactions (Stone *et al.*, 1954; Huck and Karweil, 1955; Jüntgen, 1964). The unusual kinetic parameters often determined for coal-pyrolysis reactions (Table I) suggest that the term "pseudo-activation energy" is appropriate for measured E_a values.

Van Krevelen and co-workers (1951) observed that pseudo-activation energies for coal pyrolysis increase with increasing coal rank (see Fig. 1). They concluded that the chemical structure of coal undergoes important changes during coalification, and that in the higher ranks of coal only very strong bonds remain. The anomalously low pseudo-activation energies for pyrolysis of coals containing 20–30% volatile matter (see Fig. 1) were

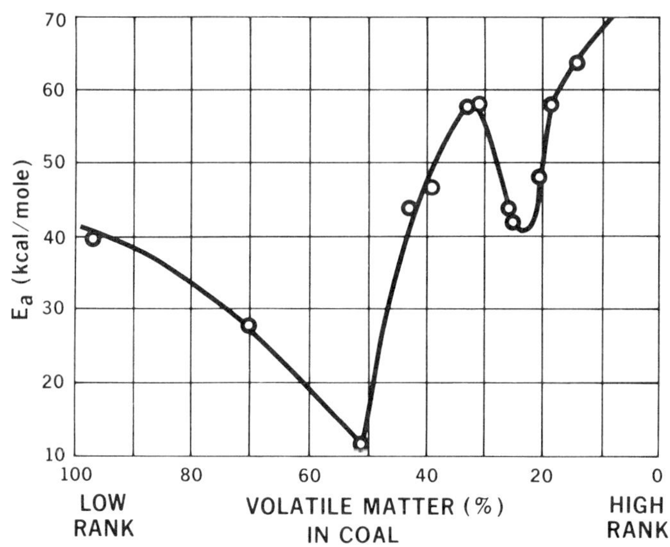

FIG. 1. Pseudo-activation energies for coal pyrolysis as a function of coal rank (determined from volatile-matter content). From van Krevelen *et al.*, 1951. *Reprinted with permission.*

attributed to changes in coal plasticity during high-temperature coking, a phenomenon which might not occur during natural coalification. Jüntgen (1964) and Berkowitz (1967) made similar observations.

Despite the relatively abundant work on the kinetics of coal pyrolysis, kinetic calculations were not utilized by coal geologists until Karweil (1956), with his classic nomograph (Fig. 2), established the first means of calculating and predicting the rank of a coal from a knowledge of its burial and thermal histories. Karweil's method of utilizing time and temperature represented a quantum leap forward in quantifying the process of thermal transformation of organic matter in sediments, and it is a testimony to his insight that essentially no improvement was made on his method over the next decade and a half.

TABLE I. *Measured kinetic parameters for evolution of hydrocarbons from fossil organic matter.*

Process	Material	Parameter Measured	Temperature Range (°C)		E_a (kcal mol^{-1})	A (s^{-1})	Reference
Kerogen pyrolysis	Recent kerogen	Colour change	150–410°	I	15–115*	—	Peters *et al.*, 1977
Kerogen pyrolysis	Recent kerogen	Colour change in oil-generative window	350°	I	45	—	Peters *et al.*, 1977
Kerogen pyrolysis	Recent kerogen	*n*-alkane generation	150–410°	I	31	—	Ishiwatari *et al.*, 1976
Oil-shale pyrolysis	Utah GRS?	Liquid evolution	275–365°	I	41.5	8.1×10^{11}	Maier and Zimmerley, 1924
Oil-shale pyrolysis	Colo. GRS	Liquid + gas evolution	350–437°	I	62	2.8×10^{18}	Hubbard and Robinson, 1950
Oil-shale pyrolysis	Colo. GRS	Liquid + gas evolution	437–525°	I	25.4	1.4×10^{7}	Hubbard and Robinson, 1950
Oil-shale pyrolysis	Colo. GRS	Liquid + gas evolution	250–465°	I	45.2	6.3×10^{9}	DiRicco and Barrick, 1956
Oil-shale pyrolysis	GRS	Methane evolution	185–400°	I	40.25	1×10^{9}	Hoering and Abelson, 1963
Oil-shale pyrolysis	Colo. GRS	Bitumen evolution	<427°	PT	40–41.7	—	Hill and Dougan, 1967
Oil-shale pyrolysis	Colo. GRS	Oil evolution	<427°	PT	42.5–48.5	—	Hill and Dougan, 1967
Oil-shale pyrolysis	Colo. GRS	Methane evolution	316–440°	I	39	—	Weitkamp and Gutberlet, 1970
Oil-shale pyrolysis	Colo. GRS	Methane evolution	316–440°	PT	29	—	Weitkamp and Gutberlet, 1970
Oil-shale pyrolysis	Colo. GRS	Liquid + gas evolution	316–440°	I	49	—	Weitkamp and Gutberlet, 1970

Oil-shale pyrolysis	Colo. GRS	Liquid + gas evolution	316–440°	PT	25–56*	—	Weitkamp and Gutberlet, 1970
Oil-shale pyrolysis	Colo. GRS	Bitumen evolution	150–350°	I	19	—	Cummins and Robinson, 1972
Oil-shale pyrolysis	Colo. GRS	Liquid + gas evolution	350–487°	I	42.4	2.0×10^{10}	Braun and Rothman, 1965
Oil-shale pyrolysis	Colo. GRS	Liquid + gas evolution	487–525°	I	10.6	14.4	Braun and Rothman, 1975
Oil-shale pyrolysis	Colo. GRS	Kerogen disappearance	350–600°	PT	25–65*	—	Arnold, 1975
Oil-shale pyrolysis	Colo. GRS	Oil evolution	350–510°	PT	47.7	8.3×10^{11}	Shih and Sohn, 1980
Oil-shale pyrolysis	GRS	Oil evolution	375–500°	I	19.86–27.38	1×10^{-3}–1.5×10^{-3}	Noble *et al.*, 1981
Oil-shale pyrolysis	Austr. Torbanite	Oil evolution	350–400°	I	48.5	1.5×10^{12}	Cane, 1950
Coal pyrolysis	Boghead	Devolatilization	300–350°	I	14.6–23.4*	—	Shapatina *et al.*, 1950
Coal pyrolysis	Boghead	Devolatilization	350–400°	I	28.2–41.0*	—	Shapatina *et al.*, 1950
Coal pyrolysis	Peat	Devolatilization	300–350°	I	16.9–39.4*	—	Shapatina *et al.*, 1950
Coal pyrolysis	Peat	Devolatilization	350–400°	I	13.7–39.4*	—	Shapatina *et al.*, 1950
Coal pyrolysis	Lignite	Devolatilization	300–350°	I	5.3–32.6*	—	Shapatina *et al.*, 1950
Coal pyrolysis	Lignite	Devolatilization	350–400°	I	6.8–13.0*	—	Shapatina *et al.*, 1950
Coal pyrolysis	Bituminous	Devolatilization	300–350°	I	3.9–37.5*	—	Shapatina *et al.*, 1950
Coal pyrolysis	Bituminous	Devolatilization	350–400°	I	3.8–29.8*	—	Shapatina *et al.*, 1950

TABLE I. (*continued*)

Process	Material	Parameter Measured	Temperature Range (°C)		E_a (kcal mol^{-1})	A (s^{-1})	Reference
Coal pyrolysis	Lignite	Devolatilization	350–550°	PT	6†	1.7×10^{-3}	van Krevelen *et al.*, 1951
Coal pyrolysis	Bituminous	Devolatilization	350–550°	PT	21–29†	67–1500	van Krevelen *et al.*, 1951
Coal pyrolysis	Semi-anthracite	Devolatilization	350–550°	PT	32†	1.7×10^{3}	van Krevelen *et al.*, 1951
Coal pyrolysis	Bituminous	Devolatilization	425–538°	I	34.6	—	Stone *et al.*, 1954
Coal pyrolysis	Bituminous	Devolatilization	410–510°	I	26.2–6.7‡	—	Stone *et al.*, 1954
Coal pyrolysis	Bituminous	Fluidity changes	394–448°	I	47–55	3.8×10^{11}–4.7×10^{14}	Fitzgerald, 1956
Coal pyrolysis	Bituminous	Degasification	300–760°	I	52–59	8.3×10^{13}	Chermin and van Krevelen, 1957
Coal pyrolysis	Bituminous	Bitumen evolution	325–370°	I	30	6.7×10^{8}	Oxley and Pitt, 1958
Coal pyrolysis	Bituminous	Methane evolution	450–950°	PT	42.9–59.1	1.05×10^{8}–1.15×10^{9}	Jüntgen, 1964
Coal pyrolysis	Bituminous	Ethane evolution	?	PT	42.1	1.7×10^{9}	Hanbaba and Jüntgen, 1969
Coal pyrolysis	Lignite	Devolatilization	200–850°	PT	6.1–8.8	—	Mochalov and Gryaznov, 1969
Coal pyrolysis	Bituminous	Devolatilization	200–850°	PT	10.1–14.3	—	Mochalov and Gryanov, 1969
Coal pyrolysis	Bituminous	Methane evolution	?	PT	53	1.7×10^{12}	van Heek *et al.*, 1971

Coal pyrolysis	Bituminous	Methane evolution	350–800°	PT	42.1	1.7×10^{9}	Jüntgen and Klein, 1975
Natural coalification	Bituminous	Devolatilization	60–80°	PT	8.4	3.6×10^{-12}	Huck and Karweil, 1955
Natural coalification	Bituminous	Devolatilization	140–160°	PT	0.9	0.3	Karweil, 1975
Natural oil generation	Kerogen	Oil evolution	60–127°	PT	11–14	—	Connan, 1974
Oil cracking	Crude oil	Light-HC evolution	267–372°	I	49	3.0×10^{11}	McNab *et al.*, 1952

* Increasing with increasing maturity
† See Fig. 1
‡ Decreasing with increasing maturity.

GRS = Green River shale
I = Isothermal
PT = Programmed-temperature

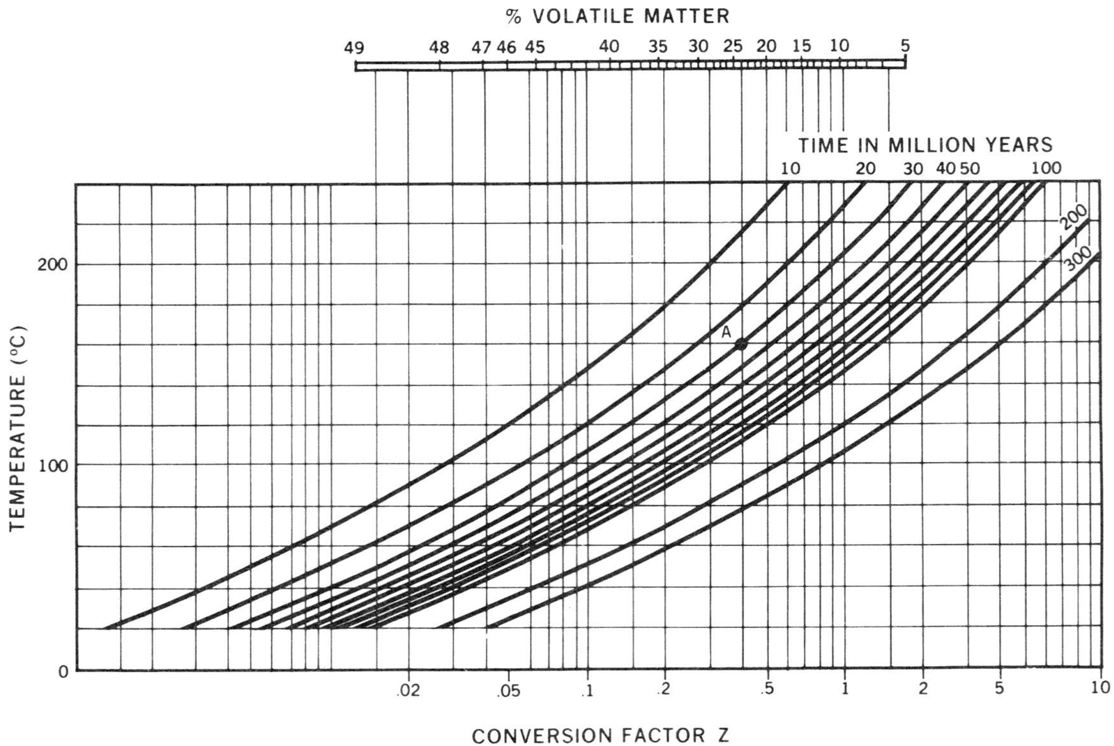

FIG. 2. Karweil's nomograph (1956) relating coal rank (determined by volatile-matter content) to length of time spent at a certain palaeotemperature. Each line represents a different maximum palaeotemperature. The *x*-coordinate of any point along any time line represents the rank of a coal seam whose maximum palaeotemperature is given by the *y*-coordinate of the same point. Example: point *A* has spent 30 million years at 160 °C and contains 24% volatile matter (corresponding to the medium-volatile bituminous rank). *Reprinted with permission in modified form.*

Both Karweil (1956) and Teichmüller and Teichmüller (1966) recognized that the relative importances of time and temperature in coalification change as temperature rises, although their conclusions on that issue were apparently in direct conflict. Karweil stated that the temperature effect was greater at high temperatures, whereas the Teichmüllers argued that time is not important at low temperatures. Application of kinetic principles clearly supports Karweil's view; the Teichmüllers' statement is best taken as an observation that at very low temperatures little coalification occurs, regardless of the time available (Teichmüller and Teichmüller, 1979).

III. Oil-Shale Retorting

It was recognized long ago that heating of oil shale in the absence of oxygen produces an organic liquid somewhat similar in chemical composition to crude oil. Crum-Brown (cited by Steuart, 1912) originally defined kerogen as the oil-producing organic material in oil shale. Engler (1913) proposed that during retorting kerogen decomposes thermally in a two-step process:

$$\text{kerogen} \rightarrow \text{bitumen} \rightarrow \text{oil}.$$

The intermediate "bitumen" consists of fragments that have been broken off from the larger kerogen molecules, but which are not identical in composition to oil molecules. Engler noted, as Bergius did simultaneously for coal pyrolysis, that time and temperature could be substituted for each other, and that the effect of temperature was greater than that of pressure. The work of McKee and Lyder (1921) and Franks and Goodier (1922) confirmed these results. The latter also noted that the chemical composition of bitumen obtained during oil-shale pyrolysis depends upon pyrolysis temperature.

Maier and Zimmerly (1924) carried out the first detailed kinetic analysis of oil-shale pyrolysis on Green River shale (see Table I). They determined that the decomposition of oil-shale kerogen was an irreversible first-order reaction, and therefore supported Engler's conclusion that the rate of pyrolysis was a function of both time and temperature. Together with Franks and Goodier (1922), they also made the important statement that there was no definite temperature threshold for the pyrolysis, but rather a gradual increase in rate with increasing temperature.

Carlson (1937), Hubbard and Robinson (1950) and Hoering and Abelson (1963) eventually corroborated the findings of Maier and Zimmerley. Hubbard and Robinson also proposed a slight modification of Engler's earlier mechanism; that gas is formed along with bitumen in the initial decomposition reaction:

$$\begin{array}{ccccl} \text{kerogen} & \rightarrow & \text{bitumen} & \rightarrow & \text{oil} \\ & & + & & + \\ & & \text{gas} & & \text{gas} \\ & & & & + \\ & & & & \text{carbon} \\ & & & & \text{residue.} \end{array}$$

By the mid-1960's other workers were beginning to find that pyrolysis of oil shale, like that of coal, was a much more complex process than earlier workers had appreciated. Allred (1966) reinterpreted Hubbard and Robinson's (1950) data and performed additional experiments to show that the decomposition mechanism was dependent upon pyrolysis temperature, and that the reaction order was therefore variable. Fausett *et al.* (1974) found that a complex mechanism involving both first- and second-order reactions gave an even better interpretation of Hubbard's and Robinson's data. Pseudo-activation energies calculated by Allred and by Fausett *et al.* ranged from 20 to 50 kcal mol^{-1} (84 to 209 kJ mol^{-1}).

Hill and Dougan (1967), in contrast, studied oil-shale pyrolysis at much lower temperatures, and found the mechanism to be simpler than at the higher temperatures employed by most earlier workers. The reactions they investigated were first order in the earlier stages, but gradually gave way to zero-order processes in the later stages. They argued that at high temperatures bitumen production occurred so rapidly that diffusional escape from the rock matrix was not possible, thus resulting in polymerization. The intermediate polymers then decomposed further to give the complex kinetic dependence noted by Allred (1966), Fausett *et al.* (1974) and Arnold (1975). Weitkamp and Gutberlet (1970) also emphasized the importance of diffusion in both the mechanism and energetics of oil-shale pyrolysis. It became clear from their work and that of Hill and Dougan (1967) that extrapolation of pyrolysis data on oil shales to reactions in natural settings would be difficult.

In 1972 Cummins and Robinson bridged this gap by carrying out oil-shale pyrolysis experiments at much lower temperatures (150–350°C) than previous workers had. They showed that at those temperatures the pyrolysis was adequately described by a first-order kinetic expression having a pseudo-activation energy of 19 kcal mol^{-1} (78 kJ mol^{-1}).

Braun and Rothman (1975) helped to simplify the mechanism of oil-shale pyrolysis even further by reinterpreting Hubbard and Robinson's data again. After considering hitherto-neglected induction times in the heating experiments, they concluded that oil-shale pyrolysis could indeed by fairly represented by the two sequential reactions originally proposed by Engler (1913). The pseudo-activation energy for bitumen formation was calculated to be about 10 kcal mol^{-1} (42 kJ mol^{-1}), whereas that for bitumen decomposition was 42.5 kcal mol^{-1} (178 kJ mol^{-1}).

By the mid-1970's, therefore, work on oil-shale-pyrolysis kinetics had come virtually full-circle, and was again fundamentally in agreement with Engler's early analysis. Pyrolysis experiments carried out at low to moderate temperatures are adequately described by first-order kinetics (Shih and Sohn, 1980; Noble *et al.*, 1981). As van Krevelen *et al.* (1951) had observed for coal pyrolysis (see Fig. 1), there was evidence that pseudo-activation energies increase as retorting proceeds (Weitkamp and Gutberlet, 1970; Arnold, 1975). It was not by any means clear, however, that such changes were related to chemical reactions; they may have been linked to physical changes in the oil shale during retorting.

Maier and Zimmerley's work (1924) had suggested a decrease in pyrolysis rate with increasing pressure, although all their experimental pressures were far too low to simulate natural environments. Noble and co-workers (1981) showed conclusively that rates of oil-shale pyrolysis are reduced under higher pressures.

No particular attempt has ever been made to compare coalification processes (natural or artificial) with oil-shale retorting. Although Bertrand and Renault (1893) recognized the algal origin of sapropelic coals, and that algal coals and oil-shale kerogens behave similarly upon retorting, rank changes in algal coals were not studied, nor was their relationship to oil-shale kerogens explored, until much later (e.g. Schopf, 1949; Cane, 1950; Kasatochkin, 1959; Bradley, 1970; Cane and Albion, 1973). There was, furthermore, no concept of thermal maturation of oil-shale kerogen analogous to coalification. Because of the poor empirical correlation between oil-shale occurrences and petroleum deposits, it was thought that oil-shale kerogen could not be transformed into bitumen or hydrocarbons in natural settings (Takahashi, 1922; Cox, 1946; Krejci-Graf, 1963). Carlson (1937) proposed that slow decomposition of oil-shale kerogen in natural settings might be responsible for the bitumen present in oil shales, but for many years no one seems to have accepted or pursued his suggestion.

Kinetic data compiled from laboratory retorting experiments were therefore not applied to the behaviour of oil-shale kerogen under natural conditions. As our understanding of the similarities among coalification, oil-shale retorting and petroleum generation has grown, data from artificial-coalification and retorting studies have instead been applied directly to understanding the chemistry and energetics of petroleum formation in the subsurface.

IV. Petroleum Formation

Very little progress in understanding the origin of petroleum was made prior to the 1960's. An organic origin had been suggested by early chemical analyses of petroleum constituents (e.g. Marcusson, 1908; Zelinsky, 1927), and was confirmed to the satisfaction of most by Alfred Treibs (1936). There was no consensus, however, about *how* petroleum forms from organic matter. Microbial activity at shallow depths of burial was proposed by some workers (e.g. Sickenberger, 1891; Morrey, cited in Bownocker, 1903, pp. 313–314; Hammar, 1934; Zobell, 1945) as responsible for transformation of various functionalized organic molecules into hydrocarbons. The data of Trask (1934), however, indicated an absence of hydrocarbons in recent sediments. In the 1950's indigenous hydrocarbons were finally identified for the first time in recent sediments (Smith, 1954), but these mixtures were soon shown to be quite distinct from petroleum hydrocarbons (Orr and Emery, 1956; Stevens *et al.*, 1956). In contrast, bitumens similar in composition to petroleum were found in all lithified shales and limestones studied by Hunt and Jamieson (1956). It thus became clear to most workers that petroleum does not form immediately after sedimentation, but rather at greater depths of burial.

A. Oil generation at depth

The mechanism by which oil is generated remained obscure, however, because the relationships among coal, oil-shale kerogen, petroleum, and disseminated organic matter in ordinary sedimentary rocks (also termed "kerogen" by Takahashi (1922), Hunt and Jamieson (1956) and Forsman and Hunt (1958), and "kerabitumen" by others, including King *et al.* (1963)) were not fully appreciated for a long time. Rogers (1860) had proposed that oil and gas in Pennsylvania were derived from coal-bearing strata at a certain coal rank. Potonié (1910) suggested that "sapropelites" might be sources of oil when subjected to elevated pressures, temperatures, or both, and Anderson and Pack (1915) offered a similar model for formation of California oils from diatoms. Many early workers had specifically proposed that kerogen is somehow converted to bitumen, which in turn gives rise to petroleum (Engler, 1913; McCoy, 1919; McKee and Lyder, 1921), although others thought either that kerogen was derived from bitumen (Cunningham-Craig, 1916), or that there was no genetic relationship at all between kerogen and bitumen in sediments (Takahashi, 1922).

Karrick (1926) had suggested that virtually all shales could, like oil shales, generate some bitumen upon retorting, but because retorting of ordinary shales is economically unimportant, the significance of this idea for oil generation in natural settings went unappreciated until much later. Numerous investigators (e.g. Roberts, 1938) suggested that elevated subsurface temperatures caused the "distillation" of oil from source rocks, but were vague about the details of this process. In any case, as Treibs (1934) and Dunning and Moore (1957) pointed out, the presence of carboxylated porphyrins in many oils indicates that the temperatures at which oil forms and is reservoired are not extremely high.

The mechanism by which oil forms in shales remained poorly understood for a long time, despite the suggestion by McNab *et al.* (1952) that petroleum is produced by thermal decomposition of kerogen. Even in the late 1950's this model was still not universally accepted (Hunt and Jamieson, 1956; Meinschein, 1959).

An important advance occurred when Forsman and Hunt (1958) and Forsman (1963) distinguished between coal-type kerogens and oil-shale-type kerogens. More recent work has proved that oil is principally derived from kerogens which are similar in origin and chemical composition to oil-shale kerogens.

With little doubt the previous great reluctance among early coal workers and petroleum geochemists alike to draw close parallels among coals, oil shales and finely disseminated organic matter in sediments (Potonié (1910) was exceptional in comparing sapropels in coals and in bituminous shales and limestones) retarded the development of petroleum geochemistry for many years. For example, Krejci-Graf (1963) thought that oxidizing conditions during deposition led to formation of coals and oil shales, whereas reducing conditions produced the sapropelites that he believed to be the sources for petroleum.

B. Role of kerogen

By the 1960's evidence had begun to accumulate that some dispersed kerogen in fine-grained sedimentary rocks is indeed transformed into bitumen by thermal processes. Increases in bitumen contents of rocks under the influence of elevated subsurface temperatures were reported by numerous workers (e.g. Louis, 1966; Louis and Tissot, 1967; Albrecht and Ourisson, 1969; Shibaoka *et al.*, 1973). Philippi (1965) in particular provided strong empirical evidence for the role of thermal kerogen decomposition in petroleum formation in his classic paper on generation of petroleum in Southern California. Louis and Tissot (1967) noted that the formation of hydrocarbons occurred subsequent to bitumen formation, as Engler (1913) had observed for retorting of oil-shale kerogen.

As many analytical methods commonly employed by coal scientists were slowly adapted for studying disseminated kerogens, the kerogen-transformation model was gradually confirmed. Correia (1967), Staplin (1969), and Bordenave and co-workers (1970), following the lead of coal workers (Schopf, 1948; Wilson, 1961), noted that spores and pollen grains in shales gradually darken with increasing thermal exposure, and developed scales for estimating thermal-maturity levels from colours of pollen grains. Vitrinite reflectance measurements, first successfully applied to coals in the 1930's, were found to be applicable to disseminated kerogens as well as to coals (Teichmüller, 1958), but were not applied routinely until somewhat later (Teichmüller, 1971). As they had for coals (see Fig. 3), elemental analyses confirmed that chemical transformations accompany thermal maturation of kerogens (Hlauschek, 1950; Forsman and Hunt, 1958; Forsman, 1963; King *et al.*, 1963; Durand and Espitalié, 1973; Tissot *et al.*, 1974). Krejci-Graf (1963) noted a parallel between changes in elemental composition of kerogens during oil generation and in coals during destructive distillation. The coal-maceral concept (White, 1908; Potonié, 1910) was finally applied to kerogens in 1958 (Forsman and Hunt), and coal evolution diagrams (van Krevelen, 1950) such as Fig. 3 were used by 1973 to describe kerogen transformations (Durand and Espitalié, 1973).

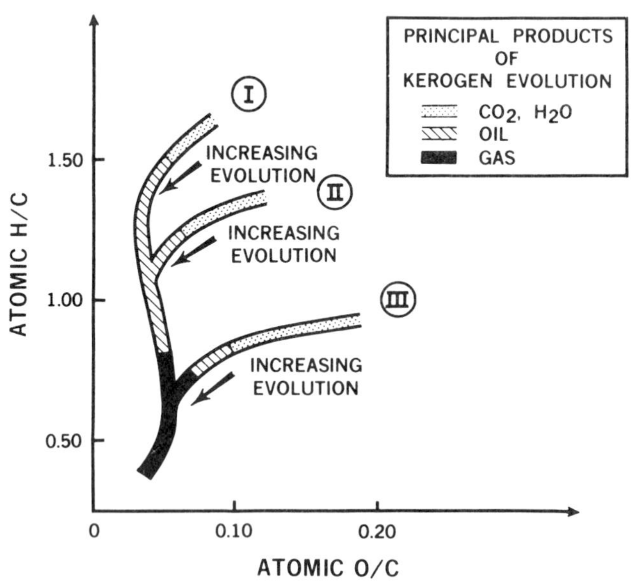

FIG. 3. Van Krevelen diagram showing elemental-compositional differences among kerogen types (I, II and III) and changes in H/C and O/C ratios with maturation for each kerogen type. After Tissot *et al.*, 1974. *Reprinted with permission.*

Pyrolysis studies have verified that dispersed organic matter in sedimentary rocks is capable of generating hydrocarbons by thermal decomposition. Even recent sediments, which do not contain true kerogen, yield hydrocarbons upon pyrolysis (e.g. Whitehead and Breger, 1950; Hoering and Abelson, 1963). Studies of dispersed kerogens (Hoering and Abelson, 1963; McIver, 1967; Bordenave *et al.*, 1970; Barker, 1974; Tissot *et al.*, 1974) have proved that virtually all kerogens break down thermally to yield hydrocarbons, the nature and quantity of product being functions of kerogen type (Dow, 1977). Barker (1974) and Peters *et al.* (1977) established for kerogen pyrolysis, as earlier workers had for coal pyrolysis and oil-shale retorting (see Fig. 1), that pseudo-activation energy increases and remaining hydrocarbon generative capacity decreases with increasing rank. Bordenave *et al.* (1970) showed by pyrolysis experiments that the oil-generative capacity of highly mature kerogens was exhausted. By the 1960's Hoering and Abelson (1963), Abelson (1967) and McIver (1967) could assert that kerogen was the most probable source of petroleum.

C. Kinetics

The work of Philippi (1965) showed that both time and temperature are important in oil generation, and that they are at least to some degree interchangeable. Oil formed at lower temperatures in the Los Angeles Basin than in the Ventura Basin because longer times were available for cooking the Los Angeles Basin source rocks. Philippi's approach was mainly qualitative, however; he did not attempt to describe the relationship between time and temperature other than to suggest that the rate of oil generation approximately doubled with every 10°C rise in temperature.

Shortly afterward the French Petroleum Institute began to develop theoretical models for petroleum-generative processes (Tissot, 1969; Deroo *et al.*, 1969). They assumed essentially the same mechanism for petroleum formation that Engler (1913) had, and employed pseudo-activation energies similar to those found by Cummins and Robinson (1972) for low-temperature pyrolysis of Green River oil shale: 10 kcal mol^{-1} (42 kJ mol^{-1}) for bitumen formation and 15 to 20 kcal mol^{-1} (63 to 84 kJ mol^{-1}) for hydrocarbon formation. *A*-factors were very low (10^2 to 10^8 per M.Y.) (10^{-11} to 10^{-5} s^{-1}). Tissot (1969) also noted that petroleum-forming reactions encompassed a range of activation energies, and that the prevalent type of bond cleavage is therefore a function of temperature. Connan's studies (1974) also showed that pseudo-activation energies of 11 to 14 kcal mol^{-1} (46 to 59 kJ mol^{-1}) gave adequate correspondence with measured data on petroleum formation (see Table I).

As many workers noted, however, these pseudo-activation energies were far lower than could be explained on purely chemical grounds. Other investigators have come to very different conclusions on the basis of kinetic measurements of kerogen transformation at high temperatures. Ishiwatari *et al*. (1976) found that the pseudo-activation energy for *n*-alkane formation during kerogen pyrolysis was 31 kcal mol^{-1} (130 kJ mol^{-1}). This high value is probably a consequence of the homogeneity of the reactions. Peters *et al*. (1977) calculated pseudo-activation energies of 40 to 50 kcal mol^{-1} (167 to 209 kJ mol^{-1}) for the darkening of kerogen within the oil-generation window. These results may indicate that kerogen-colour changes are not necessarily correlative with bitumen formation, but rather depend, as Snowdon (1979) suggested, on the temperatures at which the reactions occur. Abelson (1967) was concerned that the discrepancy between predicted and measured kinetic parameters for oil-shale pyrolysis was evidence for the inapplicability of laboratory experiments in studying petroleum formation. Snowdon (1979) has echoed and amplified these reservations.

It has been proposed that catalysis by minerals, particularly clays, was responsible for the low measured pseudo-activation energies (e.g. Brooks, 1948; Dobryansky, 1963; Hedberg, 1964; Louis, 1966; Louis and Tissot, 1967; Vassoevitch *et al*., 1967; Tissot, 1969; Shimoyama and Johns, 1971; Welte, 1972). Geocatalysis models have always been beset by two difficulties, however: the improbability that large, relatively immobile kerogen molecules could have appreciable surface interaction with catalytic mineral surfaces, and the inability of known catalysts to lower activation energies as much as the empirical and experimental data on oil generation would require (Hoering and Abelson, 1963). Recent work (e.g. Horsfield and Douglas, 1980) indicates that if catalysis is a factor in oil generation, it is much more important in cracking the initial bitumen than in the breakdown of kerogen.

In 1975, a much more plausible explanation was offered for the unusual kinetic parameters obtained from laboratory pyrolysis experiments of coals, oil shales and kerogens, and from empirical data on oil generation in natural settings (Jüntgen and Klein, 1975; Tissot and Espitalié, 1975). Both groups observed that pseudo-activation energies representing a group of parallel reactions can be very different than the actual activation energy for any one of the parallel reactions. Jüntgen and Klein showed clearly (see Fig. 4) that a group of parallel reactions having normal kinetic parameters (activation energies from 48 to 62 kcal mol^{-1} (200 to 260 kJ mol^{-1}); *A*-factor = 10^{15} min^{-1} (10^{13} s^{-1})) occurring in their natural sequence during pyrolysis at gradually increasing temperatures can be described quite accurately as though they were actually a single reaction having kinetic parameters (E_a = 20 kcal mol^{-1} (84 kJ mol^{-1}); $A = 10^4$ min^{-1}

($2 \times 10^2\ s^{-1}$)) that are physically impossible for the first-order thermal decomposition of organic matter.

Jüntgen's and Klein's contribution had the potential to be a revolutionary advance for petroleum geochemistry, because it offered a much more viable alternative than geocatalysis in explaining the low apparent-activation energies and pre-exponential factors that were often reported in pyrolysis work on fossil fuels (e.g. Huck and Karweil, 1955; Cummins and Robinson,

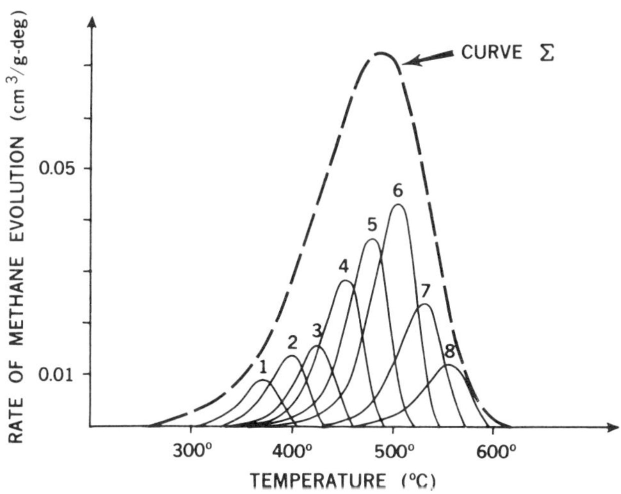

FIG. 4. Rate of methane evolution from coal for eight parallel reactions as a function of pyrolysis temperature during programmed-temperature pyrolysis. Each numbered curve represents one of the parallel reactions. A Gaussian distribution is assumed for the initial concentrations of the eight starting materials. Activation energies assumed, in kcal mol^{-1} (kJ mol^{-1}), are: (1) 48 (201); (2) 50 (209); (3) 52 (217); (4) 54 (226); (5) 56 (234); (6) 58 (242); (7) 60 (251); (8) 62 (259). *A*-factors are 10^{15} min^{-1} ($10^{13}\ s^{-1}$) for each reaction. Curve Σ, representing the sum of the eight parallel reactions, has a pseudo-activation energy of 20 kcal mol^{-1} and an *A*-factor of 10^4 min^{-1} ($2 \times 10^2\ s^{-1}$). After Jüntgen and Klein, 1975. *Reprinted with permission.*

1972; Braun and Rothman, 1975) and from empirical studies on petroleum formation (Connan, 1974). Unfortunately, Jüntgen's and Klein's work has largely been ignored by petroleum geochemists because it dealt with methane formation from coals, rather than with petroleum formation, and because it was published in German.

Not everyone is in agreement about the effects of time on oil-generative processes, however. Some opinions that time is of minimal importance are discussed below under "Models for thermal generation of petroleum". Recently Price (1982) showed that in undisturbed areas vitrinite reflectance correlates better with present temperature than with burial time. He also noted that very young samples from a geothermal area plotted along with much older samples, and thus concluded that maturation occurs in a geological instant.

Some of Price's reasoning appears to be circular, however. By choosing samples from tectonically undisturbed regions he is essentially requiring that present-day temperatures be the maximum temperatures achieved. A good correlation between R_0 and present-day temperatures would therefore be expected. Furthermore, as we shall see later, using sediment age as the cooking time is a poor approximation. Price thus represents a minority viewpoint; the great majority of workers believe in a traditional kinetic description for oil-generative reactions.

D. Kerogen type

Although earlier workers (e.g. Snider, 1934; Forsman and Hunt, 1958; Forsman, 1963) had suggested that the kinetics of oil generation are different for different types of kerogen, Louis and Tissot (1967) were perhaps the first to attribute the different temperature ranges for oil generation in different basins at least in part to different types of organic matter. Welte (1972) and Connan (1974) suggested that different types of organic material have different pseudo-activation energies for oil formation.

In 1975 Tissot and Espitalié described each of the three principal kerogen types by a distribution of activation energies ranging from 10 to 80 kcal mol^{-1} (42 to 334 kJ mol^{-1}) (see Fig. 5). They noted that Type I kerogen has the highest "average" pseudo-activation energy because it contains mainly cross-linked aliphatic chains that require cleavage of strong carbon–carbon bonds in order to produce bitumen. Type II kerogen has a lower average pseudo-activation energy, because it comprises a substantial proportion of weaker C–N, C–S and C–O bonds.

According to Tissot and Espitalié, and Vandenbroucke *et al.* (1976), Type III kerogen begins to generate hydrocarbons even earlier than does Type II kerogen, but reaches its maximum generation rate rather late, as does Type I kerogen (see Fig. 6). Because all the experiments that led to these conclusions were carried out at elevated temperatures during programmed-temperature pyrolysis, application to petroleum-generative processes in natural settings requires extrapolation of the results.

Other workers (e.g. Kartsev *et al.*, 1971) have come to opposite conclusions about oil-generation thresholds for different kerogen types. Powell *et al.* (1978) showed that amorphous and resinous (Type II) kerogens begin to generate oil at R_0 values of 0.5%, whereas terrestrial (Type III) kerogens commence later, at $R_0 = 0.7\%$. Welte and Yükler (1981) agreed with

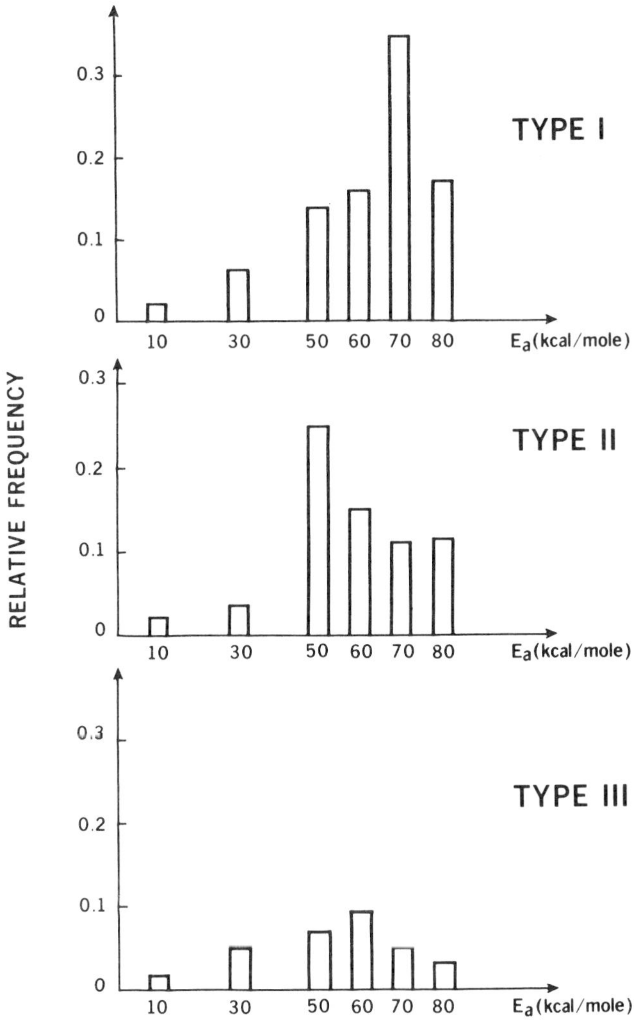

FIG. 5. Distribution of activation energies within each of the three principal kerogen types (see Fig. 3), according to Tissot and Espitalié, 1975. Relative numbers of bonds having each activation energy are represented along the vertical axis. *Reprinted with permission.*

Powell and co-workers. This apparent contradiction of the results of Tissot and Espitalié (1975) could be explained by different definitions of "threshold." Although Fig. 6 shows that in an absolute sense generation begins in Type III kerogen earlier than in Type II, if the threshold for "effective" or "measurable" generation is raised slightly (to value G in Fig.

6, for example), this latter threshold is attained earlier in Type II kerogen. The concept of a threshold for "effective" generation has often been proposed in recent years (e.g. Momper, 1978), and is probably valid.

Leythaeuser *et al.* (1980) claimed, on the basis of changes in fluorescence intensity, that the threshold for the onset of oil generation increases in going from Type I kerogen to Type II to Type III. The discrepancy between their results and those of other workers may be a result of different methodologies, and may in fact represent an important advance, because

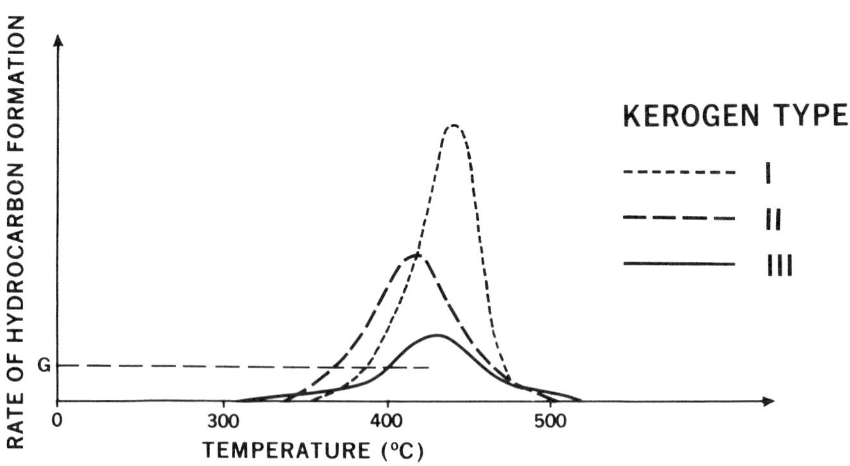

FIG. 6. Rate of hydrocarbon formation by programmed-temperature pyrolysis of kerogens of Types I, II and III. Time and temperature both increase to the right. From Tissot and Espitalié, 1975. *Reprinted with permission.*

fluorescence increases are probably more directly related to petroleum formation than are changes in reflectance of vitrinite particles (Radke *et al.*, 1980).

Despite the gradual accumulation of evidence suggesting that the kinetics of the various kerogen types are distinct, most workers (e.g. Dow, 1977; Waples, 1980) have ignored, either explicitly or implicitly, differences in oil-generation rates. This approach is undoubtedly an oversimplification that should be eliminated by further research. Definitive data on this problem are badly needed now.

E. Pressure

The role of pressure in oil generation has never been rigorously considered. McCoy proposed long ago (1919) that pressure helps to transform solid kerogen to liquid petroleum. Subsequent discussions took coalification as a model, and assumed that if pressure aided coalification (an assumption that was probably incorrect), it should also facilitate oil generation (Bordenave *et al.*, 1970). This simplistic approach was superseded by more sophisticated considerations of the effects of pressure on petroleum formation. It was assumed that petroleum is formed by thermal decomposition of large kerogen molecules, and therefore that the most important consideration was possible retardation of kerogen decomposition by the buildup of internal pressure from volatile light hydrocarbons. Low pressures were therefore believed to favour kerogen decomposition (Louis and Tissot, 1967), but were not thought to be nearly as important as temperature.

In recent years, as microfracturing has come into favour as a possible important mechanism for primary migration (Tissot and Welte, 1978, pp. 282–295, and references cited therein), buildup of internal pressure during oil generation has come to be recognized as a necessity. The question of possible retardation of oil generation by high pressures has thus become moot. Efforts to find an influence of thrust pressure on oil generation have not been successful (Kalkreuth, 1979). The main remaining interests in pressure influences on petroleum formation seem to be whether generation is retarded in over-pressured shales (Cecil *et al.*, 1977), how pressure influences subsurface temperatures (Welte and Yükler, 1981), and what precise conditions are required for initiation of microfracturing.

V. Destruction of Hydrocarbons

From an explorationist's point of view it has always been more important to know where oil might be found than to understand its origin. In 1915, in an effort to avoid drilling in areas that could not possibly yield liquid hydrocarbons, White proposed his famous "carbon-ratio theory," which stated that commercial oil deposits would not be found where fixed-carbon contents of coals exceeded 65 or 70%. His hypothesis was based on empirical studies of liquid hydrocarbon occurrences. Because in White's opinion fixed-carbon contents of coals were dependent upon frictional heat generated during thrusting, oil occurrences were limited by orogenic activity. White and other workers subsequently enlarged and modified the original scheme (White, 1921, 1935; Fuller, 1919; Russell, 1925, 1927; Hume, 1927; Heck, 1943; Teichmüller, 1958, 1974; Brooks, 1970).

By the 1930's it had been hypothesized by many geologists that light crude oils are formed by thermal cracking of initially heavy oils (e.g. Fuller, 1919; White, 1921, 1935; Pratt, 1934; Barton, 1934, 1937). This hypothesis, which became a part of the carbon-ratio theory, was not without its opponents, however (e.g. Washburne, 1919; Tarr, cited by Lockwood, 1925), in part because the correlation between fixed-carbon content and oil gravity was very imperfect in some areas (McCoy, 1921; Hume, 1927; Russell, 1927; Jones, 1928). Dorsey (1927) acknowledged that the carbon-ratio theory worked, but suggested that facies variations in depositional environments accounted better for its efficacy than did thrusting.

These objections proved minor, however, and in the intervening years additional evidence has been amassed in support of the carbon-ratio theory. Many workers (e.g. Dobryansky, 1963; King *et al.*, 1963; Landes, 1967; Albrecht and Ourisson, 1969; Milner *et al.*, 1977) have shown that liquid bitumens and petroleums change in composition and gravity as a result of thermal-cracking reactions in the subsurface, and can eventually be destroyed in extreme cases. Because the thrust-pressure explanation for coalification has now been discarded, the basic correlation that White originally implied between sediment temperature and oil preservation is instead seen as a function of normal subsurface heating over geological time.

Methods of measuring coal rank have changed since White's time, and fixed-carbon contents are now determined much less frequently. Adoption of new methods of determining coal rank has permitted extension of the carbon-ratio theory to sediments that lack coal beds. Spore colour has been used to define oil deadlines (Correia, 1967), but today vitrinite reflectance has supplanted other techniques by virtue of its great convenience (Bostick and Damberger, 1971). Various workers have proposed oil preservation deadlines at vitrinite reflectance values (in oil) from 1.0% (Bostick and Damberger, 1971) to 1.35% (Cornelius, 1975a). If a dry-gas preservation deadline exists, it probably lies at vitrinite reflectance values of about 3.0% to 3.2% (Bostick and Damberger, 1971; Dow, 1977), although the exact value is poorly known. Dry-gas preservation may in fact be more dependent upon the presence of oxidizing agents like elemental sulphur that could help destroy methane (Orr, 1974; Hunt, 1975; Milner *et al.*, 1977; Barker and Kemp, 1982) than on temperature alone.

It has sometimes been popular to fix hydrocarbon-preservation deadlines in terms of subsurface temperature (e.g. Landes, 1967; Hunt, 1975; Milner *et al.*, 1977). Such schemes have the virtue of simplicity because subsurface temperatures are readily predictable from measured or estimated geothermal gradients, but because they ignore the effects of time, these models are subject to considerable error when reservoirs of greatly different ages are compared. Destruction of hydrocarbon mixtures by cracking reactions has long been known to follow first-order kinetics with measured pseudo-activation energies of about 50 kcal mol^{-1} (about 200 kJ mol^{-1}) (McNab *et al.*, 1952). It is therefore necessary to consider both time and temperature effects in establishing hydrocarbon-preservation deadlines. Some models that partially address this question are discussed in the next section.

VI. Models for Thermal Generation of Petroleum

Through the use of spore colouration, vitrinite reflectance and a host of other less precise or less common techniques, it has become possible to estimate the degree to which the petroleum-generative process has progressed in strata from which samples are available (Héroux *et al.*, 1979). The same parameters are also employed to determine whether preservation deadlines for liquid or gaseous hydrocarbons have been reached. Whether or not these various techniques are really adequate measures of hydrocarbon generation and destruction is being debated. Most measure changes that are assumed to occur simultaneously with petroleum formation but which are distinct processes chemically. Their kinetic dependences may therefore be very different from those of petroleum formation and destruction. Although these accomplishments have been very important to petroleum explorationists working in areas where samples are readily available, they have been of limited value in frontier areas. Furthermore, measured maturity values give no direct information about timing of oil generation, a question of critical importance in many cases.

It is therefore highly desirable to be able to predict the progress of thermal evolution of sediments through time. The earliest model for thermal maturation, developed in 1956 by Karweil, was designed to predict coal rank. Models for petroleum generation were first published a decade later (Louis and Tissot, 1967). Since then the sophistication and value of newer models have gradually increased through time, although progress has been a little erratic. Models for coalification and petroleum generation are fully compatible with each other; each assumes that time, temperature, or both are important factors, and that effects of pressure are negligible. Some models have also taken into consideration variations in the type of organic matter present. Time–temperature models can be divided into several groups, based on the ways they take time and temperature into consideration (Nagornyi and Nagornyi, 1974; Karpov *et al.*, 1975; Wright, 1980). Each type of model will be discussed below.

A. Time effects

1. *Negligible effect of time*

The simplest models, promoted particularly by Neruchev and Parparova (1972a, b), assume either implicitly or explicitly that time is not a factor in oil generation. This assumption allows the establishment of strict minimum temperatures, below which petroleum formation does not occur. Numerous workers (e.g. Cox, 1946; McNab *et al.*, 1952; Vassoevitch *et al.*, 1967; Kartsev *et al.*, 1971; Pusey, 1973; Cornelius, 1975a; Milner *et al.*, 1977) have assumed an oil-generation threshold temperature of 60 to 65°C, although some have also admitted that time does play a role. Pusey (1973), for example, claimed that his temperature threshold was applicable to rocks heated for periods ranging from 1 million to 200 million years. Albrecht and Ourisson (1969) and Wright (1980) proposed a slightly higher threshold temperature (70°C), whereas Kontorovich *et al.* (1967), Lopatin (1971, 1976), Lopatin and Bostick (1973), and Buntebarth (1979) believed that the temperature threshold lay at or below 50°C. Landes (1967) has often been cited (e.g. Teichmüller, 1974) as favouring a minimum temperature of 90°C for oil generation, but the cited reference contains no such statement. Rogers (1860), Brooks (1970), Lopatin and Bostick (1973), Bartenstein and Teichmüller (1974), Demaison (1975), and Price (1982) have claimed that coal rank (or kerogen maturity) is a maximum-reading thermometer, a statement tantamount to saying that only temperature affects maturity and claiming that any temperature below the maximum temperature is completely inconsequential. Nagornyi and Nagornyi (1974) asserted that "geological time does not limit the coalification process".

Many early authors either did not state the basis for making such judgments, or used reasoning that is not defensible today. Cox (1946), for example, assumed that temperatures in petroliferous basins rarely exceed 100°C, with gradients averaging about 1°C per 100 ft (about 3°C per 100m), and that petroleum is formed at depths of about 5000 feet (about 1500m). He did not explain his choice of 5000 feet for oil generation, but given the poor understanding in 1946 of the mechanism of petroleum formation, of migration distances, and of the oil potential deep within superficially explored basins, his conclusions were based on little more than guesswork. Given the limited data and knowledge of oil-generative processes, It is surprising that the early workers came as close to the truth as they did.

Such pronouncements about absolute temperature thresholds are of some value in the broad sense of establishing an approximate boundary for oil generation (Gretener, 1981), but are of little value as predictive tools, because there are so many exceptions. For example, one of the earliest studies of oil generation (Philippi, 1965) showed clearly that young petroliferous basins generate oil at higher temperatures (115–140°C) than the rule of thumb proposed by earlier workers would predict. It became clear very early, by comparing the temperatures of oil generation in California from Philippi's study with those from the much older Paris Basin

(60–65°C) (Louis and Tissot, 1967), that time is a factor to be reckoned with in developing models for oil generation.

2. *Geological age as cooking time*

Long before the first models for maturation of organic matter in sediments were constructed, it was recognized by most workers that time is an important factor in such transformations. Connan's empirical study (1974) on the kinetics of oil generation used formation ages as their cooking time, and the maximum palaeotemperature as the cooking temperature. He was well aware that this assumption introduced errors into his method, but hoped that the errors in each case history would cancel each other.

Connan's model works adequately if the histories of the basins under consideration are simple and similar. In complex tectonic regimes, however, especially those involving Palaeozoic rocks, the deficiencies of the model became apparent (Waples, 1976).

Young *et al.* (1977) have suggested that the geological age of the source beds for oils can be estimated from gasoline-range hydrocarbon compositions of the oils. Their equations take both time and temperature into consideration, but do not distinguish kinetically between reactions that form hydrocarbons and those that destroy them. Furthermore, their method of estimating the age of an oil is only valid if the source and reservoir beds are of the same age and are stratigraphically close. It is possible using their model to determine which oils have migrated vertically, but in such cases ages of oils cannot be determined unless the timing of migration is known.

The pseudo-activation energies calculated by Young and co-workers for hydrocarbon cracking are exceedingly low (3–6 kcal mol^{-1} (12–25 kJ mol^{-1}) for clastic reservoirs and 3–20 kcal mol^{-1} (12–84 kJ mol^{-1}) for carbonate reservoirs) *A*-factors are also very low. Failure to consider the kinetics of hydrocarbon-forming and -destroying reactions separately may have been responsible for these unusual parameters.

3. *Effective heating time*

Lopatin (1969a, b) showed clearly, through the use of burial-history plots (Fig. 7), that heating effects are not constant throughout a rock's history. For example, the oldest rocks depicted in Fig. 7 (300 M.Y.: lowermost burial history line) spent their first 80 M.Y. at temperatures far below their maximum palaeotemperature. The cooking that occurred during the first 80 M.Y. is therefore much less important than that which took place later.

In an effort to take this factor into consideration, Lopatin (1969b) was the first to attempt to define an "effective heating time" for oil generation. He observed a good correlation between measured thermal maturity and the length of time spent at temperatures above 100°C. Cornelius (1975a, b) distinguished between "exposure time" (time spent at the maximum palaeotemperature) and "cooking time" (time spent reasonably near the maximum palaeotemperature), but did not define clearly his methodology for utilizing these time spans to calculate thermal maturity.

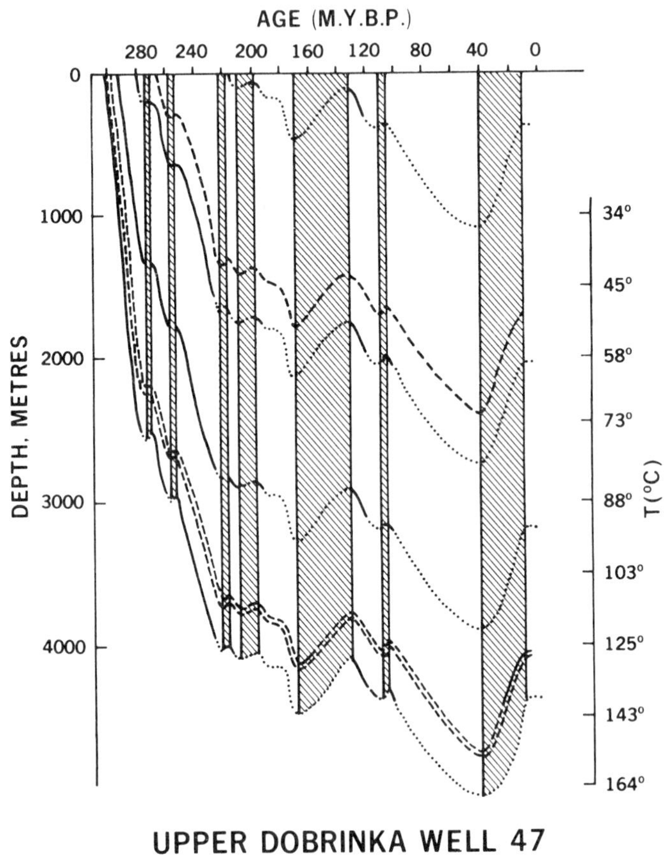

FIG. 7. Burial-history plot for Palaeozoic strata from upper Dobrinka Well 47. Horizontal scale is in millions of years; vertical scale is in metres. Each line represents the complete burial history of a single horizon, from the time of deposition (depth = 0) to the present day (time before present = 0). From Lopatin, 1969a. *Reprinted with permission.*

Hood and co-workers (1975) defined the effective-heating time of a rock as that period spent within 15°C of the rock's maximum palaeotemperature. They created a scale of thermal maturity called "level of organic metamorphism" (LOM) (see Fig. 8). They also prepared a diagram that permitted prediction of LOM values using only the effective time and maximum palaeotemperature (Fig. 9). Their choice of a 15°C range for effective heating was a reflection of the very rapid increase in rate of oil generation

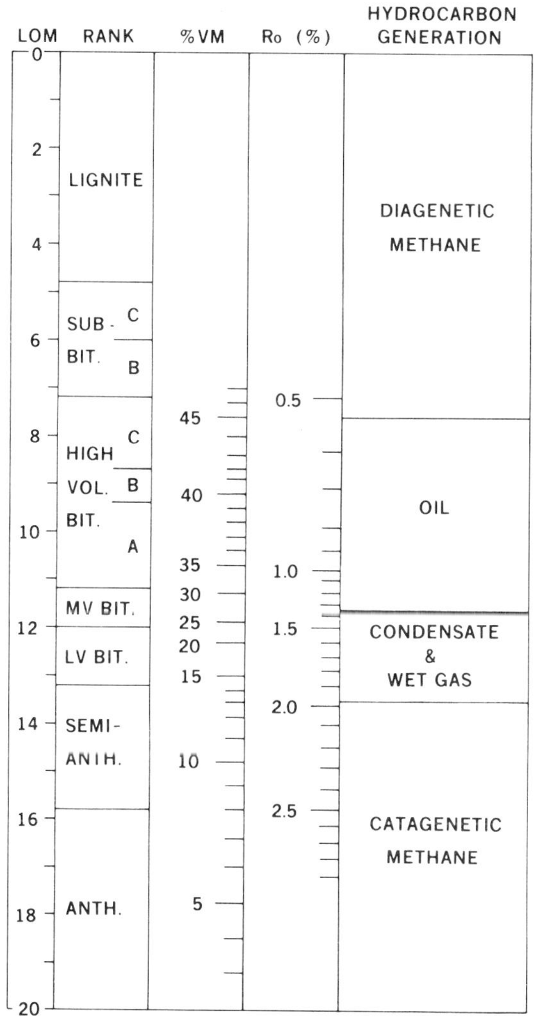

FIG. 8. Level of organic metamorphism (LOM) as defined by Hood *et al.* (1975). *Adapted with permission.*

with increasing temperature. Because of the simplifying assumptions employed in calculating effective heating time, Hood and co-workers cautioned against putting too much faith in calculated LOM values. They

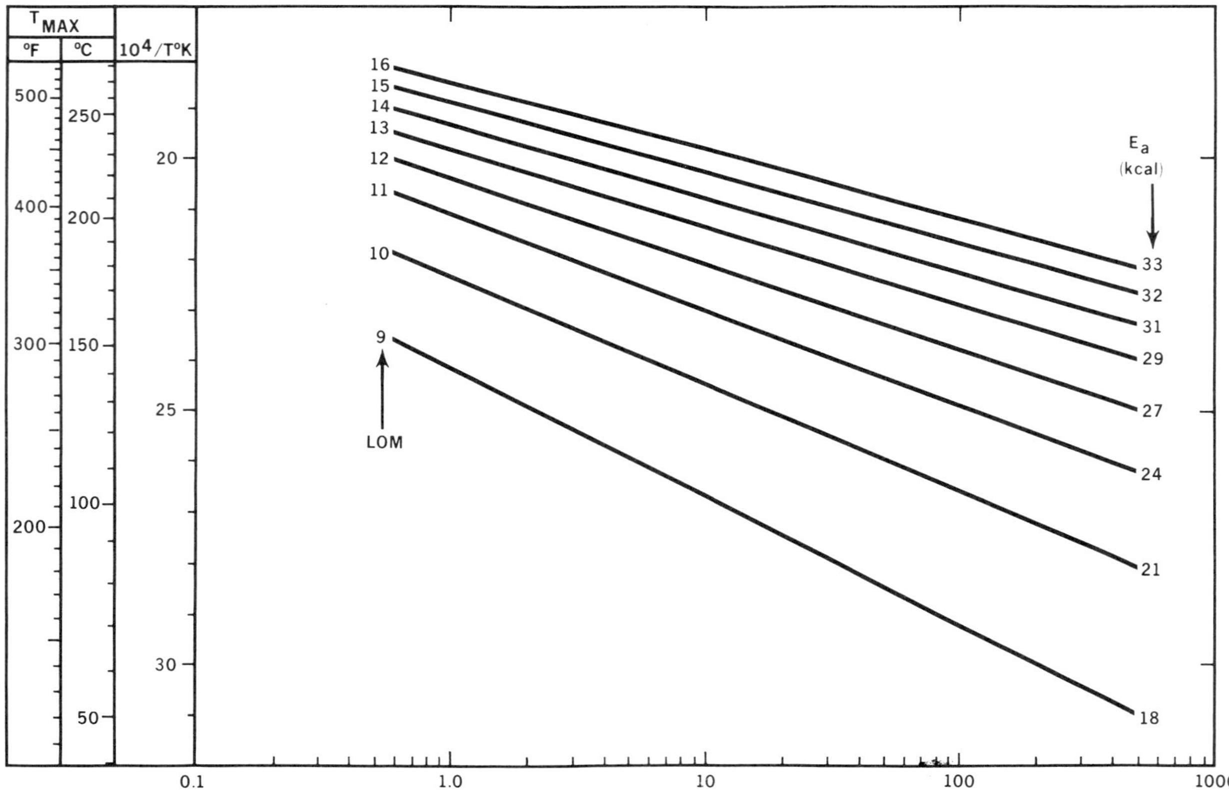

FIG. 9. Diagram for calculating level of organic metamorphism (LOM) for maximum palaeotemperature (expressed three ways on the vertical scale) and effective heating time (time spent within 15 °C of maximum palaeotemperature) on the horizontal scale. Note that pseudo-activation energies increase as LOM increases. From Hood *et al.*, 1975. *Reprinted with permission.*

preferred to limit their predictions to extrapolations within sections where some measured data already existed. Tests of their model in both Palaeozoic and Mesozoic basins gave reasonably good correspondences with measured data.

The main weakness with the "effective-heating time" model, of course, lies in its inability to consider a rock's complete thermal history. In many cases, particularly in old rocks, important thermal maturation may have occurred at temperatures more than 15°C below the maximum palaeotemperature. The next generation of models was able to take a rock's entire thermal history into account.

4. *Total-thermal-history models*

Karweil's model (1956) was capable of taking the complete burial and thermal history of a coal into account in calculating its maturity. He explained how one can sum the effects of several time–temperature regimes and calculate a coal's cumulative maturity using the conversion factor Z (see Fig. 1). Karweil's scheme was developed long before thermal conversion of kerogen to oil was understood, however, and was intended to predict coal rank. It has therefore never been popularly applied in predicting oil generation. Karweil's method has been criticized for overestimating the effect of time (Nagornyi and Nagornyi, 1974; Karweil, 1975; Kettel, 1981).

In 1969 Tissot published the first mathematical model for oil generation that used the entire thermal history of a sediment in predicting its oil-generative history. The model he used was quite sophisticated; it employed several reactions, linked both in parallel and in series, to define the oil-generative process. Kinetic parameters were selected to give the best fit with empirical data.

Tissot's work gave an excellent correspondence between measured and predicted values for twenty-three wells from the Paris Basin. The mathematical apparatus he used was rather formidable, however, and little effort was expended to make his model available to geologists as a predictive tool. Tissot's model therefore has been used mainly by his French colleagues (e.g. Deroo *et al.*, 1969; Tissot *et al.*, 1975, Tissot and Espitalié, 1975; Du Rouchet, 1980).

Deroo and co-workers brought the question of timing of oil generation to the fore by modelling bitumen formation in the Paris Basin as a function of time (see Fig. 10).

Application of Tissot's model in other basins showed clearly how strongly the thermal history of sedimentary rocks influences the timing and quantity of hydrocarbons generated (see Fig. 11).

Subsequent models developed by other workers sacrificed some of

Tissot's rigour, but in doing so more than compensated by creating better geological frameworks. A model developed by Bostick (1973) also took into account the contributions to thermal maturation of time spent at the maximum burial temperature as well as time spent at lower temperatures. Bostick even showed that the greater the difference between the maximum temperature and the temperature of any cooler period, the smaller would

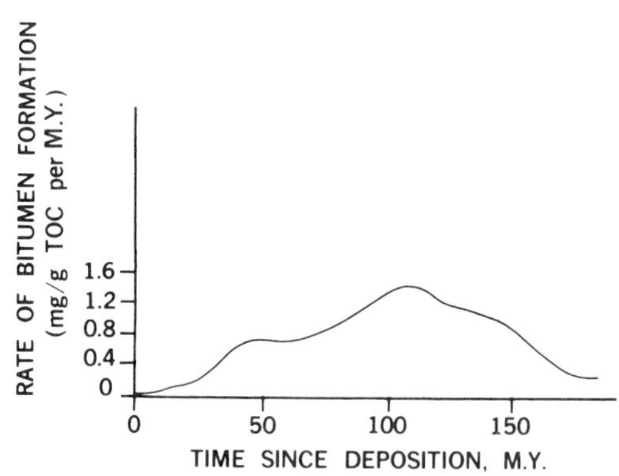

FIG 10. Rate of bitumen formation in the Toarcian Shale at Essis, Paris Basin, predicted from Tissot's model (1969). From Deroo *et al.*, 1969. *Adapted with permission.*

be the effect of the lower temperature. He produced a diagram (Fig. 12), derived from Karweil's nomograph (1956), for estimating vitrinite reflectance values from the time and temperature history, and showed a very good correlation between measured and predicted values for several samples.

Shortly after publishing the first "effective-heating time" model, and almost simultaneously with Bostick (1973), Lopatin (1971) published a model capable of considering the complete thermal history of a rock in a very simple way. Like Karweil's original nomograph (1956) for predicting coal rank, Lopatin's idea was revolutionary, because it provided a much more convenient method for combining the maturation effects of time and temperature. Bostick immediately saw the advantages of Lopatin's approach over his own (Lopatin and Bostick, 1973). Lopatin's method, which developed from Karweil's work, has been explained in detail elsewhere (Waples, 1980; Cohen and Waples 1981), but a brief synopsis is appropriate here.

Starting with a burial-history diagram and a superimposed grid of isotherms that represent the temperature history of the sedimentary section (see Fig. 13), one can calculate the thermal maturity of any rock layer at

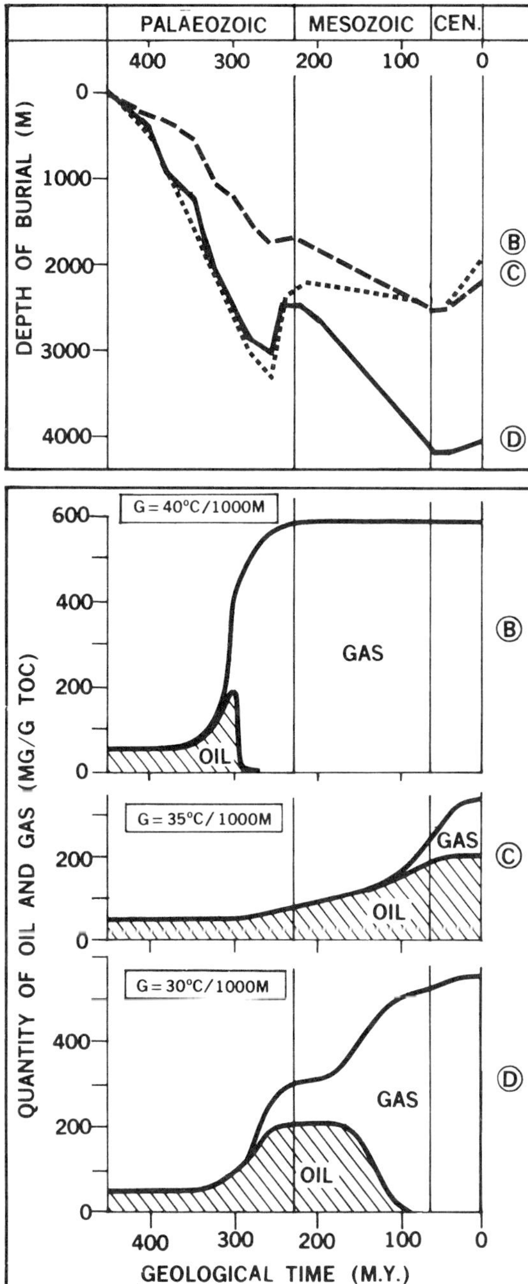

FIG. 11. Burial history and hydrocarbon generation in Palaeozoic rocks at three locations in the Illizi Basin, Algeria. Hydrocarbon generation predicted using the model of Tissot (1969). From Tissot *et al.*, 1975. *Reprinted with permission.*

the present or at any time in the past by merely summing the amount of thermal exposure attained in discrete time intervals in the past (see Waples 1980 for an example). In Lopatin's original scheme (1971) each 10°C temperature interval is assigned a temperature factor (γ) based on consideration of the kinetics of oil generation. This γ value is then multiplied by that temperature interval's time factor, which is merely the length of

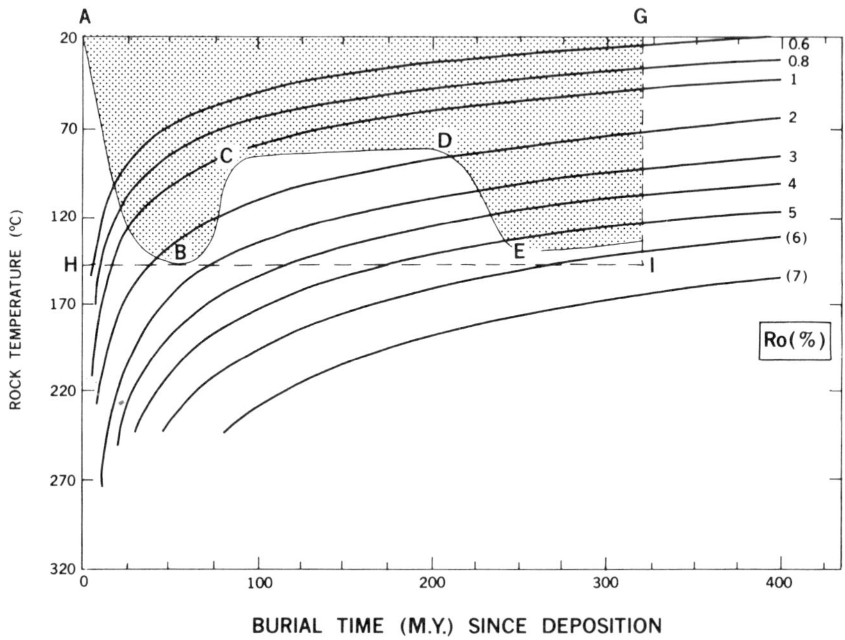

FIG. 12. Diagram for predicting vitrinite reflectance values from thermal history of sediments. Take a ratio R between the area above the sediment's thermal-history curve (stippled area) and the rectangular area AHIG created by assuming immediate burial to the sediment's maximum palaeotemperature followed by maintenance of that temperature to the present. The actual maturity is calculated by reducing the maturity that would have been achieved along burial-history curve AHI (6.3%) by the factor R. From Bostick (1973). *Adapted with permission.*

time, in million years, that the rock spent within that temperature interval. The sum of these products for all temperature and time intervals of interest gives the "Time–Temperature Index" of maturity (TTI) (Lopatin, 1971), or "Sum Heat Impulse" (Lopatin and Bostick, 1973). Mathematically stated,

$$\mathrm{TTI} = \sum_{i=1}^{n} (t_i)(\gamma_i) \tag{2}$$

where t_i is the length of time (in million years) spent by the rock in the ith temperature interval, and γ_i is the temperature factor for that temperature interval.

The value of γ determines the relative importances of time and temperature in oil generation. Following an old chemical rule of thumb, Lopatin assumed $\gamma = 2$, meaning that the rate of oil generation doubles

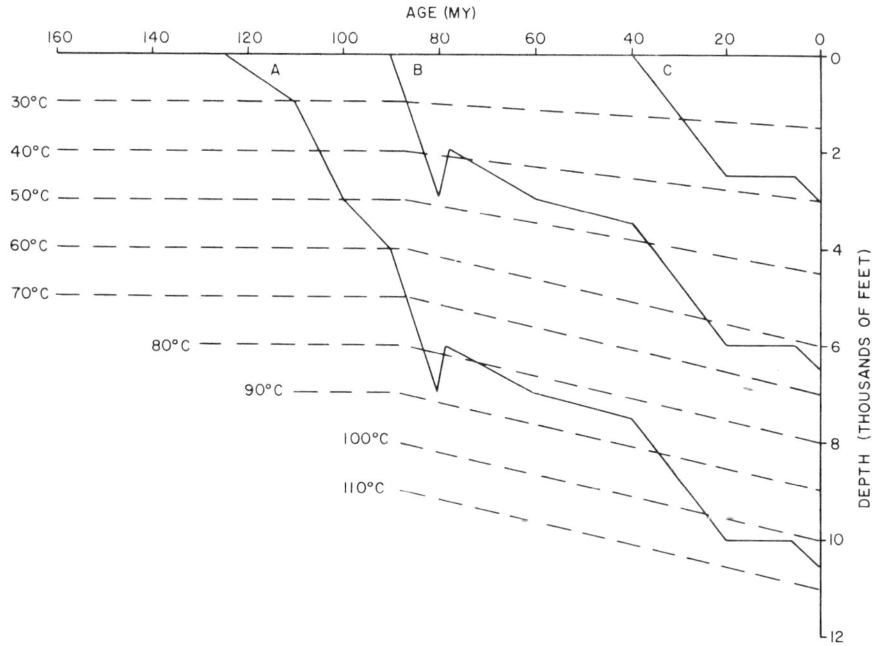

FIG. 13. Illustration of Lopatin's method for considering the complete thermal history of sedimentary rocks. Burial-history curves for three horizons (A, B, C) in a hypothetical sedimentary sequence are solid lines; superimposed temperature grid is represented by dashed lines. From Waples (1980). *Reprinted with permission.*

with every 10°C rise in temperature. To avoid dealing with inordinately large numbers, he arbitrarily assigned the 100–110°C temperature interval a relative γ value of 1. The 80–90°C temperature interval thus has $\gamma = 1/4$, whereas 120–130°C has $\gamma = 4$.

Lopatin tested and calibrated his model one one of the most difficult wells imaginable, the Münsterland-1 well in West Germany, drilled in Palaeozoic rocks burdened with a very complicated tectonic and thermal history. Although the results of his testing showed a very high internal consistency, the absolute accuracy of his calibration was questioned by several workers on the basis of his geological reconstruction and thermal history (Neruchev and Parparova. 1972a; Golitsyn, 1973; Nagornyi and Nagornyi, 1974; Karpov *et al.*, 1975; Karweil, 1975; Waples, 1980). Recalibration of Lopatin's method with larger and more reliable data sets (Waples, 1980; Kettel, 1981) has verified the validity of the model itself.

Lopatin and Bostick (1973) and Lopatin (1976) later suggested some improvements in the scheme, and the approach of Gretener (1981) and Gretener and Curtis (1982) was essentially that of Lopatin, although cosmetic changes were made. Lopatin (1976) used fewer and larger temperature intervals (15°C to 30°C instead of 10°C), and created a nomograph to minimize mathematical manipulations. The logic in making this change is discussed in the next section. Lopatin's nomograph is probably useful for hand calculations, but because both construction of the burial and thermal history of a sedimentary section and calculation of TTI values can easily be computerized (Waples, 1980), the nomograph is unnecessary. In any case, using larger temperature intervals reduces accuracy.

One weakness of Lopatin's and Karweil's total-thermal-history models in application to the problem of oil generation is that they were developed and calibrated for transformations of solid organic matter (coals and vitrinitic kerogens), rather than for generation of liquid or gaseous hydrocarbons. Waples (1980) tried to extend this methodology to hydrocarbon generation, but further work is necessary to establish more securely the correlation of TTI values with the various stages of hydrocarbon generation and destruction.

B. Temperature effects

One can choose among several ways of interrelating time and temperature in total-thermal-history models for oil generation. One obvious way to consider the effect of temperature on rates of oil generation is to maintain a constant pseudo-activation energy (e.g. Du Rouchet, 1980). As temperature increases with increasing depth of burial, the ratio of the oil-generation rate at the new, higher temperature (T_h) to that at the previous, lower temperature (T_l) decreases, as shown by dividing the Arrhenius equation at the higher temperature by that at the lower:

$$\gamma = \frac{k_h}{k_l} = \frac{A\exp(-E_a/RT_h)}{A\exp(-E_a/RT_l)}. \tag{3}$$

Simplification yields

$$\gamma = \exp\{E_a(\Delta T)/RT_lT_h\}. \tag{4}$$

If the activation energy and temperature increment (ΔT) are held constant as the temperature increases, T_l and T_h both increase, leading to a gradual decrease in γ. Because γ depends on ΔT, it is important to specify the size of the temperature intervals used. A 10°C interval has been employed by most workers thus far, and, except as noted, all discussions below of γ assume $\Delta T = 10$°C.

Golitsyn (1973), Lopatin and Bostick (1973) and Lopatin (1976) have suggested that γ decreases with increasing temperature. Golitsyn used $\gamma = 1.4$ up to 60°C, $\gamma = 1.3$ to 100°C, and $\gamma = 1.2$ to 150°C. The pseudo-activation energy required for his model would be about 7 kcal mol^{-1} (29 kJ mol^{-1}), a value much lower than that determined experimentally or empirically for oil generation by any worker. His justification for the low E_a apparently was his belief that oil-generation kinetics are diffusion-controlled. Tissot (1969), in contrast, proposed that oil generation could be satisfactorily modelled using a single pseudo-activation energy of 50 kcal mol^{-1} (209 kJ mol^{-1}). The much higher γ values (2.3 at 60°C, 1.8 at 100°C, and 1.4 at 150°C) implied by Tissot's higher pseudo-activation energy mean that in Tissot's model oil generation is much more sensitive to temperature than in Golitsyn's.

In a later publication Lopatin (1976) favoured a constant pseudo-activation energy of 10 kcal mol^{-1} (42 kJ mol^{-1}) for petroleum formation, but in contrast to Golitsyn elected to maintain a constant $\gamma = 2$. Equation (4) shows that under these conditions ΔT must increase as the reaction temperature increases. Lopatin used 15°C ΔT intervals from 50 to 80°C, gradually increasing to 30°C intervals from 170 to 230°C. There is no fundamental difference between Lopatin's and Golitsyn's approaches; in both cases E_a is constant. Whether γ or ΔT is maintained constant is a matter of computational convenience.

Other investigators have chosen to maintain a constant γ rather than E_a. If γ and ΔT remain constant, it is evident from Equation (4) that as temperature increases, the pseudo-activation energy must decrease. After testing several γ values against empirical data on thermal maturity, Waples (1980) concluded that no better value than Lopatin's original one ($\gamma = 2$) could be selected on the basis of his data. A γ value of 2 implies that pseudo-activation energies increase from approximately 12 kcal mol^{-1} (50 kJ mol^{-1}) at 25°C to 20 kcal mol^{-1} (84 kJ mol^{-1}) at 100°C to 26 kcal mol^{-1} (109 kJ mol^{-1}) at 160°C. Kettel (1981) favoured $\gamma = 1.6$ on the basis of data from four wells near the Bramsche Massif in north-west Germany, but his study lacks a strong statistical base.

On the basis of empirical and experimental evidence from coal- and oil-shale pyrolysis, the hypothesis of Golitsyn (1973) and Du Rouchet (1980) that E_a remains constant during oil generation appears untenable (Karpov *et al.*, 1975). The models of Lopatin (1971) and Waples (1980) are in some senses consistent with experimental evidence, but assert that the increase in activation energy is a function of temperature alone. All experimental evidence from pyrolysis work, however, indicates that pseudo-activation energy increases with increasing maturity, rather than with temperature. Two initially identical kerogens that are now at the same levels of thermal maturity should have the same pseudo-activation energy for further maturation, regardless of the temperatures at which maturation is occurring.

Hood *et al.* (1975) realized this, and related the increase in pseudo-activation energy to LOM values, rather than to temperature (see Fig. 9). Pseudo-activation energies range from about 15 to 26 kcal mol^{-1} (63 to 109 kJ mol^{-1}) within the oil-generation window. If the approach of Hood and his co-workers is accepted, then γ must be expressed as a function of thermal maturity. Furthermore, the exact value of γ and its dependence on maturity will be different for each kerogen type.

Golitsyn (1973) seems to have recognized the dependence of γ on maturity. Although his model was based on γ as a function of temperature, like Karweil (1956) he saw a need to treat uplifted rocks differently from those that had suffered continuous subsidence. He proposed that after uplift the effect of lowered temperatures was even less than that predicted using γ values. He therefore introduced a correction factor

$$a = T/T_{\text{max}} \tag{5}$$

that reduced the cooking effect of post-uplift temperatures even further. This correction is equivalent to saying that after the maximum temperature has been reached, the temperature dependence of maturation increases (γ increases). An increasing temperature dependence with increasing maturity means that the pseudo-activation energy also increases.

Although Golitsyn's idea is qualitatively similar to that of Hood *et al.* (1975), its application seems to be too capricious to be theoretically satisfying. For example, his correction factor is calculated using Celsius temperatures, rather than absolute temperatures. Furthermore, the magnitude of the correction factor is not a consistent function of maturity.

Recently Toth *et al.* (1982) have put forth quite a different model for the role of temperature in increasing vitrinite reflectance values. All other workers have proposed that because pseudo-activation energies for organic-maturation reactions are reasonably high (of the order of 10–50 kcal mol^{-1} (50–200 kJ mol^{-1})), the slight temperature dependence of the frequency factor A in the Arrhenius equation (Equation (1)) is negligible. If, however, the temperature dependence of A is considered in the form

$$A = \beta \exp(T/T_{\text{D}}) \tag{6}$$

then by combining Equations (1) and (6) the rate constant k can be expressed as

$$k = \beta \exp\{(-E_a/RT) + (T/T_D)\} \tag{7}$$

where β and T_D are constants. Toth and co-workers found that the best empirical fit to data from immature samples of widely varying geological ages and geographical distribution was given by an E_a value of about 0.05 kcal mol^{-1} (0.2 kJ mol^{-1}), and $T_D = 438$ K (165°C). More recent work has shown that T_D is closer to 200 K, and that imposing a threshold temperature of 22°C greatly improved the fit (Lerche, personal communication, 1982).

The pseudo-activation energy determined by Toth *et al.* is far lower than those proposed by any other workers, and, if correct, indicates that the rate-limiting step in increasing vitrinite reflectance in low-maturity rocks is some process, such as changing van der Waals forces or hydrogen bonding between molecules, that has a very low activation energy.

Mathematical reduction of Equation (7) under normal subsurface temperature conditions indicates that in fact when E_a is very small, the T/T_D term dominates the exponential expression, and that the temperature dependence of the rate constant k is very low. Lerche (personal communication) augmented the temperature dependence of vitrinite-reflectance changes by claiming that

$$R_o^{1/2} = \text{constant} + \int k(T)\,dt. \tag{8}$$

Nevertheless, the temperature dependence of reflectance increases in the model of Toth, Lerche, *et al.* is still lower than other workers have predicted or found. Further work is necessary to discover the reasons why models exhibiting both strong and weak temperature dependences seem compatible with the empirical data of different workers.

Application of the data of Toth *et al.* to oil generation is not straightforward, because there is no necessary causal link between vitrinite-reflectance changes and oil generation. Their published work was mainly concerned with reflectance changes at very low maturity levels ($R_o = 0.3$–0.6%), where oil generation has in general been minimal. Unpublished data by those same workers has shown, however, that the same kinetic analysis is valid up to at least $R_o = 2.2\%$ (Lerche, personal communication, 1982). The possible implications from their studies—that vitrinite is therefore a poor indicator of the hydrocarbon-generation window, and that the time–temperature dependence of oil generation might be quite different than predicted from other models—are thought-provoking.

One curious aspect of models of oil generation is that, as we noted earlier, many of them assume an absolute minimum temperature for oil generation, even though their kinetic foundations predict that generation should proceed at finite (albeit very modest) rates even at very low temperatures. From a theoretical point of view such absolute temperature thresholds have little to recommend them, but from a practical standpoint they are probably accurate: the amount of maturation that occurs at low temperatures in reasonable spans of geological time is of little or no consequence.

C. Modern models: weaknesses and possible improvements

The most sophisticated published model presently in common use is that developed by Lopatin (1971). Several improvements could be made, however, adapting positive features of other models, by refinement of basic parameters already in the model, or by further research.

Nagornyi and Nagornyi (1974) in particular were very critical of Lopatin's method, listing many real difficulties in its application. With careful work, however, many of their criticisms can be defused. One complaint is that Lopatin's method has not been adequately calibrated. Errors in Lopatin's original reconstruction invalidated his original calibration of TTI values to maturity. Later work (Waples, 1980) has improved the status of the calibration, but as Kettel (1981), Katz *et al.* (1982) and Magoon and Claypool (1982) have noted, more work is needed, particularly on hydrocarbon deadlines.

The success of any calibration of Lopatin's model will be critically dependent upon choosing a satisfactory mathematical relationship between the effects of time and temperature. On the basis of pyrolysis work, oil generation is adequately described by a first-order kinetic expression, with the Arrhenius equation (equation (1)) defining the rate constant k. (Nagornyi and Nagornyi (1974) and Snowdon (1979), however, do not believe that chemical kinetics can be readily applied to oil generation.) More research is needed to determine pseudo-activation energies and A-factors of the different kerogen types (Welte and Yükler, 1981). An effort should be made to integrate the model of Hood *et al.* (1975), in which E_a values change with maturity, into Lopatin's model. The main difficulty will be to make E_a dependent upon maturity rather than temperature.

All the time–temperature models proposed by the workers cited earlier seem to give good correspondences between measured and predicted maturity values. This good agreement is at first surprising, and somewhat disturbing, because it seems unlikely that each investigator has discovered a unique set of kerogens, all following the same kinetic laws, and differing from those of other workers. A more plausible explanation, as Snowdon (1979) noted, is that the temperature and time ranges over which oil generation occurs are too narrow to determine accurately its temperature dependence. The maximum range in known examples is only about 80°C (60–140°C), with the majority of cases lying between 80 and 110°C.

Most of the world's oil has been generated from rocks of Mesozoic age. In order to separate the effects of time and temperature, extreme cases must be studied carefully: very young and very old sediments, and unusually high or low temperatures. The model must be forced to conform to the data from such examples.

Extreme examples have often been discarded in the past because their results did not conform with expectations. Palaeozoic sediments are typical; poor correlations between predicted and measured values have often been explained by changing heat flows through time. Such facile explanations may actually obscure fundamental problems with some of the models.

The fact that all the models seem to be reasonably successful predictors in most instances is comforting; we can generally get close to the right answer even if our understanding is not perfect. These modest successes should not lull us into complacency about the ultimate accuracy of our models, however. Instead we should perhaps concentrate on the seeming exceptions to our models, in an effort to refine the models.

Uplift and erosion retard or stop the chemical reactions involved in coalification and petroleum formation by lowering subsurface temperatures (Teichmüller and Teichmüller, 1966). A precise knowledge of the timing, duration and amount of erosional removal is thus critical to the success of total-thermal-history models. Most important are tectonic events that occurred when the source rocks were at their maximum palaeotemperatures (Waples, 1980; Katz *et al.*, 1982); in most settings of interest to petroleum explorationists these are also the most recent and best-documented events. In some cases, however, the exact magnitude of even recent tectonic events may be poorly known (e.g. Andean orogeny in Peru) and modelling may be problematic. In such cases it is necessary to obtain measured maturity data to check the geological reconstructions. When such data are not available Lopatin's method should be used with caution (Issler, 1982).

A further complication is introduced by the phenomenon of compaction. Neither Lopatin (1971) nor Waples (1980) considered sediment compaction in their thermal models. If compaction curves were identical in all basins, and if essentially all compaction occurred prior to the onset of oil generation, neglecting compaction would be of little consequence. In comparing areas of greatly different lithologies, however, particularly those in which over-pressuring (under-compaction) occurs, it is conceivable that ignoring compaction introduces a significant error in burial history. Recently Du Rouchet (1980) and Falvey and Deighton (1982) have published maturation studies in which compaction has been considered. Further research on this problem is needed, however, and could necessitate a modification of the existing correlation between TTI values and measured maturities.

Even where precise burial-history curves can be constructed, palaeotemperatures can present problems (Gretener, 1981). In many areas it may be reasonable to extrapolate modern gradients into the past on the basis of stable heat-flow models (Tissot and Espitalié, 1975). In other areas, however, gradients certainly have changed through time (e.g. Rhine Graben (Teichmüller, 1979; Espitalié, 1979); Bramsche Massif (Kettel, 1981)). Provinces in which salt diapirism, permafrost formation or intrusive volcanism have occurred are particularly difficult to model (Gretener, 1981; Issler, 1982). Issler (1982) believes that the differences between present-day and past temperatures are often very large and essentially impossible to assess accurately. He therefore concludes that there are no TTI scales that in all cases define the oil-generation window, and that each basin must be calibrated individually. This view seems to me to be unduly pessimistic, because historical models for heat flow in sedimentary basins can be developed (e.g. Yükler *et al.*, 1978). Although accurate modelling of such changes requires a more thorough understanding of the geophysics of basin development than we are generally capable of at the present time, recent work (e.g. McKenzie, 1981; Ho and Sahai, 1982; Cohen, 1982; Toth *et al.*, 1982) is encouraging.

Modern geothermal gradients are themselves often poorly known. Most measured gradients are interpolated from only a few subsurface temperatures, which in turn are normally unequilibrated measurements corrected by standard methods that may not always be satisfactory (Landes, 1967; Demaison, 1975; Cornelius, 1975b; Hood *et al.*, 1975). In general, temperature data are by far the weakest part of any time–temperature model. Given these limitations, it is worth questioning whether improving other aspects of Lopatin's method will lead to appreciably better correlations. Much effort should be expended to improve our knowledge of modern and ancient temperatures.

Another important problem has arisen from Waples' application (1980) of Lopatin's method to oil- and gas-preservation deadlines. His implicit assumption that destruction of oil and gas follows first-order kinetics with pseudo-activation parameters similar to those for oil generation has not been adequately verified, and in fact is almost certainly not true. As we have seen, oil generation involves numerous distinct reactions with a wide range of true activation energies; it is this distribution of true activation energies that creates the pseudo-kinetic parameters explained in Fig. 4. Oil destruction, in contrast, involves numerous reactions having true activation energies that are rather similar, because they are mainly hydrocarbon-cracking reactions. The pseudo-activation energy and *A*-factor for the entire cracking process are therefore much closer to the true activation energies and *A*-factors. Data on the kinetics of relatively homogeneous systems (e.g. formation of n-paraffins from kerogen (Ishiwatari *et al.*, 1976); cracking of crude oil (McNab *et al.*, 1952)), support this interpretation. The high pseudo-activation energy for oil destruction (compared to that for oil generation) is accompanied by a much higher *A*-factor, so that kinetics of generation and destruction may have very different temperature dependences. The accurate application of Lopatin's method to oil- and gas-preservation deadlines requires a more sophisticated investigation than has yet been published.

VII. Application to Hydrocarbon Exploration

Webb (1976) applied Connan's "geological age as cooking time" model (1974) to oil generation from Palaeozoic rocks in the Oklahoma City field. His data supported the hypothesis that oil generation occurred in two pulses that were interrupted by uplift and erosion during the early Pennsylvanian. Because Connan's model was not designed to deal with timing problems, however, Webb found its application in answering questions of timing rather unwieldy.

Lopatin's method is now being rapidly adopted by many petroleum explorationists. Although the model is conceptually simple, it is nevertheless reasonably satisfying from a theoretical point of view. It lends itself well to computerization, and can be applied in both frontier and heavily explored areas where deeper, new plays are under consideration. Required input data are minimal (time stratigraphy, amounts and timing of erosional removal, and geothermal history), and can be based either on data obtained from careful analyses or on pure speculation. The quality of the predictions from the model is, of course, a direct function of the quality of input data, but even highly speculative models can be extremely valuable in frontier areas. Variation of input parameters, such as geothermal gradient or amount of erosional removal, allows one to create both optimistic and pessimistic scenarios.

There have been several published applications of Lopatin's method to date, most of which have utilized Waples' recalibration (1980) of oil- and gas-generation and -preservation deadlines. Young *et al.* (1977) used Lopatin's burial-history curve to reconstruct the thermal history of sediments in order to calculate the ages of oil-source rocks on the basis of gasoline-range hydrocarbon compositions of the oils. They did not, however, calculate TTI values.

Zieglar and Spotts (1978) used Waples' data prior to publication to explain variability of oil and gas occurrences in California's Great Valley. The vast amount of Cretaceous sediment that accumulated in the Delta depocentre near Sacramento caused hydrocarbon generation to commence there in Late Cretaceous time (see Fig. 14). At the present time the Palaeocene is in the early to middle stages of oil generation, and all younger

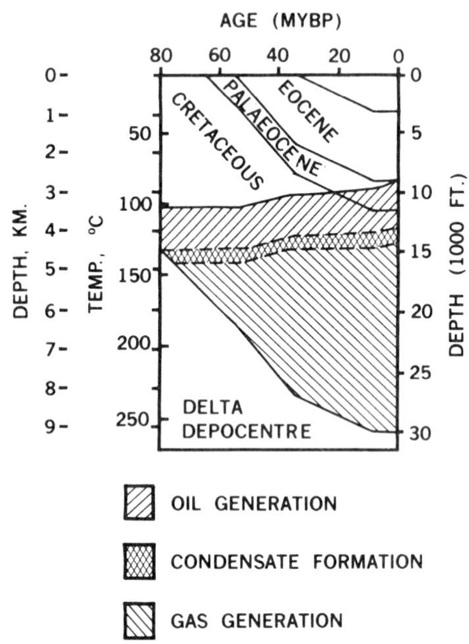

FIG. 14. Hydrocarbon-generation history in the deepest part of the Delta depocentre south of Sacramento, California. From Zieglar and Spotts, 1978. *Reprinted with permission.*

rocks are still thermally immature. Because the organic matter in the Cretaceous rocks consists principally of gas-prone Type III kerogen, liquid hydrocarbons are not important products. Therefore Cretaceous beds have been significant sources only for the gaseous hydrocarbons reservoired in Palaeogene rocks.

In contrast, the two depocentres in the southern Great Valley, which contain oil-prone Type II kerogen in the Tertiary-age source beds, have generated liquid hydrocarbons only since the middle Miocene. The dominant Tejon depocentre began to generate oil from Palaeogene source beds nearly 20 million years ago (see Fig. 15), whereas the Buttonwillow depocentre accumulated sediment more slowly. Shallower burial of the potential source rocks retarded maturation, restricting oil generation to post-Miocene

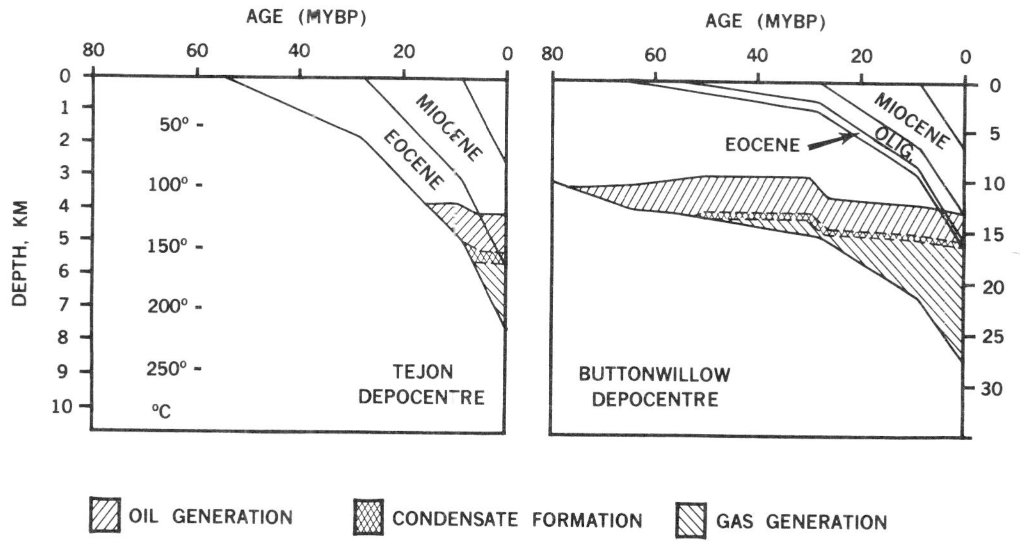

FIG. 15. Hydrocarbon-generation history in the deepest part of the Tejon and Buttonwillow depocentres, southern Great Valley, California. Horizontal and vertical scales and subsurface temperatures are the same in both diagrams. From Zieglar and Spotts, 1978. *Reprinted with permission.*

times. Neogene rocks are thermally immature at Buttonwillow but partially mature at Tejon. The authors implied that the smaller oil reserves adjacent to the Buttonwillow depocentre may be a result of less complete maturation than in the Tejon area.

Zieglar and Spotts showed the great utility of Lopatin's model in distinguishing the histories of different parts of the same region. They also noted that more burial-history curves would be needed to delineate accurately the history of each depocentre. In calculating the volumes of mature source rock in each depocentre they used a single burial-history curve for the basin centre and extrapolated the depths to maturity at that point to the basin edges. A more exact approach would involve making separate reconstructions for several points on a cross-section and tying the present-day maturities (or those at any time of interest in the past) together as iso-TTI lines. Böttcher *et al.* (1949), Teichmüller and Teichmüller (1951, 1968), Levenshteyn *et al.* (1963), Shibaoka *et al.* (1973) and Galloway *et al.* (1982) have placed measured maturity lines (from moisture and volatile-matter contents of coals and vitrinite reflectance data) on cross-sections. The application of Levenshteyn and co-workers is particularly interesting, because the cross-section they used was not from the present day, but immediately prior to an ancient tectonic event. Their analysis thus implicitly dealt with the timing of maturity compared to timing of structure formation.

MacMillan (1980) applied Lopatin's method to five Colorado basins known to contain good oil-source rocks. He determined geothermal gradients from corrected borehole-measured temperatures and extrapolated the modern gradients into the past. Burial-history curves were generalized by the author from published data, and potential hydrocarbon-source rocks were identified by analyses of depositional environments.

MacMillan's reconstructions involved a single generalized stratigraphy in each basin. He recognized the limitations that such an approach placed on the immediate applicability of his data, and noted that detailed reconstructions for specific locations within each basin would be required for more precision. Nevertheless, he was able to predict which potential source rocks had generally achieved thermal maturity. The timing of hydrocarbon generation was also examined, because, according to MacMillan, timing may have been critical in enhancement of secondary porosity by organic fluids.

Magoon and Claypool (1982) applied Lopatin's method to the Inigok-1 well on Alaska's North Slope, in an effort to determine timing of oil and gas generation. They verified the accuracy of their geological reconstructions by comparing measured maturity data with predicted maturities using a slight modification of Waples' recalibration (1980) of Lopatin's TTI values. After considering three possible temperature histories for the sediments at Inigok (see Fig. 16), they concluded that Case III best represents the geothermal history, and were able to rationalize this conclusion on the basis of geological data.

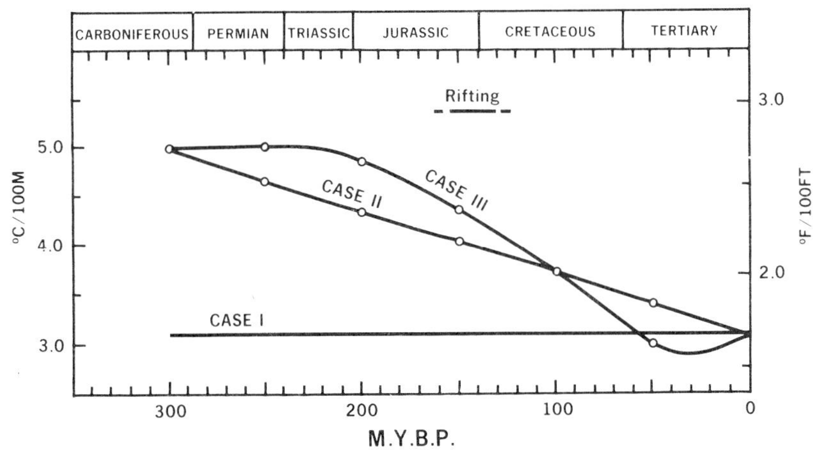

FIG. 16. Three possible scenarios for evolution of the geothermal gradient through time at the Inigok-1 well, North Slope, Alaska. Case I represents a constant geothermal gradient; Case II represents a gradual decrease in gradient since the Carboniferous, and Case III is approximately a combination of I and II. From Magoon and Claypool, 1982.

Using the best temperature history and a burial-history diagram (Fig. 17), Magoon and Claypool showed that maturation of the lowermost potential source (Shublik Fm.) commenced about 80 M.Y.B.P. and is still continuing today in the uppermost potential source rock (Pebble Shale Unit). The Shublik Fm. is now in the early gas-generative phase.

Moshier and Waples (1982) used Lopatin's method to evaluate the Lower Cretaceous Mannville Group shales as possible source rocks for Alberta's vast heavy oil deposits. They calculated TTI values for the Mannville at numerous points within the tar-sands drainage area, and thus were able to superimpose TTI lines on cross-sections through the drainage area (see Fig. 18). By making burial-history profiles for several sites, they calculated the volume of sediment at various levels of thermal maturity more accurately than Zieglar and Spotts (1978) had for the Great Valley. Combining these data with the initial generative capacity of the Mannville shales (estimated from geochemical data), they were able to show that the generative capacity of the region, including the thrusted area in the Foothills, was far less than the volume of oil in place (1300 to 2600 thousand

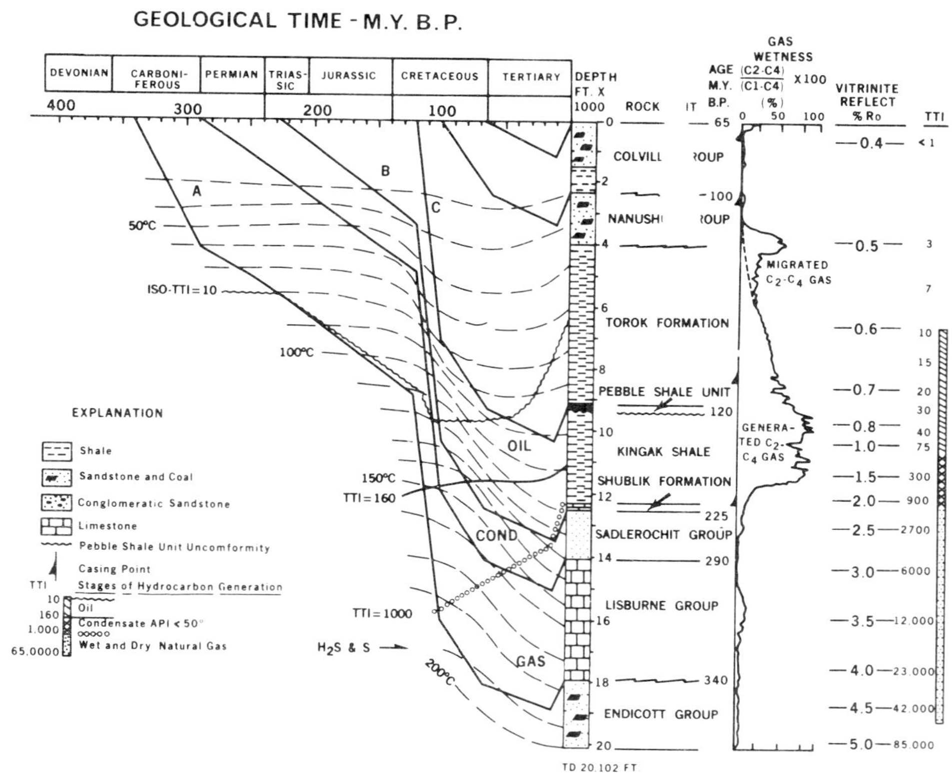

FIG. 17. Hydrocarbon-generation history and measured geochemical data at Inigok-1, North Slope, Alaska. From Magoon and Claypool, 1982.

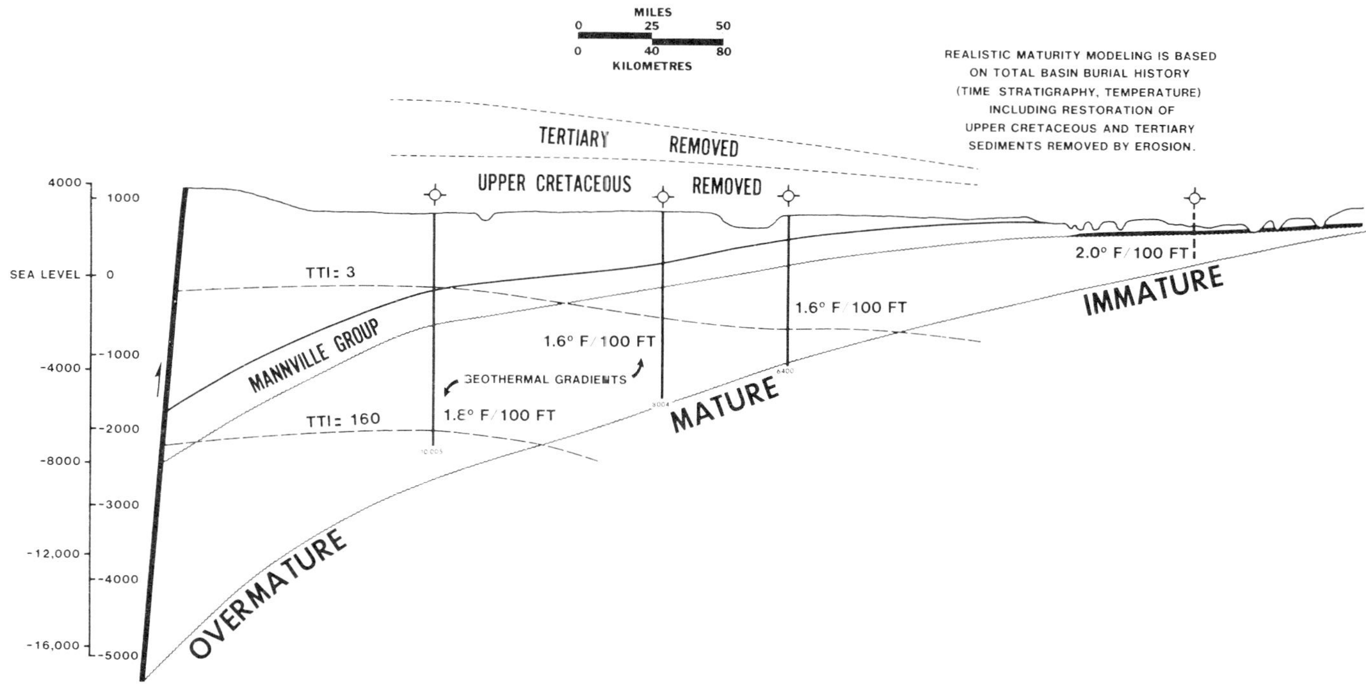

FIG. 18. Cross-section from foothills tc tar sands, Alberta, showing present-day TTI values through the basin.

million barrels). They thus concluded that the Mannville was inadequate as the sole, or even dominant, source for Alberta's heavy oils.

Bachman and co-workers (1983) have predicted the maturation history of a Neogene forearc basin the Philippines, and found a good correlation with measured reflectance data. The temperature and depth conditions required for maturation are very similar to those for sediments of equivalent age in the Los Angeles and Ventura basins (Philippi, 1965). Falvey and Deighton (1982) have studied the thermal history of sediments in the Perth basin, Australia, where comparison of measured and calculated reflectance data shows excellent agreement.

Issler (1982), in contrast, found that thermal maturities of sediments from both the Scotian Shelf (Atlantic coast of Canada) and the Canning Basin of Australia could not be predicted accurately using the same time–temperature function. He attributed this discrepancy to an inadequate understanding of palaeotemperatures.

Lopatin's method has also been used as one small part of comprehensive models of basin evolution (Welte and Yükler, 1981; Sluijk and Nederlof, 1982). In Welte and Yükler's model the entire sedimentary history of a basin is modelled, including heat flow, sediment-accumulation rates, physical properties of sediments and organic matter type and quantity. Lopatin's method is used to predict the timing of transformation of the organic matter to hydrocarbons. Sluijk and Nederlof's model focuses on both generation and preservation of hydrocarbons, using Lopatin's method both to predict kerogen maturity, and to predict gas/oil ratios as cracking proceeds.

VIII. The Future of Thermal Modelling

Over the last few years time–temperature models for oil generation have been widely used, and have become accepted exploration tools within many oil companies. This trend is likely to continue at an accelerated pace in the future. There will probably be some important changes in both the models and in their applications, however.

A. Refinement of present models

Lopatin's method is probably good enough and workable enough to be the basis for future modelling work. Several possible improvements in Lopatin's scheme have already been mentioned. Some of the refinements most likely to be brought to fruition in the next few years are listed below:

1. Some relatively minor recalibrations of the correlation between measured maturity parameters and TTI values will be made.

2. Mathematical expressions interrelating the effects of time and temperature on the cracking of bitumen and petroleum will be developed. Integration of these formulae with those for oil generation will give a much firmer basis for estimating oil- and gas-preservation deadlines.
3. Some major recalibration of oil and gas deadlines will probably occur as more data become available.
4. Better methods of predicting palaeotemperatures from geophysical models will increase confidence in our knowledge of the temperature histories of rocks. Calibration of thermal models in areas where tectonic histories are well understood will permit evaluation of palaeotemperatures in complex areas by forcing predicted maturity data to correspond to measured data (Espitalié, 1979; Magoon and Claypool, 1982).
5. Better methods for correcting measured, unequilibrated borehole temperatures will be developed.
6. Other mathematical expressions relating time and temperature during oil generation will be developed. Without improvements in our knowledge of thermal histories, however, these new models are unlikely to be significantly more accurate than Lopatin's original one.
7. Computerization (e.g. Issler, 1982) will facilitate TTI calculations and will also permit the use of much smaller temperature intervals in the calculations.

B. Novel applications

Zieglar and Spotts (1978), Waples (1980) and Magoon and Claypool (1982) have already noted that Lopatin's model is extremely useful in determining the history of hydrocarbon generation. This knowledge is valuable in comparing timing of oil generation with timing of trap formation, trap tilting, trap breaching, and reservoir diagenesis.

Recently Moshier and Waples (1982) and Sluijk and Nederlof (1982) have placed calculated isomaturity lines on cross-sections in order to make regional predictions about maturity levels (see Figs. 18 and 19). In Fig. 19, for example, there is a mature pile of sediment that generates oil near the centre of the basin. In following the most natural migration path (in this case a laterally continuous sand) the oil moves out of the mature sediment into cooler sediment along the structurally higher basin margin. This rapid and efficient movement out of the zone of generation permits preservation of oil, for if it remained behind it would quickly be cracked to gas. It is thus necessary to have good migrational lines of communication between the generation zone and the preservation zone. Any factors that hinder removal of the oil from the generative zone (such as sealing faults or laterally discontinuous conduits) will thus greatly reduce the oil-source potential of a region. Regional analyses of this type are likely to become very common and important in the future.

It is also instructive to follow basin development and maturation through

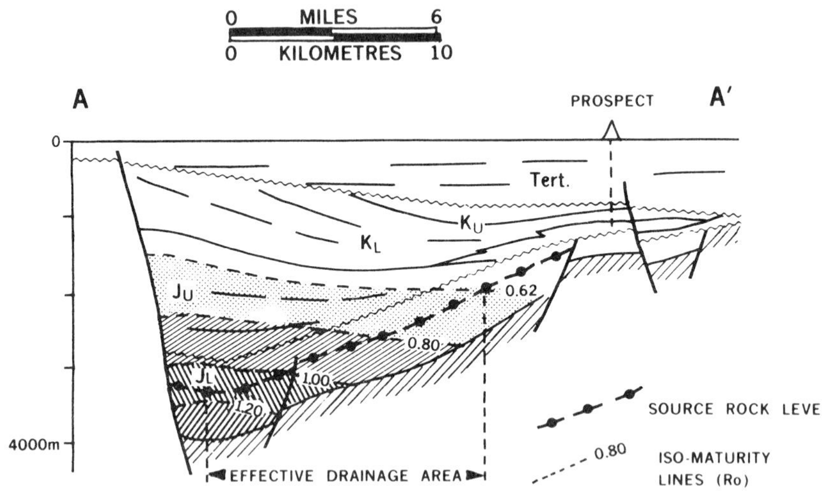

FIG. 19. Isomaturity lines (represented by TTI values) on a geological cross-section. From Sluijk and Nederlof (1982). *Reprinted with permission.*

time by means of isomaturity lines superimposed on cross-sections at different times in the past. In the near future it is likely that it will be possible on interactive computer terminals to model basin development through time and watch maturation occur on the CRT. Three-dimensional modelling on the CRT may even be feasible. It may also be possible to show the competition between generation and cracking, and thus to gain a better understanding of the delicate balance that petroleum reservoirs strike between oil generation and destruction.

The future of total-thermal-history models is bright. Aided by computers and computer-graphics capabilities, Lopatin's method (or some slightly modified version) will become an important, routinely used tool among explorationists.

Acknowledgements

I thank Jacque Kyle for technical assistance and Sue Dykes for typing the manuscript. Linda Wilson, Lois Black, Dorsalean Allen and Noelle Kehoe drafted the figures.

References

Abelson, P. H. (1967). *In* "Researches in Geochemistry", Vol. 2 (Abelson, P. H., Ed.), 63–86. John Wiley, New York.

Albrecht, P. and Ourisson, G. (1969). *Geochim. Cosmochim. Acta* **33,** 138–142.

Allred, V. D. (1966). *Chem. Eng. Prog.* **62,** 55–60.

Anderson, R. and Pack, R. W. (1915). *U.S. Geol. Surv. Bull.* **603.**

Arnold, C., Jr. (1975). *Am. Chem. Soc. Div. Petr. Chem. Preprints* **20** (1), 228–234.

Bachman, S. B., Lewis, S. D. and Schweller, W. J. (1983). *AAPG Bull.*, **67,** 1143–1162.

Barker, C. (1974). *AAPG Bull.* **58,** 2349–2361.

Barker, C. and Kemp, M. K. (1982). *AAPG Bull.* **66,** 545.

Bartenstein, H. and Teichmüller, R. (1974). *Inkohlung und Erdöl, Fort. Geol. Rhein. Westf.* **24,** 129–160.

Barton, D. C. (1934). *In* "Problems of Petroleum Geology" (Wrather, E. E. and Lahee, F. H., Eds.), 109–155. AAPG, Tulsa.

Barton, D. C. (1937). *AAPG Bull.* **21,** 914–946.

Bergius, F. (1913). *J. Soc. Chem. Ind.* **32,** 462–467.

Berkowitz, N. (1967). *Symposium on the Science and Technology of Coal, Ottawa*, 149–155.

Bertrand, C. E. and Renault, B. (1893). *Soc. de l'Incl. Min. Bull., Compt. Rend. Mensueles des Reuniones, St. Etienne* **117,** 593–596.

Bordenave, M., Combaz, A. and Giraud, A. (1970). *In* "Advances in Organic Geochemistry 1966" (Hobson, G. D. and Speers, G. C., Eds.), 389–405. Pergamon Press, Oxford.

Bostick, N. H. (1973). *Congrès International de Stratigraphie et de Géologie du Carbonifère, Septieme, Krefeld, Compte Rendu*, Vol. 2, *Illinois State Geol. Surv. reprint ser. 1974-H,* 183–193.

Bostick, N. H. and Damberger, H. H. (1971). *Ill. State Geol. Surv., Illinois Petroleum No. 95*, 142–151.

Böttcher, H., Teichmüller, M. and Teichmüller, R. (1949). *Gluckauf* **85,** 81–92.

Bownocker, J. A. (1903). *Geol. Surv. Ohio, Fourth Series, Bull.* **1.**

Bradley, W. H. (1970). *Bull. Geol. Soc. Am.* **81,** 985–1000.

Braun, R. L. and Rothman, A. J. (1975). *Fuel* **54,** 129–131.

Briggs, H. (1931). *Chem. Ind.* **50,** 127–133.

Brooks, B. T. (1948). *AAPG Bull.* **32,** 2269–2286.

Brooks, J. D. (1970). *APEA J.* **10,** 35–40.

Buntebarth, G. (1979). *Inkohlung und Geothermik, Fort. Geol. Rhein. Westf.* **27,** 97–108.

Burchfield, J. D. (1975). "Lord Kelvin and the Age of the Earth", Science History Publ., New York.

Cane, R. F. (1950). *Proc. Second Oil Shale and Cannel Coal Conf., Glasgow*, 592–604.

Cane, R. F. and Albion, P. R. (1973). *Geochim. Cosmochim. Acta* **37,** 1543–1549.

Carlson, A. J. (1937). *Univ. of Calif. Publ. in Eng.* **3** (6), 295–342.

Cecil, B. C., Stanton, R. W. and Robbins, E. I. (1977). *AAPG-SEPM Conv. Prog. and Abstr. (Washington)*, 33–34.

Chermin, H. A. G. and van Krevelen, D. W. (1957). *Fuel* **36,** 85–104.

Cohen, C. R. (1982). *AAPG Bull.* **66,** 708–718.

Cohen, C. R. and Waples, D. W. (1981). *AAPG Bull.* **65,** 1647–1649.
Connan, J. (1974). *AAPG Bull.* **58,** 2516–2521.
Cornelius, C.-D. (1975a). *Ergänzungsband der Zeitschrift Erdöl und Kohle–Erdgas–Petrochemie, Compendium 74/75*, 70–83.
Cornelius, C.-D. (1975b). *Norg. Geol. Undersøkelse* **316,** 29–67.
Correia, M. (1967). *Rev. Inst. Fran. Petr.* **22,** 1285–1306.
Cox, B. B. (1946). *AAPG Bull.* **30,** 645–659.
Cummins, J. J. and Robinson, W. E. (1972). *U.S. Bureau of Mines Rep. of Invest.* **7620.**
Cunningham-Craig, E. H. (1916). *Chem. Trade J.* **58,** 360.
Demaison, G. J. (1975). *In* "Petrographie Petroliere de la Matiere Organique des Sediments, Relations avec la Paleotemperature et le Potentiel Petrolier" (Alpern, B., Ed.), 217–224. C.N.R.S., Paris.
Deroo, G., Durand, B., Espitalie, J., Pelet, R. and Tissot, B. (1969). *In* "Advances in Organic Geochemistry 1968" (Schenck, P. A. and Havenaar, E., Eds.), 345–354. Pergamon Press, Oxford.
DiRicco, L. and Barrick, P. L. (1956). *Ind. Eng. Chem.* **48,** 1316–1319.
Dobryansky, A. F. (1963). *Rev. Inst. Fran. Petr.* **18,** 41–49.
Dorsey, G. E. (1927). *AAPG Bull.* **11,** 455–465.
Dow, W. G. (1977). *J. Geochem. Expl.* **7,** 79–99.
Du Rouchet, J. (1980). *Bull. Centre Rech. Explor.-Prod. Elf-Aquitaine*, **4,** 813–831.
Dunning, H. N. and Moore, J. W. (1957). *AAPG Bull.* **41,** 2403–2412.
Durand, B. and Espitalié, J. (1973). *Compt. Rend. Acad. Sc. Paris, Ser. D.* **276,** 2253–2256.
Engler, K. O. V. (1913). "Die Chemie und Physik des Erdöls", Vol. 1. S. Hirzel, Leipzig.
Erdmann, E. (1924). *Brenn.-Chem.* **5,** 177–186.
Espitalié, J. (1979). *Inkohlung und Geothermik, Fort. Geol. Rhein. Westf.* **27,** 87–96.
Falvey, D. A. and Deighton, I. (1982). Recent advances in burial and thermal geohistory analysis, *APEA J.* **22,** 65–81.
Fausett, D. W., George, J. H. and Carpenter, H. C. (1974). *U.S. Bureau of Mines Rep. of Invest.* **7889.**
Fitzgerald, D. (1956). *Trans. Far. Soc.* **52,** 362–369.
Forsman, J. P. (1963). *In* "Organic Geochemistry" (Breger, I. A., Ed.), 148–182. MacMillan, New York.
Forsman, J. P. and Hunt, J. M. (1958). *In* "Habitat of Oil" (Weeks, L. G., Ed.), 747–778. AAPG, Tulsa.
Francis, W. (1961). "Coal". Edward Arnold, London.
Franks, A. J. and Goodier, B. D. (1922). *Colo. Sch. Mines Quart.* **17,** Supplement A, 3–16.
Fuchs, W. and Sandhoff, A. G. (1942). *Ind. Eng. Chem.* **34,** 567–571.
Fuller, M. L. (1919). *Econ. Geol.* **14,** 536–542.
Galloway, W. E., Hobday, D. K. and Magara, K. (1982). *AAPG Bull.* **66,** 649–688.
Gold, T. and Soter, S. (1980). *Sci. Am.* **242,** 154–162.
Golitsyn, M. V. (1973). *Izv. Akad. Nauk SSSR, Ser. Geol.* (8), 90–97.
Gretener, P. E. (1981). "Geothermics: Using Temperature in Hydrocarbon Exploration", AAPG Education Course Note Series 17. AAPG, Tulsa.

Gretener, P. E. and Curtis, C. D. (1982). *AAPG Bull.* **66,** 1124–1129.
Gropp, W. and Bode, H. (1932). *Braunkohle* **31,** 277–284, 299–302, 309–313.
Hammar, H. E. (1934). *In* "Problems of Petroleum Geology" (Wrather, W. E. and Lahee, F. H., Eds.), 35–49. AAPG, Tulsa.
Hanbaba, P. and Jüntgen, H. (1969). *In* "Advances in Organic Geochemistry 1968" (Schenck, P. A. and Havenaar, I., Eds.), 459–471. Pergamon Press, Oxford.
Hardman, E. T. (1877). *J. Roy. Geol. Soc. Ireland* **4,** 200–209.
Hawkes, H. E. (1972). *AAPG Bull.* **56,** 2268–2277.
Heck, E. T. (1943). *AAPG Bull.* **27,** 1194–1227.
Hedberg, H. D. (1964). *AAPG Bull.* **48,** 1755–1803.
Héroux, Y., Chagnon, A. and Bertrand, R. (1979). *AAPG Bull.* **63,** 2128–2144.
Hill, G. R. and Dougan, P. (1967). *Colo. Sch. Mines Quart.* **62** (3), 75–90.
Hilt, C. (1873). Z. Ver. Deutsch. Ing. **17,** 193–202.
Hlauschek, H. (1950). *AAPG Bull.* **34,** 755–781.
Ho, T. T. Y. and Sahai, S. K. (1982). *AAPG Bull.* **66,** 581–582.
Hoering, T. L. and Abelson, P. H. (1963). *Carnegie Inst. Wash. Ybk* **62,** 229–234.
Hood, A., Gutjahr, C. C. M. and Heacock, R. L. (1975). *AAPG Bull.* **59,** 986–996.
Horsfield, B. and Douglas, A. G. (1980). *Geochim. Cosmochim. Acta* **44,** 1119–1131.
Hoyle, F. (1955). "Frontiers of Astronomy". Heinemann, London.
Hubbard, A. B. and Robinson, W. E. (1950). *U.S. Bureau of Mines Rep. of Invest.* **4744.**
Huck, G. and Karweil, J. (1955). *Brenn.-Chem.* **36,** 1–11.
Hume, G. S. (1927). *Can. Min. Metal. Bull.* **179,** 325–345.
Hunt, J. M. (1975). *Petr. Eng.* **47** (3), 112, 116, 118, 120, 123, 127.
Hunt, J. M. and Jamieson, G. W. (1956). *AAPG Bull.* **40,** 477–488.
Ishiwatari, R., Ishiwatari, M., Kaplan, I. R. and Rohrback, B. G. (1976). *Nature* **264,** 347–349.
Issler, D. (1982). B.Sc. Honors Thesis, Dept. of Earth Sciences, Univ. of Waterloo, Canada.
Ivanov, T. A. (1967). "Coal-Bearing Formations", 351–366. Nauka, Leningrad.
Jones, I. W. (1928). *Econ. Geol.* **23,** 353–380.
Jones, O. T. (1949). *Geol. Mag.* **86,** 303–312, 346–364.
Jüntgen, H. (1964). *Erdöl und Kohle–Erdgas–Petrochemie* **17,** 180–186.
Jüntgen, H. and Klein, J. (1975). *Erdöl und Kohle–Erdgas–Petrochemie* **28,** 65–73.
Kalkreuth, W. (1979). *Inkohlung und Geothermik, Fort. Geol. Rhein. Westf.* **27,** 277–321.
Karpov, P. A., Stepanova, A. F., Solov'yeva, N. V., Agulov, A. P., Gozhaya, A. L., Golkov, V. A. and Chaitskiy, V. P. (1975). *Int. Geol. Rev.* **18,** 397–405.
Karrick, L. C. (1926). *U.S. Bureau of Mines Bull.* **249.**
Kartsev, A. A., Vassoevich, N. B., Geodekian, A. A., Neruchev, S. G. and Sokolov, V.A. (1971). *Proc. Eighth World Petr. Cong., Moscow* **2,** 3–11.
Karweil, J. (1956). *Z. Deutsch. Geol. Gesell.* **107,** 132–139.
Karweil, J. (1975). *In* "Petrographie Petroliere de la Matiere Organique des Sediments, Relations avec la Paleotemperature et le Potential Petrolier" (Alpern, B., Ed.), 195–203. C.N.R.S., Paris.
Kasatochkin, V. I. (1959). *In* "Genezis Tverdykh Goryuchikh Iskopaemykh", 247–267. Izd. Akad. Nauk SSSR, Moscow.

Katz, B. J., Liro, L. M., Lacey, J. E., White, H. W. and Waples, D. W. (1982). *AAPG Bull.* **66,** 1150–1152.

Kettel, D. (1981). *Erdöl Erdgas Zeitschr.* **97,** 395–404.

King, L. H., Goodspeed, F. E. and Montgomery, D. S. (1963). *Mines Branch Report R114, Dept. Mines Techn. Surv., Ottawa.*

Kontorovich, A. E., Parparova, G. M. and Trushkov, P. A. (1967). *Trudi Akad. Nauk SSSR, Sib. Otdel., Geol. i Geofiz.* (2), 16–29.

Krajci-Graf, K. (1963). *Geophys. Prospecting* **11,** 244–275.

Landes, K. K. (1967). *AAPG Bull.* **51,** 828–841.

Levenshteyn, M. L. (1963). *In* "Geologiya Mestorozhdeniy Uglya i Goryuchikh Slantsev SSSR" (Kuznetsov, I. A., Ed.), Vol. 1, 348–404. Gosudarstvennoe Nauchno-Tekhnicheskoe Izd., Moscow.

Leythaeuser, D., Hagemann, H. W., Hollerbach, A. and Schafer, R. G. (1980). *Proc. Tenth World Petr. Cong., Bucharest* **2,** 31–41.

Lijmbach, G. W. M. (1975). *Proc. Ninth World Petr. Cong., Tokyo* **2,** 357–369.

Lockwood, C. D. (1925). *Oil and Gas J.* **24** (30), 51, 104.

Lopatin, N. V. (1969a). *Izv. Akad. Nauk SSSR, Ser. Geol.* (5), 69–76.

Lopatin, N. V. (1969b). *Mosc. Univ. Vest., Ser. 4, Geol.* (1), 95–98.

Lopatin, N. V. (1971). *Izv. Akad. Nauk SSSR, Ser. Geol.* (3), 95–106.

Lopatin, N. V. (1976). *In* "Issledovaniya Organicheskogo Veshchestva Sovremennykh i Iskopaemykh Osadkov, Otdel'nye Ottiski", 361–366. Nauka, Moscow.

Lopatin, N. V. and Bostick, N. H. (1973). *In* "Priroda Organicheskogo Veshchestva Sovremennykh i Iskopaemykh Osadkov", 79–90. Nauka, Moscow.

Louis, M. (1966). *In* "Advances in Organic Geochemistry 1964" (Hobson, G. D. and Louis, M. C., Eds.), 85–94. Pergamon Press, New York.

Louis, M. and Tissot, B. (1967). *Proc. Seventh World Petr. Cong., Mexico* **2,** 47–60.

McCoy, A. W. (1919). *J. Geol.* **27,** 252–262.

McCoy, A. W. (1921). *AAPG Bull.* **6,** 541–584.

McCreath, A. S. (1879). *Second Geol. Surv. of Pennsylvania 1876–8 MM.* (McCreath, A. S., Ed.).

McIver, R. D. (1967). *Proc. Seventh World Petr. Cong., Mexico* **2,** 25–36.

McKee, R. H. and Lyder, E. E. (1921). *J. Ind. Eng. Chem.* **13,** 613–618.

McKenzie, D. (1981). *Earth Planet Sci. Lett.* **55,** 87–98.

MacMillan, L. (1980). *In* "Colorado Geology" (Kent, H. C. and Porter, K. W., Eds.), 191–197. Rocky Mountain Assoc. Geol., Denver.

McNab, J. G., Smith, P. V., Jr. and Betts, R. L. (1952). *Ind. Eng. Chem.* **44,** 2556–2563.

Magoon, L. B. and Claypool, G. E. (1982). *In* "Advances in Organic Geochemistry 1981" (Bjorøy, M. *et al.*, Eds.), 28–38. John Wiley, Chichester.

Maier, C. G. and Zimmerley, S. R. (1924). *Univ. of Utah Bull. No. 14*, 62–81.

Marcusson, J. (1908). *Chem. Zeit. Nr.* **30,** 377–378.

Meinschein, W. G. (1959). *AAPG Bull.* **43,** 925–943.

Milner, C. W. D., Rogers, M. A. and Evans, C. R. (1977). *J. Geochem. Expl.* **7,** 101–153.

Mochalov, V. V. and Gryaznov, N. S. (1969). *Khim. Tverd. Topl.* (4), 94–99.

Momper, J. A. (1978). *In* "Physical and Chemical Constraints on Petroleum Migration", AAPG Continuing Education Course Note Series 8, B1–B60. AAPG, Tulsa.

Moshier, S. O. and Waples, D. W. (1982). *AAPG Bull.* **66,** 610.

Nagornyi, V. N. and Nagornyi, Yu. N. (1974). *Solid Fuel Chem.* **8** (4), 30–36.

Neruchev, S. G. and Parparova, G. M. (1972a). *Akad. Nauk SSSR, Sibir. Otdel., Geol. I Geofiz.* (10), 3–10.

Neruchev, S. G. and Parparova, G. M. (1972b). *Akad. Nauk SSSR, Sibir. Otdel., Geol. i Geofiz.* (9), 28–36.

Nobel, R. D., Harris, H. G. and Tucker, W. F. (1981). *Fuel* **60,** 573–576.

Orr, W. L. (1974). *AAPG Bull.* **58,** 2295–2318.

Orr, W. L. and Emery, K. O. (1956). *Geol. Soc. Am. Bull.* **67,** 1247–1258.

Oxley, G. R. and Pitt, G. J. (1958). *Fuel* **37,** 19–24.

Peters, K. E., Ishiwatari, R. and Kaplan, I. R. (1977). *AAPG Bull.* **61,** 504–510.

Philippi, G. T. (1965). *Geochim. Cosmochim. Acta* **29,** 1021–1049.

Porfir'ev, V. B. (1974). *AAPG Bull.* **58,** 3–33.

Potonié, H. (1910). "Die Entstehung der Steinkohle". Gebrüder Borntraeger, Berlin.

Powell, T. G., Foscolos, A. E., Gunther, P. R. and Snowdon, L. R. (1978). *Geochim. Cosmochim. Acta* **42,** 1181–1197.

Pratt, W. E. (1934). *In* "Problems of Petroleum Geology" (Wrather, W. E. and Lahee, F. H., Eds.), 235–245. AAPG, Tulsa.

Price, L. C. (1982). *AAPG Bull.* **66,** 619–620.

Pusey, W. C. (1973). *World Oil*, April, 71–75.

Radke, M., Schaefer, R. G., Leythaeuser, D. and Teichmüller, M. (1980). *Geochim. Cosmochim. Acta* **44,** 1787–1800.

Reeves, F. (1928). *AAPG Bull.* **12,** 795–823.

Roberts, J. (1924). *Proc. S. Wales. Inst. Eng.* **40,** 97–138.

Roberts, J. (1938). *Petr. Times* **40,** 269–271, 288.

Rogers, H. D. (1843). *Proc. and Trans. Assoc. Am. Geol. and Natur.*, 433–474.

Rogers, H. D. (1860). *Proc. Roy. Phil. Soc. Glasgow* **4,** 355–359.

Russell, W. L. (1925). *Econ. Geol.* **20,** 249–260.

Russell, W. L. (1927). *AAPG Bull.* **11,** 977–989.

Schopf, J. M. (1948). *Econ. Geol.* **43,** 207–255.

Schopf, J. M. (1949). *Econ. Geol.* **44,** 68–71.

Shapatina, E. A., Kalyuzhnyy, V. V. and Chukhanov, Z. F. (1950). *Dokl. Akad. Nauk SSSR* **72,** 869–872.

Shibaoka, M., Bennett, A. J. R. and Gould, K. W. (1973). *APEA J.* **13,** 73–80.

Shih, S.-M. and Sohn, H. Y. (1980). *Ind. Eng. Chem. Proc. Des. Dev.* **19,** 420–426.

Shimoyama, A. and Johns, W. D. (1971). *Nature Phys. Sci.* **232,** 140–144.

Sickenberger (1891). *Chem.-Zeit.* **15,** 1582–1583.

Sluijk, D. and Nederlof, M. H. (1982). *In* "AAPG Memoir" (Demaison, G. J. and Murris, R. J., Eds.). AAPG, Tulsa, in press.

Smith, P. V., Jr. (1954). *AAPG Bull.* **38,** 377–404.

Snider, L. C. (1934). *In* "Problems of Petroleum Geology" (Wrather, W. E. and Lahee, F. H., Eds.), 51–66. AAPG, Tulsa.

Snowdon, L. R. (1979). *AAPG Bull.* **63,** 1128–1138.

Stadnichenko, T. (1934). *Econ. Geol.* **29,** 511–543.

Staplin, F. L. (1969). *Bull. Can. Petr. Geol.* **17,** 47–66.

Steuart, D. R. (1912). *In* "The Oil-Shales of the Lothians", *Mem. Geol. Surv. Scotland.*

Stevens, N. P., Bray, E. E. and Evans, E. D. (1956). *AAPG Bull.* **40,** 975–983.

Stone, H. N., Batchelor, J. D. and Johnstone, H. F. (1954). *Ind. Eng. Chem., Eng. Proc. Dev.* **46,** 274–278.
Stutzer, O. (1940). "Geology of Coal" (Noe, A. C., Trans.), Univ. of Chicago Press.
Takahashi, J.-I. (1922). *Sendai Tohoku Univ. Sci. Rpts., Series 3*, Vol. 1 (2), 63–156.
Tan, L.-P. (1965). *Bull. Geol. Surv. Taiwan* No. 16, 1–44.
Teichmüller, M. (1958). *Revue de l'Industrie minerale, num. Spec. (petr. charbons)* 99–113.
Teichmüller, M. (1971). *Erdöl und Kohle–Erdgas–Petrochemie* **24,** 69–76.
Teichmüller, M. (1974). *Inkohlung und Erdöl, Fort. Geol. Rhein. Westf.* **24,** 36–64.
Teichmüller, M. (1979). *Inkohlung und Geothermik, Fort. Geol. Rhein. Westf.* **27,** 19–49.
Teichmüller, M. and Teichmüller, R. (1951). *Neues Jahrb. f. Geol. u. Paläont.* **93,** 69–85.
Teichmüller, M. and Teichmüller, R. (1966). *In* "Coal Science", 133–155. American Chemical Society, Washington.
Teichmüller, M. and Teichmüller, R. (1968). *In* "Coal and Coal-Bearing Strata" (Murchison, D. and Westoll, T. S., Eds.), 233–307. Elsevier, New York.
Teichmüller, M. and Teichmüller, R. (1979). *In* "Diagenesis in Sediments and Sedimentary Rocks" (Larsen, G. and Chilingar, G. V., Eds.), 207–246. Elsevier, Amsterdam.
Tissot, B. (1969). *Rev. Inst. Fran. Petr.* **24,** 470–501.
Tissot, B. and Espitalié, J. (1975). *Rev. Inst. Fran. Petr.* **30,** 743–777.
Tissot, B. and Welte, D. H. (1978). "Petroleum Formation and Occurrence". Springer, Verlag, Berlin.
Tissot, B., Durand, B., Espitalié, J. and Combaz, A. (1974). *AAPG Bull.* **58,** 499–506.
Tissot, B., Deroo, G. and Espitalié, J. (1975). Proc. Ninth World Petr. Cong., Tokyo **2,** 159–169.
Toth, D. J., Lerche, I., Petroy, D. E., Meyer, R. J. and Kendall, C. G. St. C. (1982). *In* "Advances in Organic Geochemistry 1981" (Bjorøy, M. *et al.*, Eds.), 588–596. John Wiley, Chichester.
Trask, P. D. (1934). *In* "Problems of Petroleum Geology" (Wrather, W. E. and Lahee, F. H., Eds.), 27–33. AAPG, Tulsa.
Treibs, A. (1934). *Ann.* **510,** 42–62.
Treibs, A. (1936). *Angew. Chemie* **49,** 682–686 (in German). [Translation *in* "Geochemistry of Organic Molecules" (Kvenvolden, K. A., Ed.), 17–25. Dowden, Hutchinson, and Ross, Stroudsburg, PA.]
Trotter, F. M. (1948). *Geol. Soc. London, Abstr. of Proc.* No. 1442, 79–92.
Turner, H. G. (1925). *AIME Trans.* **71,** 127–148.
van Heek, K. H., Jüntgen, H., Luft, K.-F. and Teichmüller, M. (1971). *Erdöl und Kohle–Erdgas–Petrochemie* **24,** 566–572.
van Krevelen, D. W. (1950). *Fuel* **29,** 269–284.
van Krevelen, D. W. (1952). *Troisième Congrès pour l'Avancement des Études de Stratigraphie et du Géologie du Carbonifère, Compte Rendu, Heerlen* **1,** 359–368.
van Krevelen, D. W., van Heerden, C. and Huntjens, F. J. (1951). *Fuel* **30,** 253–259.

Vandenbroucke, M., Albrecht, P. and Durand, B. (1976). *Geochim. Cosmochim. Acta* **40,** 1241–1250.
Vassoevitch, N. V., Visotski, I. V., Guseva, A. N. and Olenin, V. B. (1967). *Proc. Seventh World Petr. Cong., Mexico* **2,** 37–45.
Waples, D. W. (1976). *AAPG Bull.* **60,** 884–885.
Waples, D. W. (1980). *AAPG Bull.* **64,** 916–926.
Washburne, C. W. (1919). *AAPG Bull.* **3,** 345–362.
Webb, G. W. (1976). *AAPG Bull.* **60,** 115–122.
Weithofer, K. A. (1916). *Neues Jahrb. f. Min., Geol. u. Paläont.*, Suppl. Vol. 41, 149–236.
Weitkamp, A. W. and Gutberlet, L. C. (1970). *Ind. Eng. Chem. Proc. Des. Dev.* **9,** 386–395.
Welte, D. H. (1972). *J. Geochem. Expl.* **1,** 117–136.
Welte, D. H. and Yükler, M. A. (1981). *AAPG Bull.* **65,** 1387–1396.
White, D. (1908). *Econ. Geol.* **3,** 292–318.
White, D. (1913). *U.S. Bureau of Mines Bull.* **38,** 91–130.
White, D. (1915). *J. Wash. Acad. Sci.* **5,** 189–212.
White, D. (1921). *AIME Trans.* **65,** 176–198.
White, D. (1925). *AIME Trans.* **71,** 253–281.
White, D. (1935). *AAPG Bull.* **19,** 589–617.
Whitehead, W. L. and Breger, I. A. (1950). *Science* **111,** 335–337.
Wilson, L. E. (1961). *Tulsa Geol. Soc. Digest* **29,** 131–140.
Wright, N. J. R. (1980). *J. Petr. Geol.* **2,** 411–425.
Yi-gang, Zh. (1981). *J. Petr. Geol.* **3,** 427–444.
Young, A., Monaghan, P. E. and Schweisberger, R. T. (1977). *AAPG Bull.* **61,** 573–600.
Yükler, A., Cornford, C. and Welte, D. (1978). *Geol. Rund.* **67,** 970–979.
Zelinsky, N. D. (1927). *Ber.* **60B,** 1793–1800.
Zieglar, D. L. and Spotts, J. H. (1978). *AAPG Bull.* **62,** 813–826.
Zobell, C. E. (1945). *Science* **102,** 364–369.

A Review of Models Used in Petroleum Resource Estimation and Organic Geochemistry

M. Arif Yükler* and Fritz Kokesh
Phillips Petroleum Company, Bartlesville, Oklahoma, U.S.A.

I. Introduction

The systematic search for economic oil and gas reserves started with Colonel Drake's discovery of oil at Titusville, Pennsylvania, U.S.A., in 1859. Cycles of imbalance between demand and supply of hydrocarbons have been experienced ever since. The 1973 oil embargo led to a worldwide oil shortage. Conservation measures adopted by the developed countries in response to this embargo and worldwide recession have led to the present "oil glut". These cycles of imbalance between demand and supply require efficient methods for the assessment of hydrocarbon resources. Because of the complex interrelationships that need to be evaluated and the multi-

* Currently with IES MBH., Jülich, West Germany.

ADVANCES IN PETROLEUM GEOCHEMISTRY Vol. 1.
ISBN 0-12-032001-0

plicity of approaches which exist, this paper will attempt to briefly review and evaluate those which bear on the subject of quantitative oil- and gas-reserve assessment.

In the five years following Colonel Drake's discovery, the idea was clearly formulated that hydrocarbons impelled by the forces of buoyancy accumulate on the crests of anticlines or, in general, in the highest local positions of a structure to which petroleum can migrate. Petroleum was thought to move upward in a water-saturated static medium where pressures were hydrostatic. By 1890, the "anticlinal theory" or the "gravitational theory" had become firmly established as the controlling principle of oil accumulation (Howell, 1934; Hubbert, 1953). Studies on multiphase flow systems of gas, oil and water resulted in the "hydraulic theory" at the beginning of the twentieth century (Munn, 1909; Shaw, 1917; Mills, 1920; Rich, 1921, 1923, 1931, 1934; Illing, 1938, 1939). Hubbert (1940, 1953) provided the mathematical basis of the hydraulic theory. These theories addressed only the question of location of traps. The anticlinal theory assumed hydrostatic conditions, a special case of hydrodynamics. On the other hand, the hydraulic theory considered the migration of petroleum together with the general hydrodynamic conditions.

As new instruments were developed, geophysics became an important tool in petroleum exploration based on the anticlinal theory. Geophysical methods have been developed, applied and improved, and can now measure the gravimetric, magnetic and electric properties of the subsurface. These techniques led to new hydrocarbon finds. In the 1920's, first refraction and then reflection seismics were invented. These were major advances in the scientific search for petroleum. In the past twenty years, remote sensing techniques have been developed that use airborne instrumentation to measure the physical properties of the subsurface. These techniques have proved valuable, especially in studying areas that have hostile terrain or environments.

The above methods are concerned only with the most probable location of petroleum traps; however, one has to determine not only "where", but also "what, when, and how much" to understand the generation, migration and accumulation of hydrocarbons. It is now widely acknowledged that petroleum originates from finely disseminated organic matter buried within sediments that have been subjected to temperatures above 50°C. The criteria for the assessment of source rocks are: (i) the amount of organic matter, both soluble (bitumen) and insoluble (kerogen); (ii) the type of organic matter; (iii) the maturity level of organic matter; and (iv) the expulsion efficiency (Tissot and Welte, 1978).

Organic geochemistry is now highly developed, and applicable to problems of the generation, migration and accumulation of petroleum. New rapid analytical methods have been developed to fulfill the main requirements of geochemical petroleum exploration, such as assessment of source-rock potential, organic maturity and source-rock/oil correlation (Welte *et al.*, 1980). Integration of the geochemical information with the geological, hydrodynamic and geothermal processes enable explorationists to make successful quantitative predictions of petroleum accumulations (Welte and Yükler, 1981).

Methods for assessment of amounts of undiscovered oil and gas accumulations in a given basin are many and varied. These can be summarized as: (i) descriptive (using classical geology); (ii) statistical; and (iii) deterministic methods. In this review we limit our discussions to statistical and deterministic methods.

This paper reviews the concepts and models developed to help solve this complex problem. We also briefly discuss geothermics to illustrate its importance in geochemical modelling. A successful model can only be developed with the integration of all the geological, geochemical, hydrodynamic and geothermal processes, and should also be continuously improved as better understanding of the subsurface and surface processes develops.

II. Methods of Estimating Hydrocarbon Resources

Many different methods of estimating hydrocarbon resources before and after drilling have been used to answer the two basic questions "is there any hydrocarbon potential in the area?" and "how much?". Different model types are used for different levels of geological, geophysical, hydrodynamic, geothermal and geochemical information and for different purposes (see Fig. 1). Physical models are used to study one variable at a

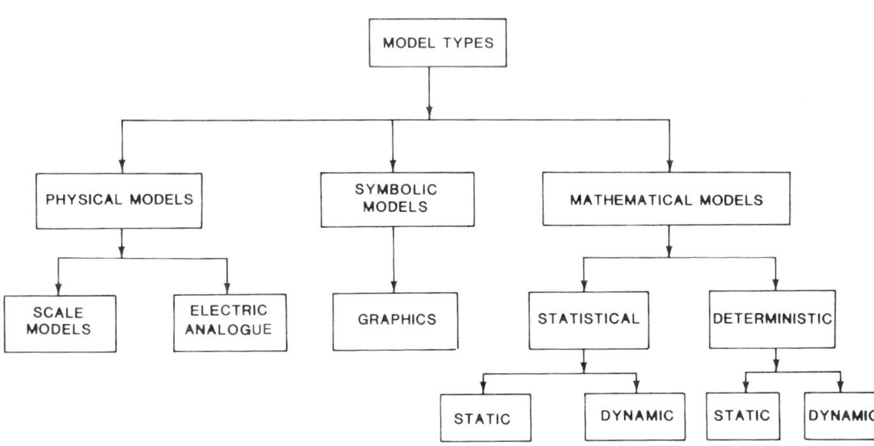

FIG. 1. Various model types.

time, for example faulting under different stress distribution, groundwater movement with varying recharge and discharge, etc. Symbolic models are used for graphical presentation of the interrelationships among the components of a system as maps, cross-sections and block diagrams. Symbolic models may only be used to describe simulation. Mathematical models, however, are constructed to simulate complex processes with one or more variables and thus they are essential in the assessment of hydrocarbon resources. Mathematical models are applied either as statistical models to summarized observational data or as deterministic models, assuming that the processes studied are deterministic. The amount and quality of data are critical in choosing a particular method of estimation.

A. Statistical methods

The methods commonly used in the petroleum industry are static in nature and do not directly analyse the dynamic processes of the system. The normal approach is first to construct assessment probability of expectation curves (see Fig. 2). These curves give the probability of finding hydrocarbon accumulations of various volumes. Probability curves are then computed using a Monte Carlo simulation (White and Gehman, 1979; Nederlof, 1980). Nederlof's model uses Bayesian statistics, which differ from classical statistics in the interpretation of the parameter under investigation. Bayesian statistics assume that the parameter has a probability distribution, whereas classical statistics assume that the parameter has a single value within a confidence interval or based on a hypothesis test. In classical statistics every distribution has a frequency interpretation; in Bayesian statistics the parameter may have a frequency distribution *or* an informal and subjective distribution based on the experience of the investigator.

Various methods of "risk analysis" have been used to improve the assessment of probability curves (Gehman *et al.*, 1975; Megill, 1977). The most widely used procedure concentrates on the various geologic risk factors such as source rock, reservoir and trap conditions that are essential for the occurrence of an accumulation. One other important factor is the recovery potential. No assessment is complete without some kind of risk analysis (White and Gehman, 1979). The most commonly used assessment methods as summarized by White and Gehman are geological analogy, delphi, areal and volumetric yields, geochemical yields, field number and size, and summations and extrapolations. Each of these probabilistic methods is described briefly in the following paragraphs.

Geological analogy has been used in one form or another in almost every model. The differences stem from the number of geological controls that are used to assess hydrocarbon accumulations. Studies based on a single geological control, such as source rocks (Conybeare, 1963), reservoir beds (Zhdanov, 1962) or trap conditions, have been replaced by those using multiple geological controls with the aid of sophisticated computer techniques (Nederlof, 1980; Miller, 1981). Other geological analogies are based on general genetic basin types (Weeks, 1952; Klemme, 1971, 1975; McCrossan and Porter, 1973; Bally, 1975; Pitcher, 1976; Warren, 1979).

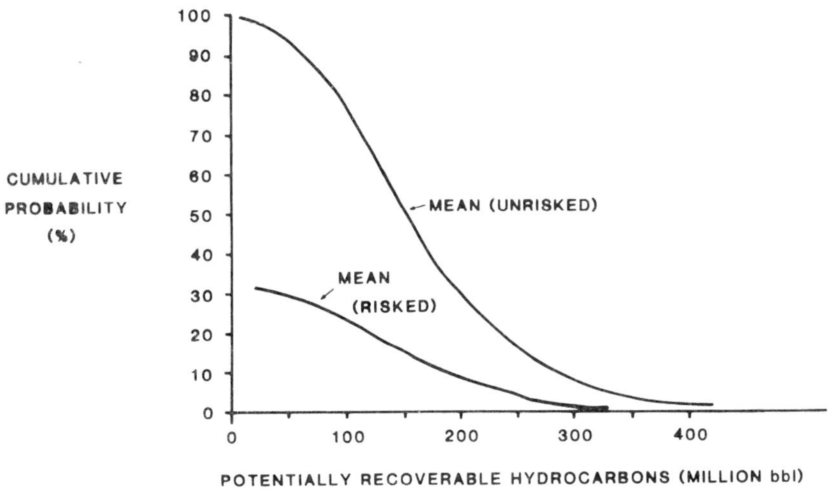

FIG. 2. Mean unrisked and risked assessment probability curves.

The major problem is that the number of basin types can approach the number of basins studied (Jones, 1975). There also are computer studies based on similarities between structural patterns using computerized statistical mapping techniques (Hambleton *et al.*, 1975; Davis, 1977; Harbaugh *et al.*, 1977).

In the delphi approach, a group of experts reviews all the geological data and each member constructs his own probability distribution of potential resources. Then all the curves are averaged (Miller *et al.*, 1975). This is a highly subjective method.

The areal-yield and the volumetric-yield methods endeavour to determine the expected volume of hydrocarbons by multiplying the total basin area or potentially productive basin area by a predetermined productivity factor. The volumetric-yield method handles the problem in three dimensions by considering the average net pay thickness. Each factor is then entered as a range of values in a Monte Carlo method (Weeks, 1949; Stoian, 1965; Walstrom *et al.*, 1967; Smith, 1968; Megill, 1971; Jones, 1975; Newendorp, 1975; Roadifer, 1975).

Geochemical material balance is a special form of volumetric yield based on the generation, expulsion, migration and accumulation of hydrocarbons. The amount and type of kerogen in a drainage area defines the initial hydrocarbon potential. The present hydrocarbon potential is determined by the amount of petroleum generated from the kerogen, its expulsion and migration efficiency and, finally, the amount accumulated and preserved in reservoirs within the system. The present hydrocarbon potential has been defined differently by different researchers depending on the techniques used, as discussed in this chapter. The last parameter involved, recovery factor, is the fraction of the accumulated hydrocarbons that can be recovered from a given reservoir. (Halbouty and Hardin, 1959; Neruchev, 1962; Conybeare, 1963; McDowell, 1975; Semenovich *et al.*, 1977; Tissot and Welte, 1978.)

The field distribution method is used to estimate the number and size of potential fields that can be found in a given area. Atwater (1956) applied this method to offshore Louisiana with data obtained from onshore fields. Other applications are given by Arps and Roberts (1958), Belov (1960), Roy *et al.* (1975), Ivanhoe (1976), Energy, Mines and Resources Canada (1977), Semenovich *et al.* (1977), Nehring (1978), Attanasi *et al.* (1980) and Ejedawe (1981). A summation of prospects or plays is then used to assess larger areas by summing the assessment of smaller areas using the Monte Carlo summation technique (White and Gehman, 1979).

Amounts of hydrocarbons discovered are commonly plotted against time, drilling activity and number and size of fields discovered. Davis (1958), Hubbert (1962, 1967, 1974) and Menard and Sharman (1975) have used this information to predict future discoveries based on various exploration routines.

The usefulness of the above statistical methods is strongly limited by the uniqueness of basins (Jones, 1975). The importance of particular geological and geochemical control factors varies from one basin to another. If the effects of one or more negative factors are being overlooked, the effectiveness of the positive or favourable control factors can be completely meaningless, and models are then only interesting mathematical exercises. To overcome this limitation other techniques have been developed and applied in petroleum exploration.

Additional techniques that can be grouped under non-geological or economic appraisal methods include systematic grid drilling and petroleum exhaustion maps. Systematic grid drilling is an old but simple method and is applicable to petroleum, uranium, gold, silver and other ores. Griffiths (1966) described applications of the technique in Kansas, Pennsylvania and Ohio.

Two similar methods for calculating an area exhausted by exploratory drilling have been developed by Singer (1976) and Schuenemeyer and Drew (1977). Drew *et al.* (1979) applied this type of analysis to the Denver Basin and presented physical exhaustion maps that showed the degree that each point in the basin had been physically exhausted of target petroleum deposit of different sizes.

B. Deterministic models

The simulation of sedimentary basin development was established in the 1960's (Sloss, 1962; Harbaugh, 1966; Bonham-Carter and Sutherland, 1968) with the advent of large and fast computers. Several books were published at this time on systems analysis and simulation concepts (Chorafas, 1965; Krumbein and Graybill, 1965; Leontief, 1966; Kendall, 1968; Griffiths, 1967; Chorley and Haggett, 1967; Harbaugh and Merriam, 1968; Harbaugh and Bonham-Carter, 1970).

The initial models were basically process-response models (Krumbein, 1964; Whitten, 1964) and they were followed by simple static deterministic models which evolved into more complex static models and then into dynamic models (Harbaugh and Bonham-Carter, 1970). Later, researchers started to develop efficient and fast algorithms for bookkeeping purposes in a one-, two- or three-dimensional framework to handle mass-balance formulae (Ojakangas, 1970).

Complete basin-evolution models were the next step attempted by various researchers. Early models attempted to determine the petroleum potentials of a given sedimentary basin based on structural and stratigraphic relations (Ojakangas, 1970). Other models used only the geochemical information (Tissot, 1969); Tissot and Espitalié, 1975). The temperature history of a sedimentary basis is the most important physical parameter in uniting the sedimentary system with the petroleum potential. The most popular physical method is based on the plate-tectonics concept (Sleep, 1971; McKenzie, 1978; Royden *et al.*, 1980). Geochemical parameters such as vitrinite reflectance and chemical fossils also have been used to determine temperature history (Lopatin, 1971; Hood *et al.*, 1975; Waples, 1980; Mackenzie *et al.*, 1981a). Model studies by Yükler *et al.* (1978) have shown that all the processes operating in a sedimentary basin are interrelated and thus should be considered together dynamically. Yükler and Welte (1980) developed the first complete deterministic basin model that determines geological history, palaeohydrodynamics, palaeotemperatures and hydrocarbon generation, and migration and accumulation. This three-dimensional model has been successfully applied to four basins (Welte and Yükler, 1981).

Dynamic interpretation of the geological and geochemical processes is essential to our understanding of the mechanisms that finally lead to hydrocarbon accumulations. Unfortunately, most researchers have used the word "dynamics" only for time-dependent events, without considering the forces that cause these events. Consequently, only the data have some continuity in time but the basic concept always remains static. Other researchers enumerate the processes and mechanisms operating in a basin—as they see them—and then prophesy that the problem can be solved using the computer. Identifying the processes is just a part of the problem.

Deterministic models have a great advantage over statistical models in not being limited by the uniqueness of basins. Both types of model have common sources of error, namely, uncertainties in the parameters and simplification of the basins by the mathematical techniques used (see Fig. 3). There are several ways of handling uncertainties in hydrocarbon resource estimations. The most common technique is to perform repeated experiments and to draw conclusions from the distribution of the results of these experiments. Different statistical methods have been developed and routinely used to evaluate the uncertainty. Another approach is to construct a mathematical model which simulates the complex processes with all the dependent and independent variables. This model is then run with a wide range of values of the parameters. Statistical methods are applied to the results of these simulated experiments (Walstrom *et al.*, 1967). Sensitivity analysis is a powerful tool in determining uncertainties in deterministic models (Yükler, 1976; Yükler, *et al.*, 1978; Welte and Yükler, 1981).

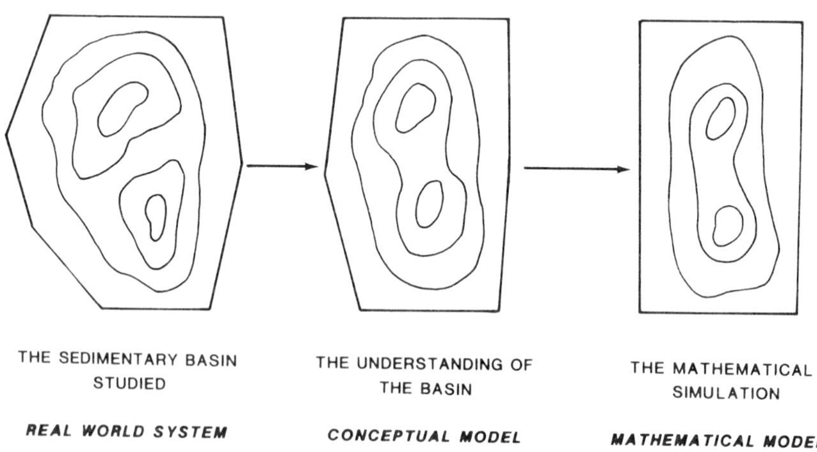

FIG. 3. Steps in the construction of a mathematical model.

III. Applications of Geochemical Models in Petroleum Exploration

In recent years organic geochemistry has been increasingly used as a tool in petroleum exploration (Tissot and Welte, 1978; Hunt, 1979). Its greatest power is realized when both geochemical and geological considerations are integrated (Conybeare, 1965; Philippi, 1965; Welte, 1965, 1972; Jones, 1975; McDowell, 1975; Tissot, 1977; Yükler *et al.*, 1979; Welte and Yükler, 1980, 1981; Welte *et al.*, 1980; Nakayama and van Siclen, 1981).

Quantitative organic geochemical models are used to address the following questions:

1. What stratigraphic intervals have source-rock potential?
2. How much oil and gas has been generated in each source-rock interval?
3. When was it generated?
4. How much was expelled?

In conjunction with geological and hydrodynamic models, the additional questions of the direction of migration and the locations of possible accumulations and post-generation changes in petroleum can be addressed.

A summary of approaches to these geochemical questions is given in Fig. 4.

To answer the question—"How much oil and gas has been generated?"—we must know the ultimate petroleum potential of a source rock given adequate conditions of time and temperature. We will express this ultimate potential as the product of two factors:

$$\text{ultimate source potential} = \text{kerogen quantity} \times \text{kerogen quality}$$

The quantity of kerogen or insoluble organic matter is conveniently measured by total organic carbon (TOC) analyses. Quality can be defined as the petroleum potential expressed on a per-TOC basis and can be measured directly by the pyrolysis yield (see Horsfield, this volume) or hydrogen-to-carbon ratio (Durand and Monin, 1980) of an immature sample, and can be estimated indirectly from visual properties of kerogen (Alpern, 1980; Masran and Pocock, 1981) or the composition of an extract or kerogen pyrolysate (Larter and Douglas, 1980).

To know "how much" was generated we must also know what portion of the ultimate source potential has been realized, that is, we must answer the question—"What is the maturity of the source interval?". There are two quite different approaches. According to the first, it is possible to evaluate the maturity of a source rock by using some conveniently measured property that has been correlated with the maturity of the bulk kerogen. Vitrinite reflectance (Teichmüller and Teichmüller, 1981) and spore colouration index (Gutjahr, 1966) are two examples of such properties, which can be called "Indicator Properties".

To overcome errors and shortcomings inherent in uses of such Indicator

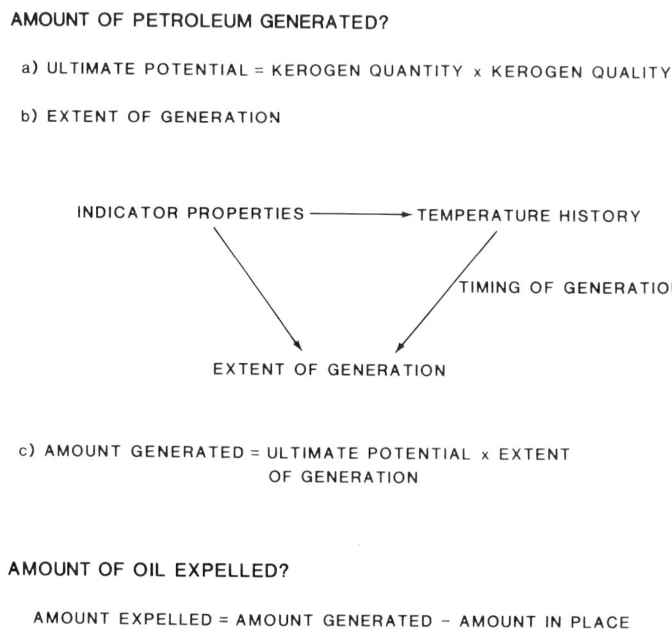

FIG. 4. Summary of the geochemical approaches to determining the amount of petroleum generated and expelled from a source rock.

Properties, maturity can be evaluated by a two-step process. The first step is a reconstruction of the temperature history, followed by the application of a kinetic model for petroleum generation to calculate the extent of generation of oil and gas through time. This process is complex and requires the use of computers.

The difference between the amount generated and the amount present provides an estimate of the "amount expelled". As shown in Fig. 4, this is based on a comparison of the amount of oil that would be in place without expulsion with the amount of thermally- or solvent-extractable organic matter, and can be greater than, less than, or equal to zero, depending on net migration. The quantity of mobile gaseous organic matter is relevant, but it is difficult to obtain good samples, and thus it is not routinely measured.

This leaves the question of the timing of generation. Quantitative geochemical exploration models capable of describing the timing of generation are necessarily based on a kinetic model for petroleum generation. This model consists of a rate law that describes the dependence of the rates of formation of petroleum (or individual kerogen components or classes of compounds) on concentrations (reaction order) and on physical conditions—time, temperature, pressure, etc. The temperature and pressure dependence of the rate constants in the rate law is also required. For the case of temperature dependence the Arrhenius equation commonly is used. When the temperature (and pressure) histories are included in the model, the extent of generation through time can be calculated.

Geochemical models can provide useful exploration information that is otherwise unattainable, but it must be recognized that there is no unique approach or answer to any of the geochemical questions that are considered in this review. Since our topic concerns integrated models of petroleum generation and accumulation, we will stress those particular approaches that (in our experience or opinion) are most powerful as components of such models.

A. Indicator Properties

Properties other than the extent of petroleum generation itself used for assessing source rock maturity are Indicator Properties. In 1915, White described the use of the rank of coal beds to assess the maturity of associated source rocks. The correlation between coal rank and vitrinite reflectance (McCartney and Teichmüller, 1972) allowed the application of the method to coaly inclusions in sedimentary rocks (Castano and Sparks, 1974). Spore colour (Gutjahr, 1966), and odd–even predominance of *n*-alkanes (OEP) (Bray and Evans, 1961) are among a multitude of other proposed Indicator Properties (Heroux *et al.*, 1979).

The implicit assumption in the use of an Indicator Property is that there is a one-to-one correspondence between the property and the extent to which the bulk of the kerogen has reacted to produce petroleum. That is, the Indicator Property is assumed to track maturity. This implicit assumption is the basis of charts that purport to show unique relationships between different Indicator Properties and the extent of petroleum generation, for example, see Fig. 1 of Heroux *et al.*, (1979). Based on physical chemical principles, it is likely that such relationships depend strongly on temperature history.

The assumption is more likely to be valid in the case where the Indicator Property is a manifestation of the same material whose maturity is being monitored. Thus, vitrinite reflectance is more likely to be a valid measure of the maturity of spores than of vitrinite (Bernard *et al.*, 1981a). But even for two properties of the same material, the general validity of the correspondence cannot be assumed, because not all properties of a given material are affected identically by conditions like time and temperature. If the scope of our interest includes the thermal conversion of oil to gas, as well as its formation from oil, then there are additional requirements for the behaviour of indicators that make their general validity even less likely.

Often, a specific value of vitrinite reflectance is assumed or claimed to be associated with the onset of petroleum generation. A simple argument can demonstrate that this is not so. Connan (1974) empirically established the relationship shown in Fig. 5 between the time and temperature required to reach the "threshold of intense oil generation". This relationship has a form like that for a kinetically first-order reaction with an Arrhenius activation energy of 13.8 kcal mol^{-1} (57.7 kJ mol^{-1}) (Connan, 1974) to 15.3 kcal mol^{-1} (64.0 kJ mol^{-1}) (Waples and Connan, 1976). By contrast, Karweil (1955) and Huck and Karweil (1955) found an E_a of 8.4 kcal mol^{-1} (35.1 kJ mol^{-1}) described the temperature dependence of coalification—and hence vitrinite reflectance. The two lines in Fig. 5 have been arbitrarily made to intersect near the temperature midpoint for Connan's data. This figure shows that a particular value of vitrinite reflectance will not be found at the onset of generation. Because of the differences in slope of the two lines, the value of vitrinite reflectance that near the centre of temperature range is associated with the onset of oil generation can be reached before generation when heating is at low temperatures for long time, or after generation when heating is at high temperatures for short times.

Fig. 6 is based on a similar sort of plot by Barnard *et al.* (1981b), where lines for several different values of vitrinite reflectance (R_o) and several spore colour indices (SCI) are included. Note that the lines for spore colour are not parallel to lines for vitrinite reflectance. Therefore, these two indicators change differently in response to time and temperature. Focusing on the line for SCI = 7.5, note that at high heating rates (high temperature/short time) the line is closest to $R_o = 0.5$, but at the slow heating rate extreme the line corresponds to $R_o > 1.0$.

Ratios of particular compounds or isomers also are commonly used as Indicators. Mackenzie and Maxwell (1981) showed the relationships as functions of depth between different molecular ratios and extract content. In subsequent laboratory experiments Mackenzie *et al.* (1981a) established that the sterane aromatization reactions are more temperature sensitive (i.e. have a larger activation energy) than alkane isomerization. Therefore, the relative rates at which the different molecular ratios change will depend on temperature history. In particular, a geologically high heating rate could lead to sterane changes occurring before alkane isomerization—the opposite of the original observation.

For reasons like these, we feel that the timing and extent of petroleum generation can be established much more accurately by palaeotemperature reconstruction and use of kinetic models of petroleum generation than they

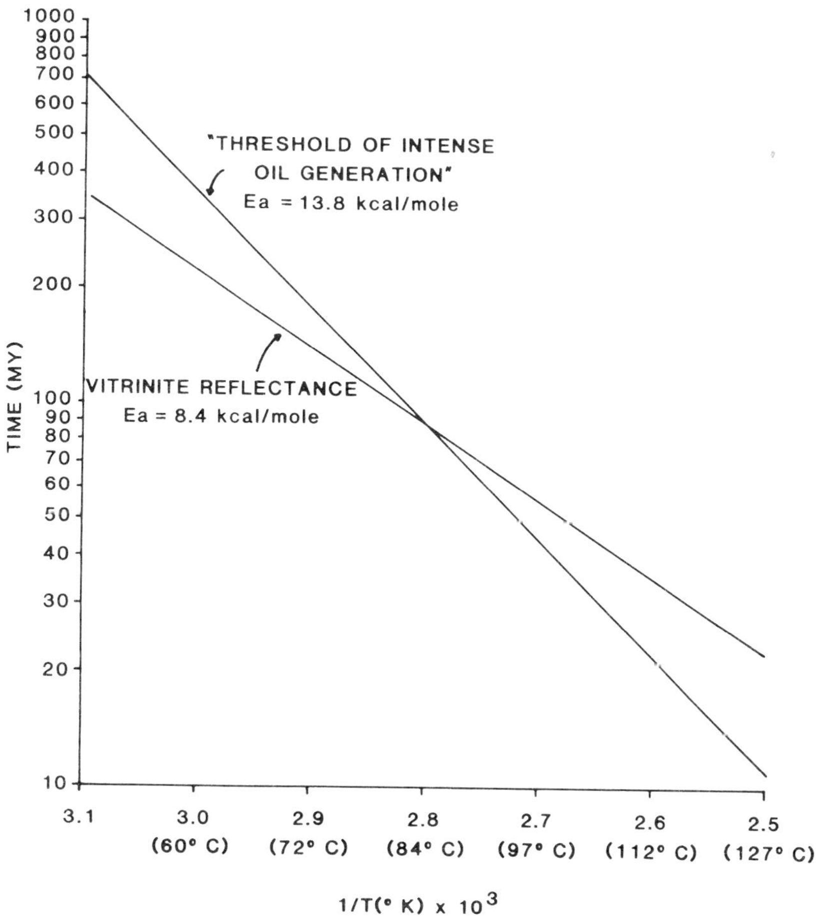

FIG. 5. Relative sensitivity of the threshold of intense oil generation and vitrinite reflectance to time and temperature, based on differences in activation energies.

can be by the simple use of Indicator Properties. This involves using an Indicator Property to reconstruct temperature history rather than to directly estimate the extent of petroleum generation. To do so, it is necessary to relate the property to time and temperature within the dynamic framework

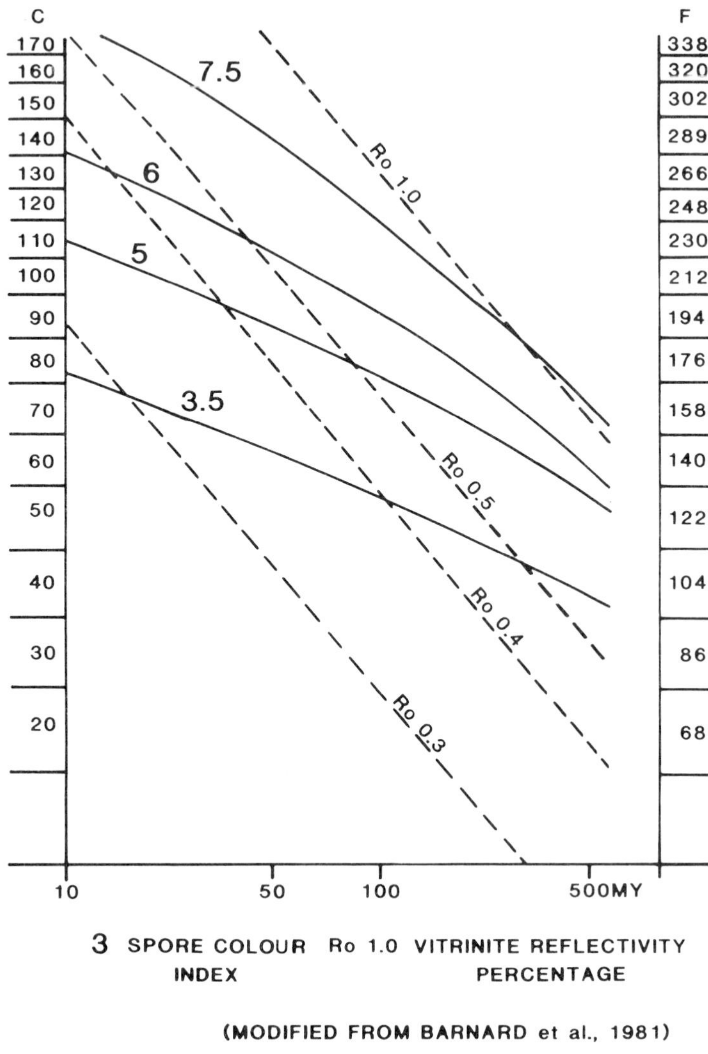

FIG. 6. Relationship of spore colour index (continuous curve) and % vitrinite reflectance (dashed curve) to time and temperature. From Barnard *et al.* (1981b). *Modified with permission.*

of a sedimentary basin, and then to use the time and temperature information to evaluate the extent of petroleum generation. This process has the added advantages of allowing the timing of generation to be determined, and of treating oil and gas generation separately, as will be shown by later examples.

B. Palaeotemperature reconstruction from geochemical concepts

There are at least two approaches to the reconstruction of palaeotemperatures. The first, which is discussed here, is the use of a property of mineral or organic matter as an indicator of the cumulative time–temperature conditions that the material has experienced. The second involves physical models that describe heat flow and thermal conductivities, and is discussed later.

As discussed by Demaison (1975), calibration of indicator-based palaeotemperature methods can be done in two ways: (i) by an empirical/statistical approach in which a property is correlated with maximum temperature; or (ii) in terms of a chemical-kinetic model of the manner in which the Indicator Property changes as a function of time and temperature. When a chemical-kinetic model is used, palaeotemperature reconstruction based on organic indicators and the calculation of the extent and timing of petroleum genesis are closely related, and the classification of a particular method into one or the other category depends on how a model is used—that is, whether palaeotemperature or extent of coalification or petroleum genesis is the dependent variable.

Methods for palaeotemperature reconstruction have been reviewed by Castano and Sparks (1974), Hood and Castano (1974), Cooper (1977), Bostick (1979) and Middleton (1982). Indicator Properties considered included vitrinite reflectance, spore colouration, kerogen electron spin resonance (ESR) and mineral alteration. Another approach is the use of oxygen isotopes as explained by Guiliani (1954, 1956), Dorman (1966), Devereaux (1967) and Servier *et al.* (1975). In this review we discuss in detail only methods based on vitrinite reflectance, with special emphasis on the method of Tissot and Espitalié (1975).

Vitrinite reflectance is very useful as an Indicator Property for palaeotemperature reconstruction (Teichmüller and Teichmüller, 1981). Vitrinite, a coal maceral, is a common component of sedimentary organic matter. The reflectance of vitrinite is a conveniently measured property, and satisfies other criteria for palaeotemperature indicators detailed by Bostick (1979). However, there are problems due to: (i) differences in vitrinite reflectance for coals and associated dispersed coaly matter (Bostick and Foster, 1975; Gill *et al.*, 1979); (ii) the presence of reworked vitrinite (Hood and Castano, 1974); (iii) differences in measurement techniques; and (iv) for miscellaneous other reasons, like those tabulated by Dow (1977).

Empirical methods involve the comparisons of measured present-day thermal gradients and vitrinite-reflectance gradients in order to recognize geological intervals or areas where palaeotemperatures exceeded those for the present day. For example, Hacquebard and Donaldson (1970) related curves of vitrinite reflectance versus depth to geothermal gradients. Dow (1977) described the use of semi-logarithmic plots of vitrinite reflectance versus depth to recognize faults, stratigraphic unconformities and thermal effects of intrusives. Ammosov (1981) described similar uses of vitrinite reflectance and other data.

The same sorts of exploration problems for which empirical approaches have been useful can be addressed using the kinetic-model based methods. These correlate the effects of time and temperature on vitrinite reflectance with equations whose form has a basis in physical chemistry. The chemical-kinetic model of coalification determined by Karweil (1955) and Huck and Karweil (1955) was modified by Bostick (1973) to produce the well known Karweil diagram that relates vitrinite reflectance to time and temperature. The Karweil model is based on a single first-order reaction, the temperature dependence of which is described by the Arrhenius equation with an activation energy of 8.4 kcal mol^{-1} (35.1 kJ mol^{-1}). For an example of how this model allows palaeotemperature to be treated quantitatively, see Kantsler *et al.* (1978).

The model of Lopatin (1971) is similar, but the reaction rate is assumed to double for a 10°C increase in temperature. The use of this doubling rule is equivalent to assuming that activation energy increases with temperature. Waples (1980) has provided additional explanation of this method. Lopatin (1980) has used his model to calculate how the extent of coalification will be affected by burial rate and geothermal gradient. Wright (1980) has done the same for the Karweil/Bostick model.

Variations on the basic ideas of Karweil and Lopatin have been reported by Stanov (1972, 1980), Hood *et al.* (1975), Karpov *et al.* (1976), Shibaoka and Bennett (1977) and Buntebarth (1978, 1982). Since vitrinite reflectance can be related to degree of coalification, in principle any kinetic model of coalification could be used for palaeotemperature determination.

Wright (1980) and Veto (1980) compared the methods of Karweil and Bostick, Lopatin, and Hood. For forty-five burial histories Veto (1980) compiled histograms of deviations for each method.

The model of Tissot and Espitalié (1975) is conceptually similar to those already mentioned, but more complex in detail and in application. The literature contains little discussion and few examples of applications of using this model. We will examine it here in detail.

The palaeotemperature method of Tissot and Espitalié (1975) is an extension of their petroleum-generation model that involves the establishment of the correspondence between R_0 and extent of reaction of Type III kerogen shown in Fig. 7. The use of the petroleum-generation model for palaeotemperature reconstruction is possible because of the special nature of the Douala Basin samples used as the prototype of Type III kerogen. The insoluble organic matter that they contain is extremely uniform geochemically and consists almost exclusively of a mixture of structured vitrinite and an amorphous kerogen for which reflectance and elemental composition parallel those of vitrinite (Durand and Espitalié, 1975). It was, therefore, reasonable to assume that the measured transformation ratio

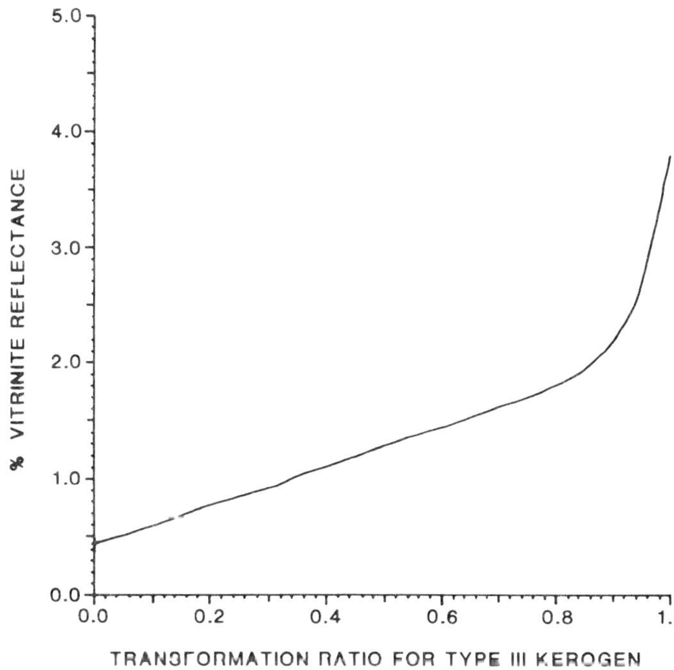

FIG. 7. Relationship between extent of reaction (transformation ratio) of Type III kerogen and vitrinite reflectance. This correspondence is based on data of Tissot and Espitalié (1975).

(see below) and vitrinite reflectance are properties of the same organic matter, and to set up a correspondence between the two properties. This special circumstance of the Douala Basin samples is in contrast to the more common situation where the organic matter is of mixed origin, in which case no correspondence is expected. The R_0 intercept in Fig. 7 is 0.43%. Vitrinite particles with values of R_0 less than 0.43% are found in immature sediments, but such particles are not yet in the catagenic stage of maturation that is described by the Tissot–Espitalié model.

Use of the model requires: (i) a temperature history with ages expressed in absolute terms (van Hinte, 1978); and (ii) a set of R_0 data. In many cases, it is adequate to base the temperature history on a burial history on which is superimposed a surface temperature and thermal gradient. The surface temperature and thermal gradient are treated as variables and are adjusted in order to optimize the agreement between the calculated R_0 values and those actually found. This process is flow-charted in Fig. 8. The "Effective Palaeothermal Gradient and Surface Temperature" are the gradient and surface temperature that produce the best fit. Adjustments

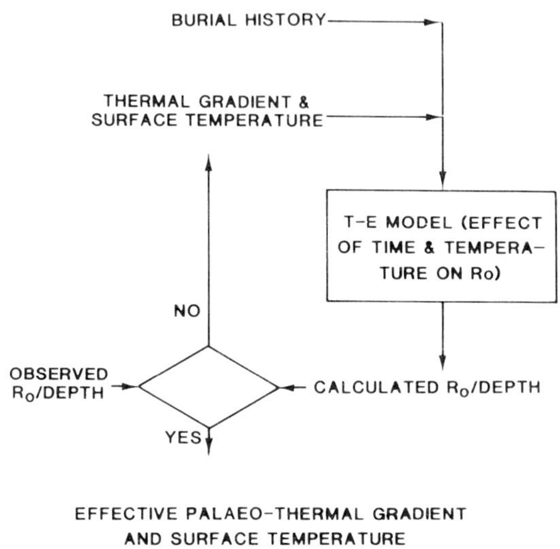

FIG. 8. Flow chart for use of the Tissot–Espitalié type R_0-based palaeotemperature method.

in the burial history can also be made in order to determine the effect on the fit.

The set of Figs. 9 to 12 illustrate a palaeotemperature reconstruction using the Tissot–Espitalié type model. Figure 9 is the burial history, which shows ages in millions of years before present (M.Y.B.P.). This burial history was prepared with allowance for changes in interval thicknesses due to compaction. Note that this particular burial history includes two episodes of erosion. Figure 10 shows observed R_0 values (percent mean vitrinite-reflectance values using oil-immersion optics) versus depth; error bars represent +/− one standard deviation. In this particular example, the present day temperatures determined during logging were especially uncertain. Uncorrected well-log temperatures correspond to a present-day thermal gradient of 1.96°F per 100 ft (3.58°C per 100 m). Two methods of correction yielded gradients of 2.55 and 3.07°F per 100 ft (4.65 and 5.60°C per 100 m). Figure 11 shows depth versus R_0 curves calculated using a surface temperature of 50°F (10°C) and each of these gradients. The measured R_0 values, also plotted in Fig. 11, fall closest to the line calculated

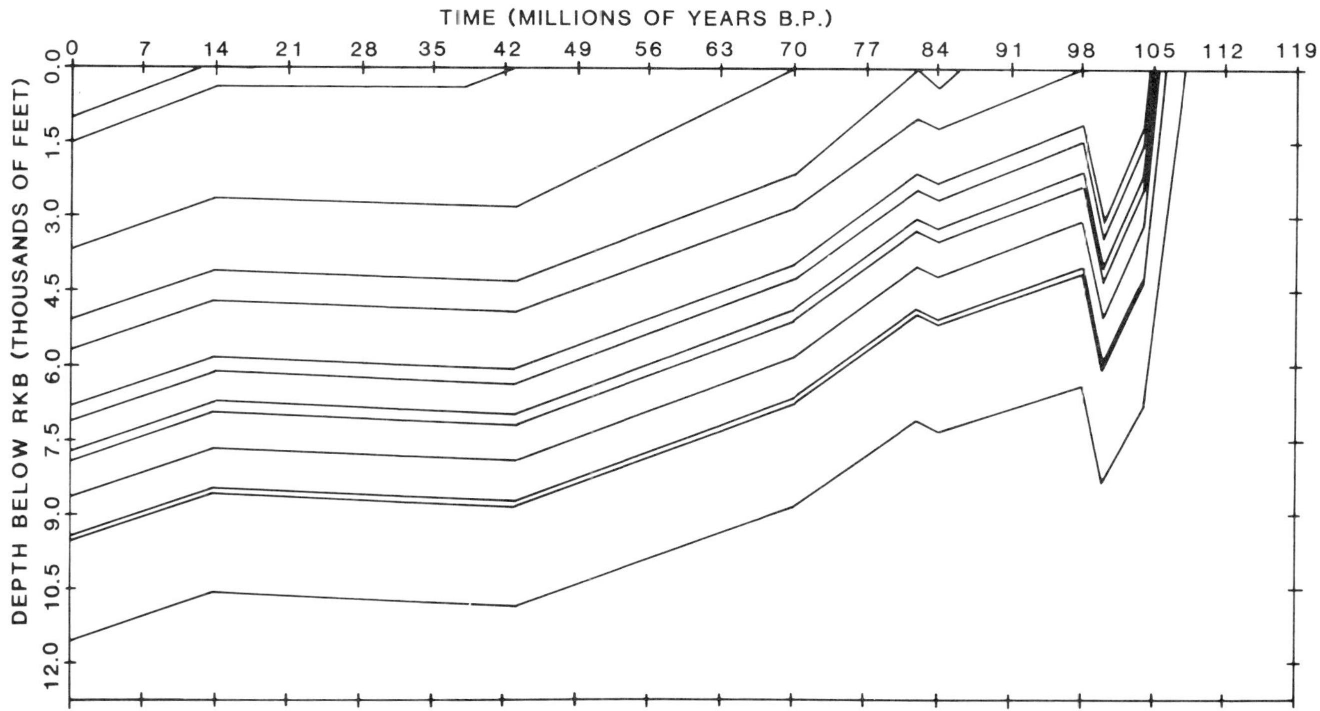

FIG. 9. A burial-history diagram.

using the gradient 2.55°F per 100 ft (4.65°C per 100 m). The fit is significantly improved using the gradient 2.40°F per 100 ft (4.37°C per 100 m), as shown in Fig. 12.

The agreement in Fig. 12 between the observed R_0 values and the calculated line based on an "Effective Palaeothermal Gradient and Surface Temperature" obviously is excellent. This application example is typical of cases where a highly precise match can be achieved between measured

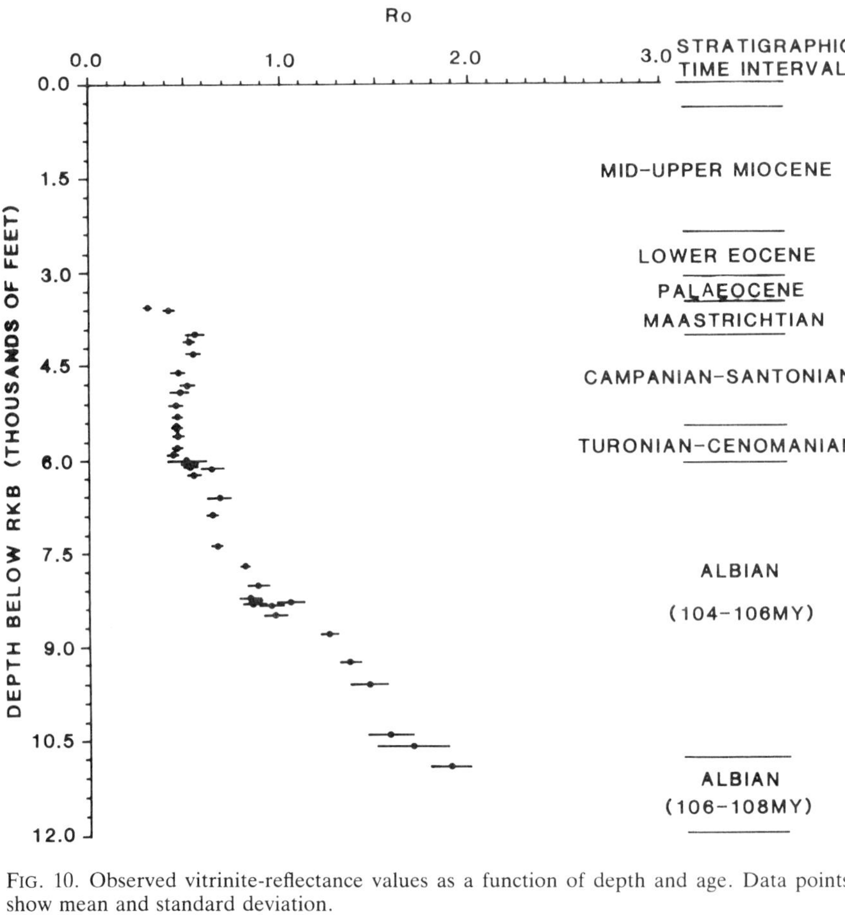

FIG. 10. Observed vitrinite-reflectance values as a function of depth and age. Data points show mean and standard deviation.

and calculated values of R_0 using a single palaeothermal gradient and surface temperature. However, one should also check the validity of the effective palaeothermal gradient using the physical constraints on a sedimentary basin, as discussed in the following section.

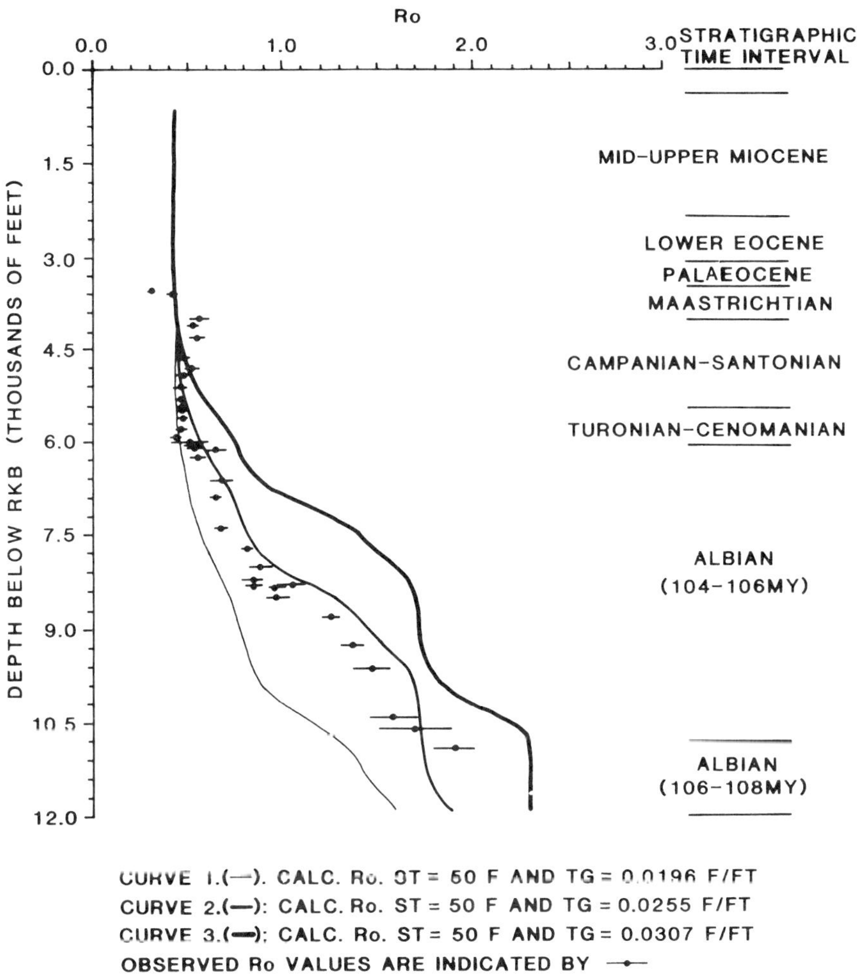

FIG. 11. Calculated vitrinite reflectance versus depth relationship, using the Tissot–Espitalié type model, the burial history shown in Fig. 9, and surface temperatures and thermal gradients obtained from well-log temperatures using three correction methods. Observed vitrinite-reflectance data are also shown.

C. Palaeotemperatures from physical concepts

The simplest method of palaeotemperature determination based on physical concepts is the geothermal-gradient method. The geothermal gradient defines the increase (or, in very few cases, decrease) of temperature with depth. It requires a minimum of two temperature values at different depths

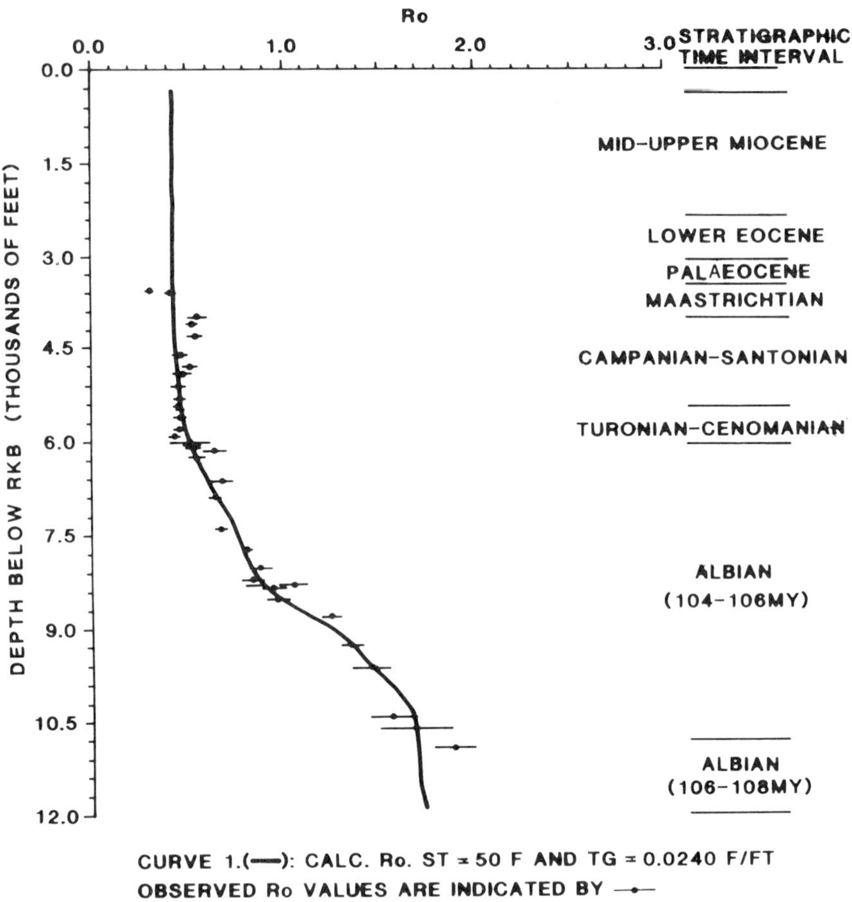

FIG. 12. Optimized fit of calculated and observed vitrinite reflectance.

and assumes that:

(i) heat flow is constant;

(ii) heat distribution occurs by simple conduction and all other modes of heat transfer (convection, radiation, etc.) are negligible;

(iii) changes in thermal conductivities are negligible or linear with depth; and

(iv) no structural deformation or magma intrusion occurs.

Consequently, palaeotemperatures of sedimentary units are computed by multiplying their depths during geological time by the assumed palaeo-geothermal gradient. The vertical movements of the sedimentary units during geological time are determined from burial histories (Schwab, 1976; Katz, 1979; Gretener, 1981).

Correct formation temperatures are critical for the determination of the correct geothermal gradients. The most reliable temperature data can be obtained from drillstem tests and from shut-in producing wells. Static formation-temperature determinations from well-logs have received great attention, due to the ease with which these data can be obtained. The main problem is that the well-log measurements are commonly recorded during short pauses in drilling operations, and thus the disturbed subsurface temperatures cannot return to their pre-drilling levels. Several methods have been developed to determine static formation temperatures from the well-logs (van Everdinger and Hurst, 1949; Edwardson *et al.* 1962; Schoeppel and Gilarranz, 1966), but none has proved to be useful under general conditions.

In an attempt to improve on the well-log method, Houbolt and Wells (1980) developed an empirical equation which relates heat flux, formation temperature and sound travel times (or sound velocity). Their equation is based on Karl's original work (1965), which showed that the ratio of acoustic velocity to thermal conductivity is constant for most water-saturated sedimentary rocks at room temperature. Thermal conductivity is strongly dependent on temperature (Karl, 1965, Shatz and Simmons, 1972), whereas sound velocities are affected to a lesser degree. Schatz and Simmons (1972) gave an empirical formula to show the relationship between thermal conductivity and temperature. Houbolt and Wells then combined these equations with the one-dimensional conductive heat-flow equation and derived their empirical equation.

The other important parameter in the determination of geothermal gradient is depth. The burial or subsidence history of sedimentary units or a sedimentary basin has to be determined to find the depth of a particular sedimentary unit as a function of depth.

Plate-tectonics studies show that the lithosphere between the surface and a depth which can be defined by the isotherm 600°C ± 100°C behaves elastically under loading (Caldwell and Turcotte, 1979). Researchers have studied the lithospheric flexure since the beginning of the twentieth century (Smoluchowski, 1909a, b; Gunn, 1937, 1944, 1947). Lithospheric flexure calculations may be based on thin-plate theory, thick-plate theory, or finite-element numerical approaches. The thin-plate theory assumes that the wavelength of the flexure is long compared with the thickness of the plate, so shear stresses due to vertical loading can be neglected compared with the bending stresses (Turcotte, 1979). The thin-plate theory is a special and simple case of the lithospheric flexure problem, and thus other more general solutions are being sought using the thick-plate theory and/or finite-element numerical approaches under various initial and boundary conditions (Smoluchowski, 1909a, b; Goldstein, 1926; Prager and Hodge, 1951; Nadai, 1963; Walcott, 1970, 1972; Hanks, 1971; Caldwell *et al.*, 1976; Parsons and Molnar, 1976; Watts and Ryan, 1976; Banks *et al.*, 1977; DeBremaecker, 1977; Kirby, 1977; Molnar, 1977; Turcotte *et al.*, 1977; McAdoo *et al.*, 1978; Melosh, 1978; Neugebauer and Spotin, 1978). The results of these studies have been applied to sedimentary basins to determine dynamic basin formation or subsidence history by Haxby *et al.*, (1976), Sleep and Snell (1976), and Beaumont (1978).

Dynamic basin formation is calculated by back-stripping sediments and adjusting the water load for palaeobathymetry and eustatic sea-level changes. The lithospheric flexure is then determined from the rigidity of the lithosphere and the rate of loading. All of these steps may have large errors in them, which may lead to erroneous results. Consequently, flexure models have only limited success in the improvement of the palaeotemperature determinations.

Thermal subsidence models have been developed to determine dynamic basin formation as a function of heat distribution. These models are based on the stretching of the lithosphere, which has often been used to account for the normal faulting and crustal thinning associated with rifting (Watts and Ryan, 1976; Parsons and Sclater, 1977; Turcotte and Ahern, 1977; McKenzie, 1978). The thinning of the lithosphere results in higher heat flux at the upper boundary of the lithosphere, which then decreases as a function of time. The subsequent cooling of the lithosphere causes subsidence by isostacy. As the lithosphere cools, the density increases and subsidence starts. The subsidence causes the formation of sedimentary basins (Turcotte and Ahern, 1977; McKenzie, 1978).

Turcotte and Ahern (1977) developed a model which assumes that the sediment and the basement subside with a constant velocity. The temperature of the sediments are then determined from a one-dimensional heat-conduction equation and a convection term using the velocity of subsidence. This application was for oceanic basins, but McKenzie (1978) proposed a model for the development and evolution of sedimentary basins based on stretching of the continental lithosphere. The model assumes a rapid stretching of the continental lithosphere, which produces thinning and upwelling of the hot asthenosphere. The sudden stretching is then followed by slow cooling of the lower part of the plate and basin subsidence by thermal contraction. Later Jarvis and McKenzie (1980) developed an improved model to investigate the effects of finite extension rates on the heat-flux and subsidence histories of sedimentary basins. In both of these latter models, the subsidence of an empty basin was considered; thus the effects of sediments on the thermal structure of the lithosphere was neglected. The importance of the sediments and their compaction on the heat distribution has been shown by Yükler *et al.* (1978) and Sclater and Christie (1980).

All the thermal subsidence models which have been presented thus far are one- or two-dimensional models that require a high degree of symmetry in a real three-dimensional system. Thermal conductivity and specific heat of sediments are commonly lumped together as thermal diffusivity, which is kept as a constant. The effects of sediment types, sedimentation rates and compaction of sediments on the heat distribution are ignored. Furthermore, heat distribution by water flow is completely ignored. McKenzie (1981) attempted to check the validity of the stretching model by comparing the computed values of vitrinite reflectance with the measured ones, but failed to match the present temperatures. A very common mistake has been made in these models by ignoring the ability of a sedimentary system to store heat under moving boundary conditions. A high heat flux does not yield high temperatures unless the system has an ability to retard heat flow. All these models may have very limited applications under very special conditions where the accumulation of heat in the sedimentary column results in linear geothermal gradients and linear maturity versus depth profiles. One should always remember that such over-simplifications may easily lead to solutions of fictitious problems, but the answers can only be related to the real system through empirical corrections.

The first terrestial heat flow measurements were by Bullard (1939) and Benfield (1939), and the first oceanic heat flow measurements were by Bullard (1954). Oceanic heat flow measurements are made more easily, and consequently these outnumber the continental measurements (Jessop *et al.*, 1976). The Deep Sea Drilling Project in particular yielded a large volume of valuable heat flow data. More work, however, is needed to determine values of palaeo-heat flow for continents and ocean floor (Beck, 1970; Jessop *et al.*, 1976; von Herzen, 1979; Gretener, 1981; Ward *et al.*, 1981). Palaeo-heat flow values are essential in the determination of thermal history in a given basin.

Heat transfer within and through sedimentary sequences takes place by mass transport (conduction), advection (based on water movement) and radiation. Palaeohydrodynamic determinations are essential in advection studies (Sharp and Domenico, 1976; Yükler *et al.*, 1978; Bishop, 1979; Welte and Yükler, 1981). Heat transfer by radiation is believed to play a significant role at temperatures above 500°C (Kappelmeyer and Haenel, 1974).

Studies by Yükler *et al.*, (1978), Yükler and Welte (1980), Welte and Yükler (1981) have shown that thermal-history models should be integrated into the geological-history, palaeohydrodynamic and geochemical models. Sediment types, sedimentation rates, compaction and water flow through and within the sediments have a great effect on the heat distribution. Tectonic events also affect the heat distribution by changing the boundaries of the sedimentary system. The reliable palaeotemperature values can only be obtained from three-dimensional dynamic models. The sedimentary system has to be defined with all the available data and the known mechanisms that operate in it. The models should be based on this system's concept (Welte and Yükler, 1980, 1981). The three-dimensional dynamic deterministic model developed by Yükler (Yükler *et al.*, 1978; Yükler and Welte, 1980; Welte and Yükler, 1981) proved to be a very valuable tool when applied to real basins.

D. Kinetic models of petroleum generation

When palaeotemperatures and pressures have been established, the next step is to determine the extent and timing of petroleum generation. Temperature information can be used in an empirical fashion to estimate the stage of petroleum generation (e.g. Philippi, 1965; Demaison, 1975; Leith and Roswell, 1979). In this review we are stressing the use of chemical-kinetic models as the basis for this evaluation of extent of generation. Such models are also capable of providing estimates of the timing of generation and of the extent to which oil has subsequently been converted to gas.

A large amount of literature exists on the kinetics of formation of hydrocarbons from geo-organic polymers. We will not review the studies concerning the processing of coals or oil shales, which make up the bulk of the published information. We will emphasize kinetic studies that have included geological reconstructions to establish the temperature history of samples that have reacted under geological conditions, and laboratory studies that have been performed for the purpose of understanding processes on a geological scale, or that have been applied to this purpose.

Perhaps the earliest application, or at least the suggestion for application, of chemical-kinetic studies to petroleum exploration was by White (1930). White (1915) is widely credited for introducing the use of rank of associated coal beds in assessing maturity of petroleum source beds. He suggested in 1930 that the results of Maier and Zimmerly (1924) for the retorting of a Utah oil shale could be used to determine the "exchange of time for temperature in petroleum generation". This is the essence of kinetic models.

In the section of palaeotemperature reconstruction from geochemical concepts, we have already mentioned the kinetic models of coalification.

Many of these are stated in terms of vitrinite reflectance levels. But if restated in terms of volatile matter content or other descriptors of coal rank, they could be used as kinetic models of petroleum genesis for coaly-type organic matter. The model of Lopatin (1971) is widely used in this way (e.g. Waples, 1980). Detailed laboratory kinetic studies of thermal decomposition of coals have been published by Hanbaba and Jüntgen (1968) and Suuberg *et al.*, (1978). Other kinetic studies of laboratory thermal decomposition of coals have been reviewed by Yellow (1965).

Tissot (1969, 1973), Tissot and Pelet (1971) and Tissot and Espitalié (1975) have described the development of a kinetic model for petroleum generation based on both geological reconstructions and laboratory studies. This model will be discussed in detail. Koncz (1977) also has used laboratory studies to obtain kinetic data for petroleum generation specifically intended to be applicable to exploration problems.

In the Tissot–Espitalié model (1975), petroleum generation is described kinetically as a two-step process:

$$\text{kerogen} \rightarrow \underset{\text{carbonaceous residue}}{\text{oil} + CO_2 + H_2O +} \rightarrow \underset{\text{carbonaceous residue.}}{\text{gas} + CO_2 + H_2O +}$$

The kerogen-to-oil step is described in terms of six independent kerogen components, but oil is treated as a single component for the oil-to-gas step. Because of the use of multiple kerogen components, the model can accommodate the formation of some gas directly from kerogen, in spite of what the equation implies. Under conditions where the rate of conversion of oil to gas is faster than the rate of formation of oil from a particular kerogen component, the gas is essentially formed directly from the kerogen component. All reaction steps are treated as kinetically first order; that is, the rate is proportional to the amount of reactant. The Arrhenius equation is used to describe the temperature dependence of the rate constant. The equation that results, for a particular kerogen, relates amounts of oil and gas to time and temperature and contains twenty-one parameters that include six "genetic potentials" of each kerogen component, seven activation energies, seven Arrhenius factors and an initial oil content.

The data necessary to develop the model and parametrize it were obtained in studies in which selected geological formations were sampled throughout the basin. For each formation, samples were collected in states from the thermally immature to the mature, and for each sampling location the burial history was reconstructed. Two kinds of geochemical analysis were performed on the samples. The first was a conventional source-rock analysis in which the extractable organic matter and organic carbon content were obtained. The second consisted of determination of weight loss or hydrocarbon yield during temperature-programmed pyrolysis in an inert atmosphere. Total hydrocarbon potential was determined by Rock-Eval type analyses. These organic geochemical characterizations were performed for kerogens of three particular geological formations and the kerogens are examples of distinct kerogen "Types".

The data generated allow the state of petroleum generation of each sample of a particular formation to be described in terms of the "transformation ratio". If X is the hydrocarbon-generating potential of a particular sample in units of milligrams of hydrocarbon per gram of organic carbon, and X_0 is its initial hydrocarbon potential (as determined from pyrolysis of an immature sample), then the transformation ratio (TR) can be written:

$$\mathrm{TR} = (X_0 - X)/X_0.$$

The transformation ratio can have values between zero and one, and is the fraction of the hydrocarbon potential of a sample that has already been realized. For samples that have not been subjected to time/temperature conditions severe enough to cause the cracking of oil to gas, and where all of the oil that has been formed has remained in place, the transformation ratio can also be evaluated using the initial potential and the extractables content, Y, which has units of milligrams of extract per gram of organic carbon:

$$\mathrm{TR} = (Y - Y_0)/X_0 \text{ (all oil has remained in place).}$$

(For the components that react with low activation energies, TR must be determined in this way, since increased temperature does not adequately accelerate reaction.)

The parameters of the model were chosen so that it can reproduce the observed effects of time and temperature on transformation ratios determined on native samples and in artificial ageing experiments. The details of precisely how this was done are not available. Tissot and Espitalié (1975) published the parameters for the Green River formation of the Uinta Basin, U.S.A., the Lower Toarcian shale of the Paris Basin, France, and for an Upper Cretaceous shale of the Douala Basin, Cameroon. Organic geochemical characterizations of these formations have been reported by Tissot *et al.* (1971, 1978), Albrecht *et al.* (1976), Durand *et al.* (1976) and Vandenbroucke *et al.* (1976); they are the prototypes of kerogen Types I, II and III respectively.

The description of the kerogen-to-oil step in terms of a set of kerogen components that react by parallel reactions allows the model to explain observations, like those of Leventhal (1976), that in laboratory stepwise pyrolyses of sedimentary rocks the amount of hydrocarbon product released at a given final temperature is limited. Kerogen isolated from recent sediments (Ishiwatari *et al.*, 1977) and coals (Berkowitz, 1960; Pitt, 1962) show similar behaviour. Kinetic studies of the pyrolysis of coals have also demonstrated the necessity of a range of activation energies to adequately describe the experimental results (Suuberg *et al.*, 1978). On the other hand, a single-step reaction mechanism seems adequate to describe the formation of oil from Green River oil shale during laboratory retorting experiments (Shih and Sohn, 1980).

Ultimately, the validity of any kinetic model is determined by the accuracy with which it describes the extent of reaction versus time and temperature. For reaction under geological conditions of time and temperature, Tissot and Espitalié (1975) published comparisons (Fig. 13) of

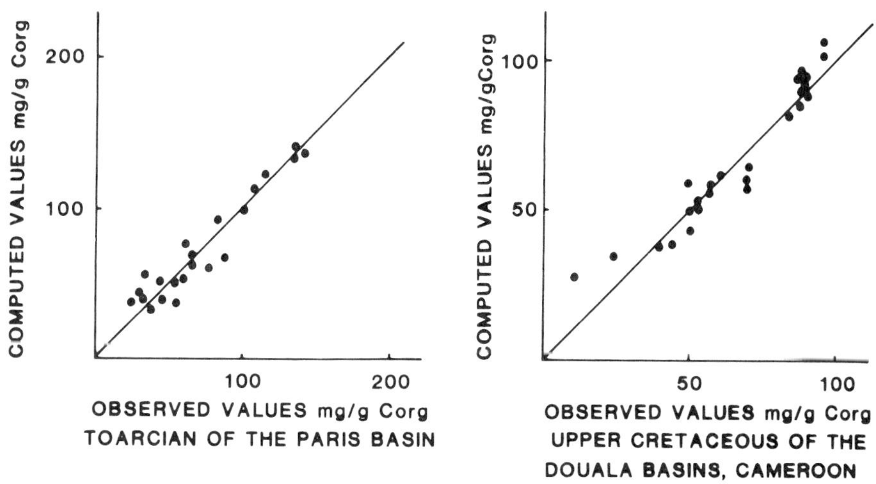

FIG. 13. Comparison of observed amount of extractable organic matter with amounts calculated from the Tissot–Espitalié model. From Tissot and Espitalié (1975). *Adapted with permission.*

the observed and calculated amounts of oil formed from Type II and III kerogens. Of necessity these include only part of the total hydrocarbon potential. The comparisons of observed and calculated extent of reaction under laboratory conditions are shown in Fig. 14 for Types II and III. In each case the agreement is satisfactory, although it must be recognized that the comparisons are most complete for laboratory conditions.

The ability of the model to reproduce hydrocarbon evolution behaviour under laboratory conditions can be considered a necessary but not a sufficient condition for its validity and application to reactions occurring under geological conditions of time and temperature. Snowdon (1979) criticized the application of kinetic parameters obtained in the laboratory at high temperatures and short times to geological problems. The potential disadvantages presented by Snowdon are well known to chemical kineticists. There are also less well known problems arising from the fact that kerogen is a solid (Yellow, 1965).

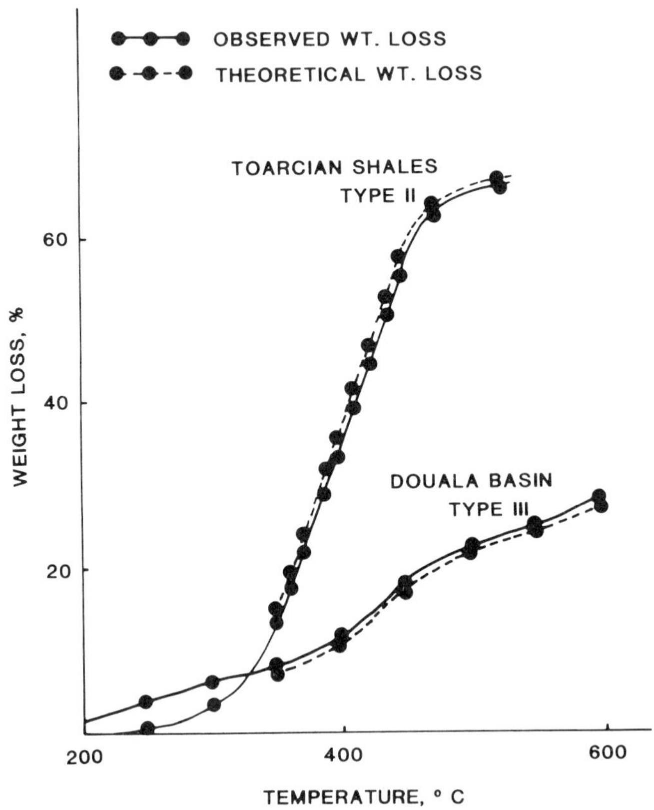

FIG. 14. Observed and calculated total weight losses in laboratory thermogravimetric (TGA) experiments. From Tissot and Espitalié (1975). *Adapted with permission.*

In defence of the use of kinetic models of petroleum generation, and of the Tissot–Espitalié type model in particular, we want to note that the parameters of the model are only partly based on laboratory experiments that involved reaction at high temperatures. The parameters are based at least in part on the analysis of samples of a given formation that have experienced different geological temperature histories. The extent to which the latter sort of data is used in the formulation of a kinetic model depends on the availability of suitable samples. In the absence of such samples, the geological result must be simulated in laboratory heating studies. Snowdon (1979) and other critics of kinetics-based models do not seem to recognize this dilemma and have not suggested an alternative approach.

Furthermore, there are similarities between the chemistry that occurs at high temperature and that under geological conditions. The strongest argument for the validity of laboratory ageing experiments comes from the parallelism in bulk changes in chemical composition and structure of kerogens under laboratory and geological conditions. Oberlin *et al.* (1974a, b) used a van Krevelen diagram, Fig. 15A, to compare the natural evolution paths of kerogens with artificial evolution paths of kerogens that were heated for extended periods at the temperatures noted. The natural and artificial evolutionary paths are very similar, and electron micro-diffraction studies showed that the progressive structural reorganization of the kerogen into graphite in the artificially aged samples parallels the reorganization in the natural samples. Similar results for kerogens from the formations studied by Tissot and Espitalié are shown in Fig. 15B, (Alpern *et al.*, 1978). Since artificial ageing duplicated the gross changes in elemental compositions of the kerogens, mass balance requires that it also duplicates the gross elemental composition of the other products. At least for Type I and II kerogens, where CO_2 and H_2O production is limited, we can infer that bulk compositions of the petroleum products will be similar under natural and artificial ageing conditions.

There are, however, a few problems concerning the model. First, weight-loss measurements serve as the basis of some of the kinetic parameters, although genetic potentials are reported as yields in units of milligrams of hydrocarbon per gram of organic carbon. It would have been preferable to directly measure hydrocarbon evolution. Secondly, many of the mathematical and experimental details of the derivation of the kinetic parameters have not been published. For a kinetic mechanism as complicated as that employed by Tissot and Espitalié, the fitting process by which the parameters are obtained is not straightforward. Finally, the fraction of the total hydrocarbon potential that can be realized as oil is extremely sensitive to kinetic parameters for the oil-to-gas step. Information on the choice of these parameters is exceptionally sketchy. Additional information on these points would be very useful and welcome.

In application of the model to real problems, it must be remembered that the published kinetic parameters refer to the kerogens found in particular geological formations. For example, the extent to which the parameters for the Douala Basin kerogen can be applied to other Type III kerogens is unknown. An even more difficult problem is encountered when

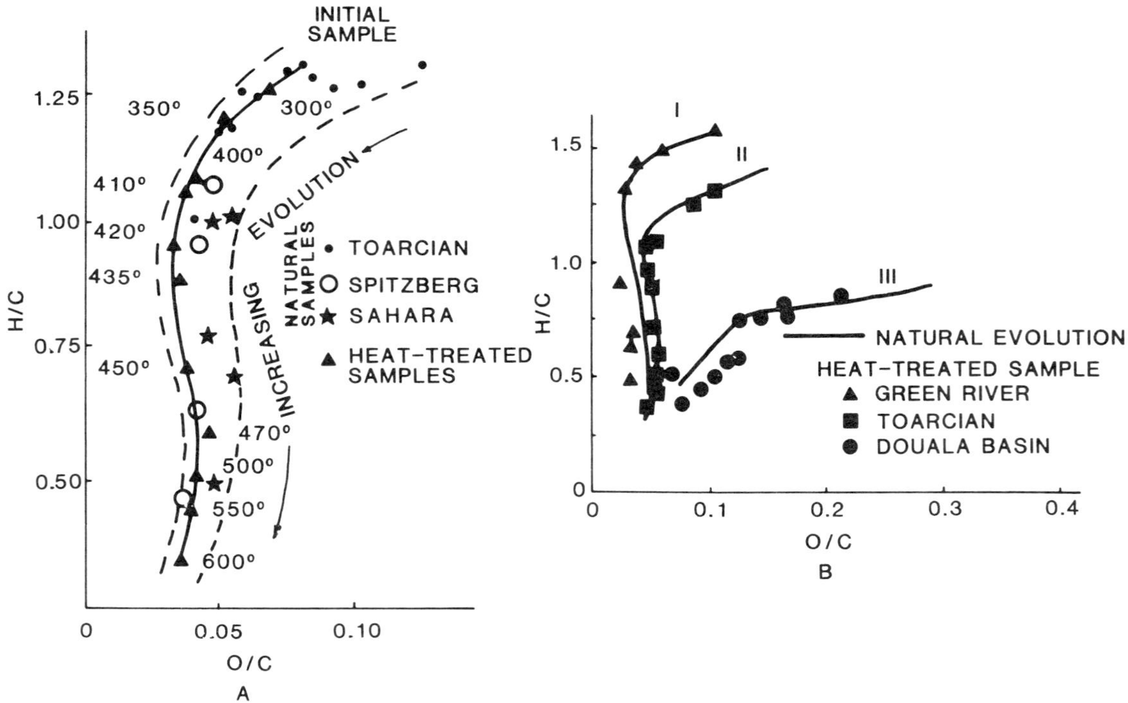

FIG. 15. Comparisons of the "evolutionary pathways" for naturally and artificially aged kerogens: (A) artificial ageing by constant-temperature pyrolysis; and (B) by temperature-programmed pyrolysis. From Oberlin *et al*. (1974) and Alpern *et al*. (1978) respectively. *Adapted with permission.*

the kerogen in the area of interest must be described as a mixture of kerogen types. It is very likely that a particular kerogen cannot be adequately described as a mixture of the three kerogens for which Tissot and Espitalié have published parameters. Dungworth (1981) has emphasized the need to determine the kinetic parameters for the particular kerogen whose behaviour will be modelled.

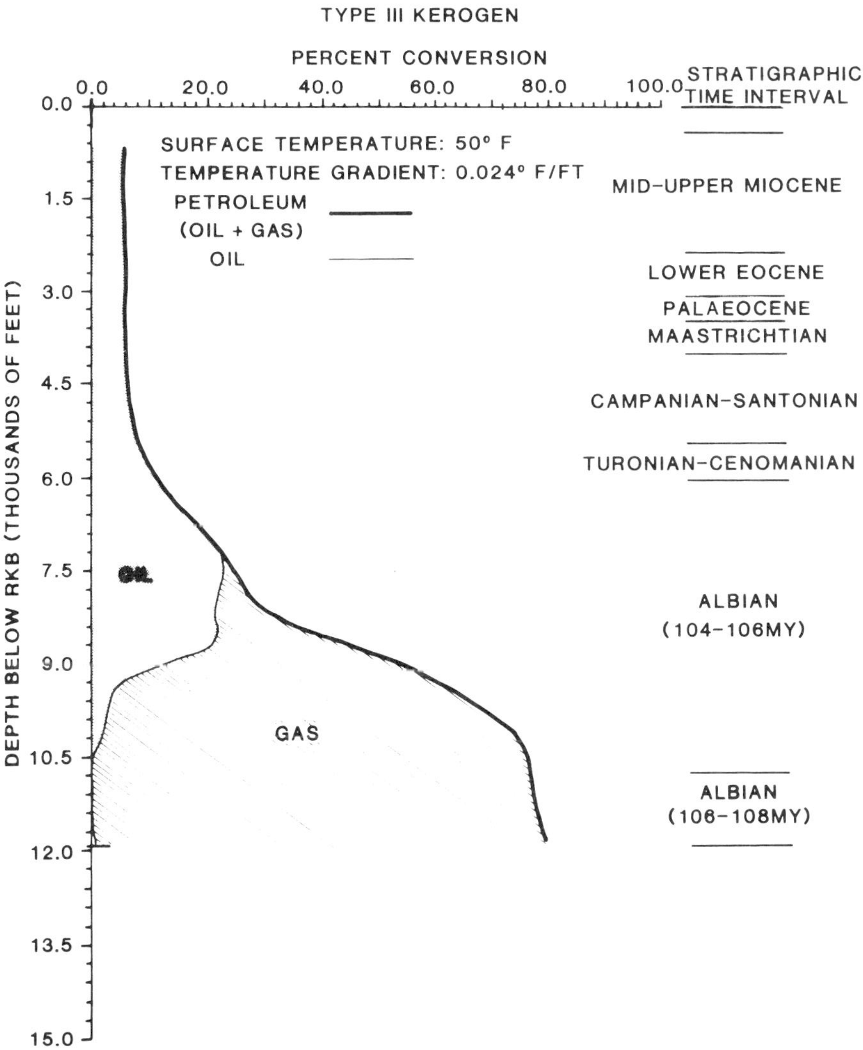

FIG. 16. Extent of conversion of kerogen to oil and gas, calculated using the Tissot–Espitalié model.

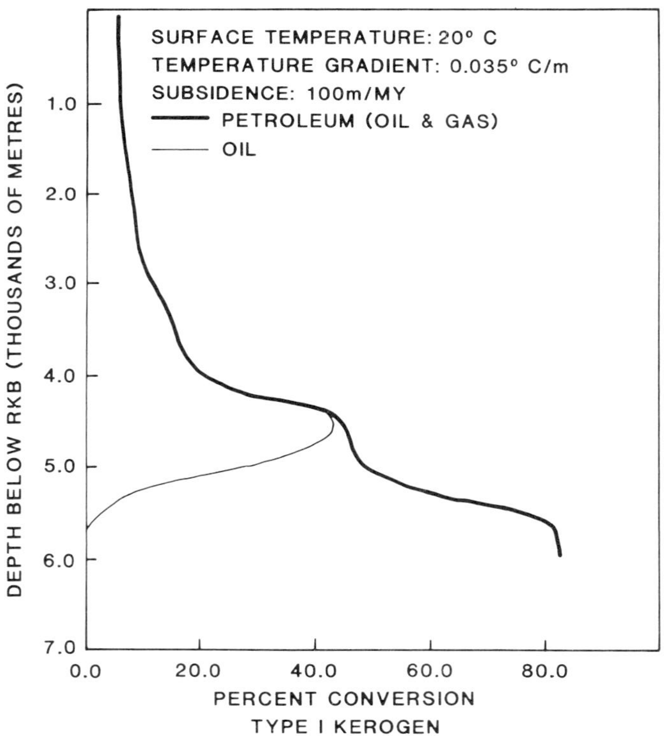

FIG. 17(*a*). Comparison of the sensitivities of kerogen Type I to time and temperature according to the Tissot–Espitalié model. The extent reaction curves were calculated assuming a constant rate of burial.

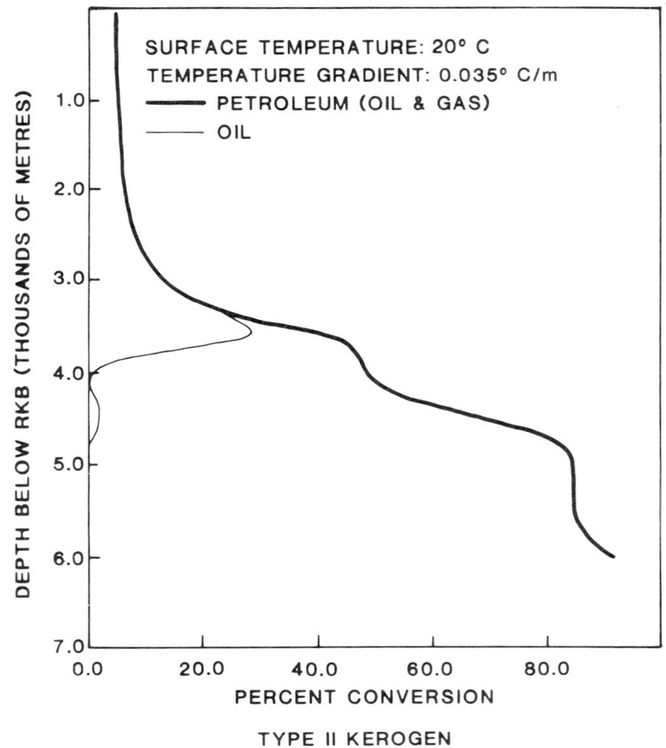

FIG 17(*b*). Comparison of the sensitivities of kerogen Type II to time and temperature.

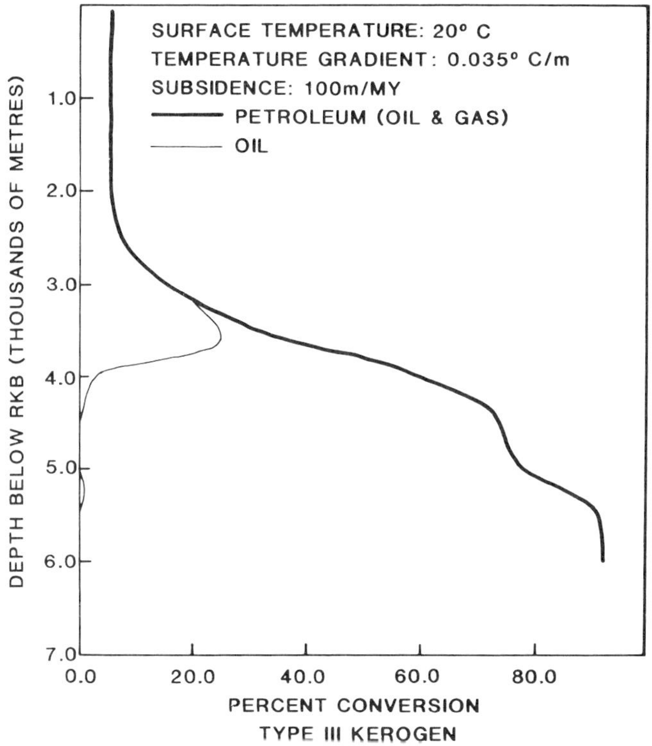

FIG. 17(*c*). Comparison of the sensitivities of kerogen Type III to time and temperature.

The Tissot–Espitalié type model and its parameters, together with a thermal history, can be used to calculate the extent of oil and gas generation as a function of depth. Figure 16, which is based on the burial history in Fig. 9, shows the extent of petroleum (oil + gas) and gas generation as a function of present-day depth for Type III kerogen. The calculation for the extent of gas formation is based on the assumption that any oil formed remains in place. If it migrated to a cooler reservoir interval then it could be (at least partially and temporarily) preserved. The extent reaction curve starts at a value greater than zero because the extractable organic matter originally present is included as "oil". (However, this extractable content is ignored in calculating the extent of kerogen degradation for palaeotemperature calculations.)

For a given temperature history, the timing of generation differs for the different kerogen types. In Figs. 17(*a*), (*b*) and (*c*), the timing of generation is shown for each prototype kerogen using an idealized temperature history in which burial occurs at a rate of 100 m per million years, the thermal gradient is 0.035°C per m and the surface temperature is 0°C. The generation curves for Types II and III are fairly similar, with generation becoming significant at about 90°C, and going through a maximum at about 125°C. For Type I generation begins at about 105°C and goes through a maximum at about 160°C.

In the next step, for each interval in the geological section for which a total organic carbon (TOC) and kerogen type information can be supplied,

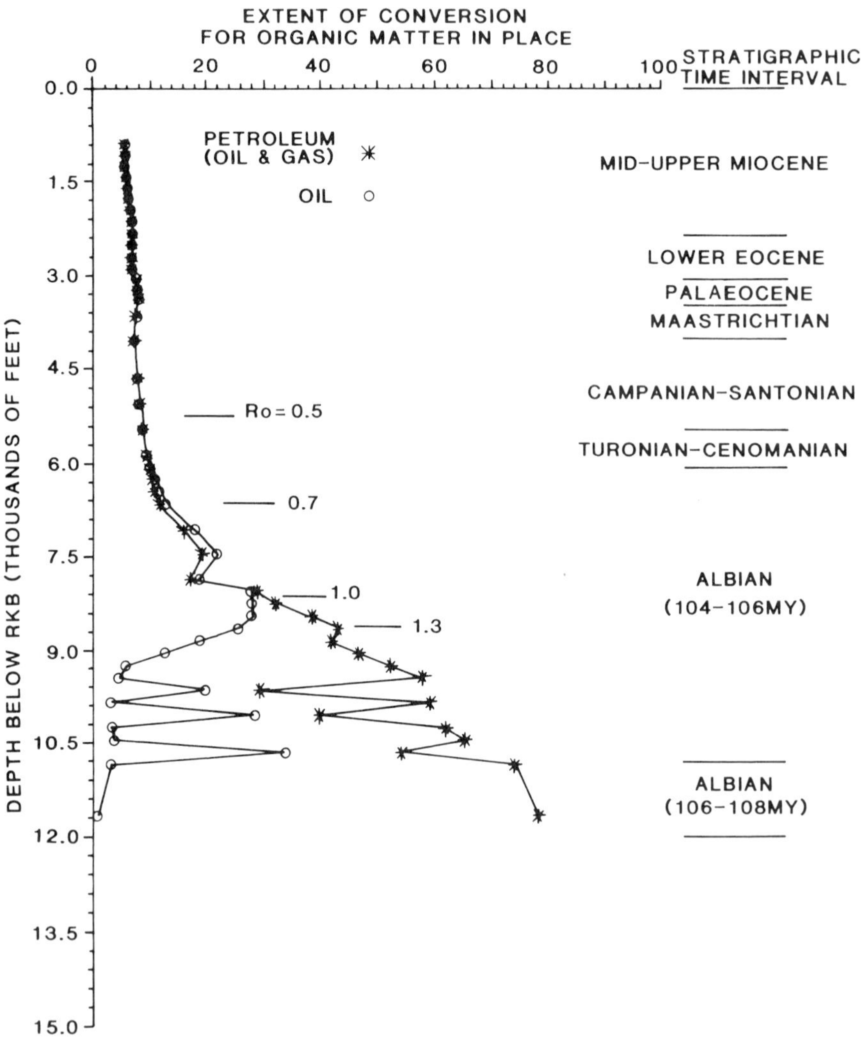

FIG. 18. Extent of generation curves for a case of changing kerogen type.

the amounts of oil and gas generated can be calculated. The specification of the kerogen type implies a genetic potential and an initial content of extractables. For intervals in which the kerogen is not equivalent to one of the prototypes, the kerogen can be treated as a mixture of the prototypes using visual kerogen, elemental composition or pyrolysis potential data

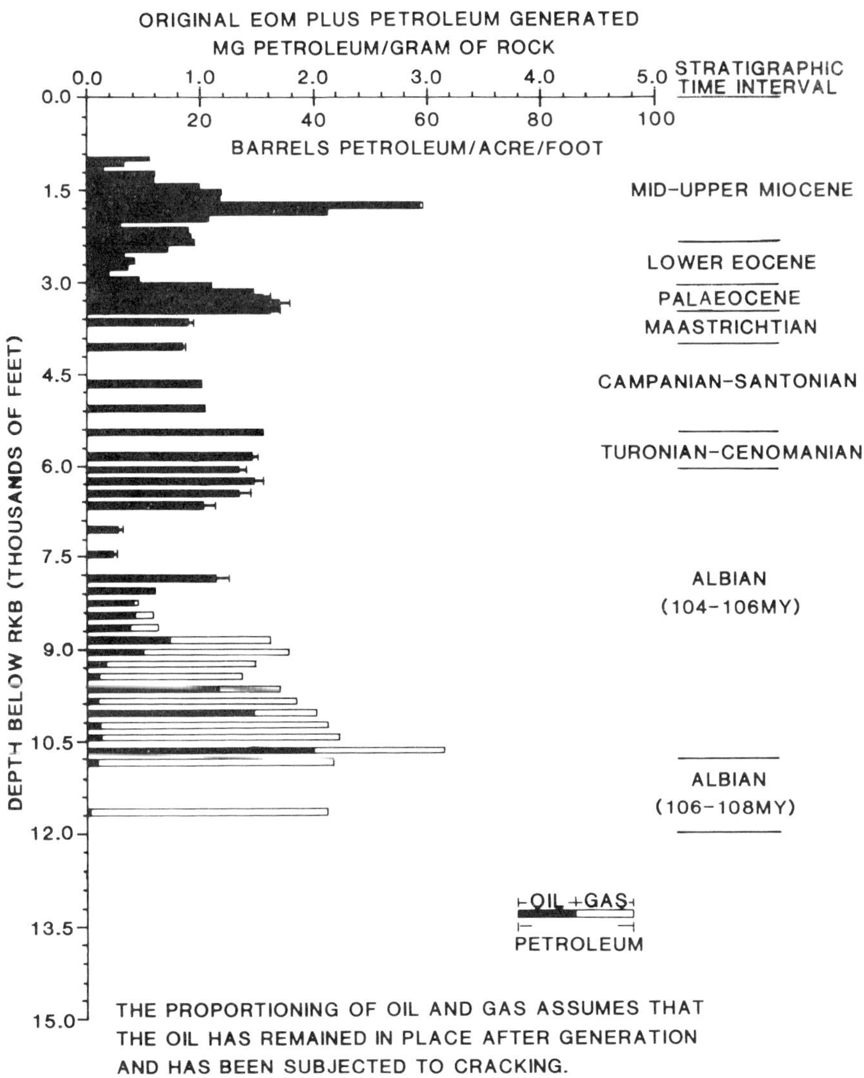

FIG. 19. Amounts of petroleum generated verses depth. These results are based on Fig. 18 and total organic carbon and kerogen type.

(Barnard *et al.*, 1981a). However, the results of this "kerogen typing" process are only approximate. The ambiguity in results from kerogen type assignment is eliminated when kinetic parameters and genetic potential are determined for kerogen from the interval of particular interest (Dungworth, 1981).

Figure 18 shows curves of extent of petroleum and gas generation for a section containing a mixture of kerogen types. (For comparison, measured R_0 values also are indicated.) The spikiness of the curves between 9000 and 12000 ft is the result of some intervals containing Type I kerogen. These maturities, together with TOC values, genetic potentials and original extractables contents, can be used to calculate the amount of oil and gas generated in each sampled interval, as shown in Fig. 19.

IV. Conclusions

This review has described some of the ways in which organic geochemical and geological information can be (and has been) integrated and applied to problems of petroleum exploration by way of quantitative models. The potential of these models is currently limited by both intrinsic and extrinsic factors.

The intrinsic limitations are those that result from the state of our geochemical knowledge. The full promise of the geochemical techniques depends on improved methods for rapid and accurate identification of kerogen types, and the relationship of kerogen type with geochemical and geological palaeoenvironment. The reliability of the answer to the question "how much?" is directly related to the kerogen type and its spatial distribution (organofacies).

Various quantitative model studies have shown that better kinetic data on kerogen degradation and petroleum generation are needed. The kinetic parameters in the literature have been found by curve fitting in a multiparameter system. It is very difficult, if not impossible, to obtain accurate values of the parameters with this technique. Sensitivity analysis has to be used in the determination of the kinetic parameters.

Kinetic models deal only with the kinetics and not the dynamics of the system, that is, kerogen degradation and petroleum generation are computed as functions of time without the forces which cause these changes. Constraints of chemical mass balance have to be integrated into the kinetic models.

The generation, migration and accumulation of petroleum takes place in the three-dimensional dynamic geological framework of a sedimentary basin. They are controlled by very complex and interrelated mechanisms such as sedimentation, tectonic movements, erosion, cementation, water flow, heat flow, mineralogic changes and chemical and physico-chemical changes. To solve this difficult problem, scientists from many fields such as geology, geophysics, geochemistry, hydrodynamics, geothermics, mathematics and computer science should work together in an integrated and well coordinated manner. The extrinsic limitations, thus, relate to problems of communications between these scientists.

Acknowledgments

We acknowledge the important contributions to ideas in this review of our colleagues J. G. Erdman, F. W. Beghtel, M. D. Brondos, W. B. Hughes, W. G. Lyon, K. S. Schorno, D. Schumacher, J. S. Shveima, M. A. Sloan, J. E. Smith and K. R. Sundberg. We also thank the management of Phillips Petroleum Company, in particular B. M. Thompson, J. R. Davis and M. N. McElroy, for permission to publish this review.

Special acknowledgments are made to W. A. Fowler, Jr., for his editorial comments, and to D. A. Juckett for his recommendations concerning the organization of this review and his invaluable assistance in the publication process.

References

Albrecht, P., Vandenbroucke, M. and Mandengue, M. (1976). *Geochim. Cosmochim. Acta.* **40,** 791–799.

Alpern, B. (1980). *In* "Kerogen—Insoluble Organic Matter from Sedimentary Rocks" (Durand, B., Ed.), 339–383. Editions Technip, Paris.

Alpern, B., Durand, B. and Durand-Souron, C. (1978). *Rev. Inst. Fran. Petr.* **33,** 867–890.

Ammosov, I. I. (1981). *Int. Geol. Rev.* **23,** 406–416.

Arps, J. C. and Roberts, T. G. (1958). *AAPG Bull.* **42,** 2549–2566.

Attanasi, E. D., Drew, L. J. and Schuenemeyer, J. H. (1980). *U.S. Geol. Surv. Professional Paper 1138-C.*

Atwater, G. I. (1956). *AAPG Bull.* **40,** 2624–2634.

Bally, A. W. (1975). *Proc. Ninth World Petr. Cong., Tokyo* **2,** 33–44.

Banks, R. J., Parker, R. L. and Huestic, S. P. (1977). *Geophys. J. Roy. Astron. Soc.*, **51,** 431–452.

Barnard, P. C., Collins, A. G. and Cooper, B. S. (1981a). *In* "Organic Maturation Studies and Fossil Fuel Exploration" (Brooks, J., Ed.), 271–282. Academic Press, London, Orlando and New York.

Barnard, P. C., Collins, A. G. and Cooper, B. S. (1981b). *In* "Organic Maturation Studies and Fossil Fuel Exploration" (Brooks, J., Ed.), 337–342. Academic Press, London, Orlando and New York.

Beaumont, C. (1978). *Geophys. J. Roy. Astron. Soc.* **55,** 471–497.
Beck, A. E. (1970). *Geophysics* **1,** 29–35.
Belov, K. A. (1960). *Petr. Geol.* **4,** 185–192.
Benfield, A. E. (1939). *Proc. Roy. Soc. London*, Ser. A, Vol. 173, 428–450.
Berkowitz, N. (1960). *Fuel* **39,** 47–58.
Bishop, R. S. (1979). *AAPG Bull.* **63,** 918–933.
Bonham-Carter, G. F. and Sutherland, A. J. (1968). *Kansas Geol. Surv. Computer Contr.* **24.**
Bostick, N. H. (1973). *Cong. Int. Stratigr. Geol. Carbonifère Compt. Rend.* **7,** 183–189.
Bostick, N. H. (1979). *SEPM Spec. Publ. No. 26*, 17–43.
Bostick, N H. and Foster J. N. (1975). *In* "Petrographie de la Matière Organique des Sediments, Relations avec Palaeo-temperature et le Potential Petrolier" (Alpern, B., Ed.), 13–25. C.N.R.S., Paris.
Bray, E. E. and Evans E. D. *Geochim. Cosmochim. Acta* **22,** 2–15.
Bullard, E. C. (1939). *Proc. Soc. London, Ser. A* **173,** 474–502.
Bullard, E. C. (1954). *Proc. Roy. Soc. London, Ser. A.* **222,** 408–429.
Buntebarth, G. *Pure Appl. Geophys.* **117,** 83–91.
Buntebarth, G. (1982). *Tectonophys.* **83,** 101–108.
Caldwell, J. G. and Turcotte, D. L. (1979). *J. Geophys. Res.* **84,** 7572–7576.
Caldwell, J. G., Haxby, W. F., Karig, W. F. and Turcotte, D. E. (1976). *Earth Planet Sci. Lett.* **31,** 239–246.
Castano, J. and D. M. Sparks, D. M. (1974). *Geol. Soc. Am. Spec. Paper No. 153*, 31–52.
Chorafas, D. N. (1965) "Systems and Simulation". Academic Press, Orlando, New York and London.
Chorley, R. J. and Haggett, P. (Eds.) (1967). "Models in Geography". Methuen, London.
Connan, J. (1974). *AAPG Bull.* **58,** 2516–2521.
Conybeare, C. E. B. (1963). *Bull. Can. Petr. Geol.* **13,** 509–528.
Cooper, B. S. (1977). *In* "Developments in Petroleum Geology" (Hobson, G. D., Ed.), 127–146. Applied Science Publishers, London.
Davis, J. C. (1977). *Mathematical Geol.* **9,** 409–427.
Davis, W. (1958). *Oil and Gas*, 24 Feb., 105–119.
DeBremaecker, J. C. (1977). *J. Geophys. Res.* **82,** 2001–2004.
Demaison, G. J. (1975). *In* "Petrographie de la Matière Organique des Sediments, Relations avec la Palaeo-temperature et le Potential Petrolier" (Alpern, B., Ed.), 217–274. C.N.R.S., Paris.
Devereaux, I. (1967). *New Zealand J. Sci.* **10,** 988–1011.
Dorman, F. H. (1966). *J. Geol.* **74,** 49–61.
Dow, W. G. (1977). *J. Geochem. Explor.* **7,** 79–99.
Drew, L. J., Schuenemeyer, J. H. and Bawiec, W. J. (1979). *U.S. Geol. Surv. Miscellaneous Investigation Series G.*
Dungworth, G. (1981). *First European Union of Geosciences*, 13–16 *April, Strasbourg, Terra Cognita.*
Durand, B. and Espitalié, J. (1976). *Geochim. Cosmochim. Acta* **40,** 801–808.
Durand, B. and Monin, J. C. (1980). *In* "Kerogen—Insoluble Organic Matter from Sedimentary Rocks" (Durand, B., Ed.), 113–142. Editions Technip, Paris.
Durand, B., Marchand, A., Amiele, J. and Combaz, A. (1967). *In* "Advances in Organic Geochemistry 1968" (Schenck, P. A. and Havenaar, I., Eds.). Pergamon Press, Oxford.

Edwardson M. J., *et al.* (1962). *J. Petr. Tech.*, April, 416–426.
Ejedawe, J. E. (1981). *AAPG Bull.* **71,** 1574–1585.
Energy, Mines and Resources Canada (1977). *Oil and natural gas resources of Canada, 1976, Ottawa, Report EP77–1.*
Gehman, H. M., Baker, R. A. and White, D. A. (1975). *In* "Probability Methods in Oil Exploration," (Davis, J. C., *et al.*, Conveners), 16–20. AAPG Research Symposium, Stanford Univ.
Gill, W. D., Khalaf, F. I. and Massoud, M. S. (1979). *J. Petr. Geol.* **7,** 39–62.
Goldstein, S. (1926). *Proc. Cambridge Phil. Soc.* **23,** 120–129.
Gretener, P. L. (1981). "Geothermics: Using Temperature in Hydrocarbon Exploration", AAPG Education Course Note Series 17. AAPG, Tulsa.
Griffiths, J. C. (1966). *Operations Research*, March–April, 189–209.
Griffiths, W. C. (1967). "Scientific Method in the Analysis of Sediments". McGraw-Hill, New York.
Guiliani, C. (1954). *Science* **119,** 853–855.
Guiliani, C. (1956). *J. Geology* **64,** 281–288.
Gunn, R. (1937). *J. Franklin Inst.* **244,** 19–53.
Gunn, R. (1944). *J. Franklin Inst.* **237,** 139–154.
Gunn, R. (1947). *Geophysics* **12,** 238–255.
Gutjahr, C. C. M. (1966). *Leidse Geologische Mededelingen* **38,** 1–29.
Hacquebard, P. A. and Donaldson J. R. (1970). *Can. J. Earth Sci.* **7,** 1139–1158.
Halbouty, M. T. and Hardin, G. C., Jr. (1959). *Proc. Fifth World Petr. Cong., New York*, **1,** 95–108.
Hambleton, W. W., Davis, J. C. and Doveton, J. H. (1975). *In* "Methods of Estimating the Volume of Undiscovered Oil and Gas Resources" (Haun, J. D., Ed.), AAPG Studies in Geology 1, 171–185. AAPG, Tulsa.
Hanbaba, P. and Jüntgen, H. (1968). *In* "Advances in Organic Geochemistry 1968" (Schenck, P. A. and Havenaar, I., Eds.), 459–471. Pergamon Press, Oxford.
Hanks, T. C. (1971). *Geophys. Jr. Roy. Astron. Soc.* **23,** 173–189.
Harbaugh, J. W. (1966). *Kansas Geol. Surv. Computer Contr.* **1.**
Harbaugh, J. W. and Bonham-Carter, G. (1970). "Computer Simulation in Geology". John Wiley and Sons, New York.
Harbaugh, J. W. and Merriam, D. F. (1968). "Computer Applications in Stratigraphic Analysis". John Wiley and Sons, New York.
Harbaugh, J. W., Doveton, J. H. and Davis, J. C. (1977). "Probability Methods in Oil Exploration". John Wiley, New York.
Haxby, W. F., Turcotte, D. L. and Bird, J. M. (1976). *Tectonophys.* **36,** 57–75.
Heroux, Y., Chagnon, A. and Bertrand, R. (1979). *AAPG Bull.* **63,** 2128–2144.
Hood, A. and Castano, J. R. (1974). *Coordinating Comm. Offshore Prospecting Techn. Bull.* **8,** 85–118.
Hood, A., Gutjahr, C. C. M. and Heacock, R. L. (1975). *AAPG Bull.* **59,** 986–996.
Houbolt, J. J. H. C. and Wells, P. R. A. (1980). *Geologie en Mijnbouw* **59,** 215–224.
Howell, I. V. (1934). *In* "Problems of Petroleum Geology" (Wrather, W. E. and Lahee, F. H., Eds.), 1–23. AAPG, Tulsa.
Hubbert, M. K. (1940). *J. Geol.* **48,** 784–944.
Hubbert, M. K. (1953). *AAPG Bull.* **37,** 1954–2026.
Hubbert, M. K. (1962). *Natl. Acad. Sci.-Natl. Research Council Pub.*, 1000-*D*.
Hubbert, M. K. (1967). *AAPG Bull.* **51,** 2207–2227.

Hubbert, M. K. (1974). *U.S. Senate Comm. on Interior and Insular affairs*, Rept. 93–40 (92–75), Pt. 1.
Huck, G., and Karweil J. (1955). *Brentstoff-Chemie* **36,** 1–11.
Hunt, J. M. (1979). "Petroleum Geochemistry and Geology". W. H. Freeman and Co., San Francisco.
Illing, V. C. (1938). *In* "The Science of Petroleum" (Dunstan, A. E. *et al.*, Eds.), Vol. 1, 209–215. Oxford Univ. Press, London.
Illing, V. C. (1939). *Inst. Petr. J.* **25,** 201–225.
Ishiwatari, R., Ishiwatari, M., Rohrback, B. G. and Kaplan, I. R. (1977). *Geochim. Cosmochim. Acta* **41,** 815–828.
Ivanhoe, L. F. (1976). *Oil and Gas J.*, 6 Dec., 154–156.
Jarvis, G. T. and McKenzie, D. P. (1980). *Earth Planet Sci. Lett.* **48,** 42–52.
Jessop, A. M., Hobart, M. A. and Sclater, J. G. (1976). "The World Heat Flow Data Collection—1975", Geothermal Series No. 5. Earth Physics Branch, Geothermal Service of Canada.
Jones, R. W. (1975). *In* "Methods of Estimating the Volume of Undiscovered Oil and Gas Resources" (I. K. Habicht, Ed.), AAPG Studies in Geology 7, 186–195. AAPG, Tulsa.
Kantsler, A. J., Smith, G. C. and Cook, A. C. (1978). *Austral. Petr. Explor. Assoc. J.* **18,** 143–155.
Kappelmeyer, D. and Haenel, R. (1974). *In* "Geoexploration Monographs" (Rosenbach, D. and Morelli, R., Eds.), Series 1, No. 4. Gebrüder Borntraeger, Paul Christian K.G.
Karl, R. (1965). *Freiberger Forschungshefte C197.*
Karpov, P. A., Stepanova, A. F. and Solov'yeva, N. V. (1976). *Int. Geol. Rev.* **18,** 397–405.
Karweil, J. (1955). *Deutsch. Geol. Gesell. Z.* **107,** 132–139.
Katz, H. R. (1979). *Roy. Soc. New Zealand Bull. No. 18*, 121–130.
Kendall, M. G. (1968). *In* "Mathematical Model Building in Economics and Industry" (Kendall, M. G., Ed.), 1–14 Hafner, New York.
Kirby, S. H. (1977). *Pure Appl. Geophys.* **115,** 245–258.
Klemme, H. D. (1971). *Oil and Gas J.*, 1 March, 85–90.
Klemme, H. D. (1975). *Bull. Can. Petr. Geol.* **23,** 30–66.
Koncz, I. (1977). *Koolaj es Foldgaz* **10,** 225–228.
Krumbein, W. C. (1964). *Northwestern Univ, Geol. Dept. Technical Report* No. 8, 1–15.
Krumbein, W. C. and Graybill, F. A. (1965). "An Introduction to Statistical Models in Geology". McGraw-Hill, New York.
Larter, S. R. and Douglas, A. G. (1980). *In* "Advances in Organic Geochemistry 1979" (Douglas, A. G. and Maxwell, J. R., Eds.), 579–583. Pergamon Press, Oxford.
Leith, M. J. and Roswell, D. M. (1979). *Geol. Soc. S. Afr. Spec. Publ.* No. 6, 205–217.
Leontif, W. (1966). "Input–Output Economics". Oxford Univ. Press, New York.
Leventhal, J. S. (1976). *Chem. Geol.* **18,** 5–20.
Lopatin, N. V. (1971). *Izv. Akad. Nauk SSSR, Ser. Geol.* (3), 95–106.
Lopatin, N. V. (1980). *Int. Geol. Rev.* **22,** 193–200.
McAdoo, D. C., Caldwell, J. G. and Turcotte, D. L. (1978). *Geophys. J. Roy. Astron. Soc.* **54,** 11–26.
McCartney, J. T. and Teichmüller, M. (1972). *Fuel* **51,** 64–68.

McCrossan, R. G. and Porter, J. W. (1973). *Can. Soc. Petr. Geol. Mem.* **1,** 589–720.
McDowell, A. N. (1975). *Oil and Gas J.*, 9 June, 85–90.
Mackenzie, A. S. and Maxwell, J. R. (1981). *In* "Organic Maturation Studies and Fossil Fuel Exploration" (Brooks, J., Ed.), 239–254. Academic Press, London, Orlando and New York.
Mackenzie, A. S., Lewis, C. A. and Maxwell J. R. (1981a). *Geochim. Cosmochim. Acra* **45,** 2369–2376.
Mackenzie, A. S., Patience, R. L. and Maxwell, J. R. (1981b). *In* "Origin and Chemistry of Petroleum" (Atkinson, G. and Zuckerman, J. J., Eds.), 1–14. Pergamon Press, Oxford.
McKenzie, D. (1978). *Earth Planet Sci. Lett.* **40,** 25–32.
McKenzie, D. (1981). *Earth Planet Sci. Lett.* **55,** 87–98.
Maier, C. G. and Zimmerly, S. R. (1924). *Bull. Univ. Utah* **14,** 62–81.
Masran, Th. C. and S. A. J. Pocock, S. A. J. (1981). *In* "Organic Maturation Studies and Fossil Fuel Exploration" (Brooks, J., ed.), 145–175. Academic Press, London, Orlando and New York.
Megill, R. E. (1971). "An Introduction to Exploration Economics". Petroleum Publ. Co., Tulsa.
Megill, R. E. (1977). "An Introduction to Risk Analysis". Petroleum Publ. Co., Tulsa.
Melosh, H. J. (1978). *Geophys. Res. Lett.* **5,** 321–324.
Menard, H. W. and Sharman, G. (1975). *Science* **190,** 337–343.
Middleton, M. F. (1982). *Geophys. J. Roy. Astron. Soc.* **68,** 121–132.
Miller, B. M. (1981). *Proc. SPE of AIME Econ. and Evaluation Symp., Dallas, 25–27 Feb.* 1981, 101–109 (*SPE*-9561).
Miller, B. M., *et al.* (1975). *U.S. Geol. Surv. Circ.* **725**.
Mills, R. van E. (1920) *Econ. Geol.* **15,** 398–421.
Molnar, P. (1977). *Geophys. J. Roy. Astron. Soc.* **51,** 701–708.
Munn, M. J. (1909). *Econ. Geol.* **4,** 141–157.
Nadai, A. (1963). "Theory of Flow and Fractures of Solids", Vol. 2. McGraw-Hill, New York.
Nakayama, K. and Van Siclen, D. C. (1981). *AAPG Bull.* **65,** 1230–1255.
Nederlof, M. H. (1980). *Proc. World Petr. Cong., Bucharest*, Sec. PD **1**(2), 1–9.
Nehring, R. (1978). *Rand Corp. Rept. R*-2284-*ClA*.
Neruchev, S. G. (1962). "Oil-Generating Suites and the Migration of Oil", 217–221. Gostoptekhizdat, Leningrad.
Neugebauer, H. J. and Spotin, T. (1978). *Tectonophys.* **50,** 275–305.
Newendorp, P. D. (1975). "Decision Analysis for Petroleum Exploration". Petroleum Publ. Co., Tulsa.
Oberlin, A., Boulmier, J. L. and Durand, B. (1974a). *Geochim. Cosmochim. Acta* **38,** 647–649.
Oberlin, A., Boulmier, J. L. and Durand, B. (1974b). *In* "Advances in Organic Geochemistry 1973" (Tissot, B. and Bienner, F., Eds.), 15–27. Editions Technip, Paris.
Ojakangas, D. R. (1970). *Kansas Geol. Surv. Computer Contr.* **49**.
Parsons, B. and Molnar, P. (1976). *Geophys. J. Roy. Astron. Soc.* **45,** 707–712.
Parsons, B. and Sclater, J. G. (1977). *J. Geophys. Res.* **82,** 803–809.
Philippi, G. T. (1965). *Geochim. Cosmochim. Acta* **29,** 1021–1049.

Pitcher, M. G. (1976). *Oil and Gas J.*, 22 March, 34–35.
Pitt, G. J. (1962). *Fuel* **41,** 267–274.
Prager, W. and Hodge P. G. (1951). "Theory of Perfectly Plastic Solids". John Wiley, New York.
Rich, J. L. (1921). *Econ. Geol.* **16,** 247–371.
Rich, J. L. (1923). *AAPG Bull.* **7,** 213–225.
Rich, J. L. (1931). *AAPG Bull.* **15,** 911–924.
Rich, J. L. (1934). *In* "Problems of Petroleum Geology" (Wrather, W. E. and Lahee, F. H., Eds.), 337–345. AAPG, Tulsa.
Roadifer, R. (1975). *In* "Probability Methods in Oil Exploration" (Davis, J. C., *et al.*, Conveners), p. 18. AAPG Research Symposium, Stanford Univ.
Roy, K. J., Procter, R. M. and McCrossen, R. G. (1975). *In* "Probability Methods in Oil Exploration" (Davis, J. C., *et al.*, Conveners), 56–60. AAPG Research Symposium, Stanford Univ.
Royden, L., Sclater, J. G. and von Herzen, R. P. (1980). *AAPG Bull.* **64,** 173–182.
Schatz, J. F. and Simmons, G. (1972). *J. Geophys. Res.* **77,** 6966–6983.
Schoeppel, R. J. and Gilarranz, S. (1966). *J. Petr. Tech.*, June, 667–673.
Schuenemeyer, J. H. and Drew, L. J. (1977). *Computers and Geosciences* **3,** 617–630.
Schwab, F. L. (1976). *Geology* **4,** 723–727.
Sclater, J. G. and Christie, P. A. F. (1980). *J. Geophys. Res.* **85,** 3711–3739.
Semenovich, V. V., *et al.* (1977). *In* "The Future Supply of Nature-made Petroleum and Gas" (Meyer, R. F., Ed.), 139–153. Pergamon Press, New York.
Servier, S. M., Douglas, R. G. and Stehli, F. G. (1975). *Bull. Geol. Soc. Am.* **86,** 1499–1510.
Sharp, J. M., Jr. and Domenico, P. A. (1976). *Bull. Geol. Soc. Am.* **87,** 390–400.
Shaw, E. W. (1917). *Econ. Geol.* **12,** 610–628.
Shibaoka, M. and Bennett, A. J. R. (1977). *Austral. Petr. Expl. Assoc. J.* **17,** 58–63.
Shih, S. M. and Sohn, H. Y. (1980). *Ind. Eng. Chem. Proc. Des. Dev.* **19,** 420–426.
Singer, D. A. (1976). *Computer and Geosciences* **2,** 149–260.
Sleep, N. H. (1971). *Geophys. J. Roy. Astron. Soc.* **24,** 325–350.
Sleep, N. H. and Snell, N. S. (1976). *Geophys. J. Roy. Astron. Soc.* **45,** 125–154.
Sloss, L. L. (1962). *J. Sed. Petrology* **32,** 415–422.
Smith, M. B. (1968). *Oil and Gas J.*, 11 March, 81–84.
Smoluchowski, M. (1909a). *Bull. Int. Acad. Sci. de Cracovie*, 3–20.
Smoluchovski, M. (1909b). *Bull. Int. Acad. Sci. de Cracovie*, 727–734.
Snowdon, L. R. (1979). *AAPG Bull.* **63,** 1128–1134.
Stanov, V. V. (1972). *Dokl. Akad. Nauk SSSR* **207,** 1200–1203.
Stanov, V. V. (1980). *Int. Geol. Rev.* **23,** 1076–1084.
Stoian, E. (1965). *J. Can. Petr. Tech.* **4,** 120–129.
Suuberg, E. M., Peters, W. A. and Howard, J. B. (1978). *Ind. Eng. Chem. Proc. Des. Dev.* **17,** 37–46.
Teichmüller, M. and Teichmüller, R. (1981). *Bull. Centre Rech. Explor.-Prod. Elf-Aquitaine* **5,** 491–534.
Tissot, B. (1969). *Rev. Inst. Fran. Petr.* **24,** 479–501.
Tissot, B. (1973). *Rev. Assoc. Fran. Tech. Petr.* **222,** 27–31.
Tissot, B. (1977). *In* "Developments in Petroleum Geology" (Hobson, G. D., Ed.), Vol. 1, 53–82. Applied Science Publishers, London.
Tissot, B. and Espitalié, J. (1975). *Rev. Inst. Fran. Petr.* **30,** 743–777.

Tissot, B. and Pelet, R. (1971). *World Petr. Congr.* **2,** 35–46.
Tissot, B. and Welte D. H. (1978). "Petroleum Formation and Occurrence. A New Approach to Oil and Gas Exploration". Springer Verlag, Berlin.
Tissot, B., Califet-Debyser, V., Deroo, G. and Oudin, J. L. (1971). *AAPG Bull.* **55,** 2177–2193.
Tissot, B., Deroo, B. and Hood, A. (1978). *Geochim. Cosmochim. Acta* **42,** 1469–1485.
Turcotte, D. L. (1979) *Adv. in Geophys*, **21,** 51–86.
Turcotte, D. L. and Ahern, J. L. (1977). *J. Geophys. Res.* **82,** 3762–3766.
Turcotte, D. L., Ahern, J. L. and Bird, J. M. (1977). *Tectonophys.* **42,** 1–28.
van Everdinger, A. F. and Hurst, W. (1949). *Trans. AIME* **186,** 305–324.
van Hinte, J. E. (1978). *AAPG Bull.* **62,** 201–222.
Vandenbroucke, M., Albrecht, P. and Durand, B. (1976). *Geochim. Cosmochim. Acta* **40,** 1241–1249.
Veto, I. (1980). *In* "Advances in Organic Geochemistry 1979" (Douglas, A. G. and Maxwell, J. R., Eds.), 163–167. Pergamon Press, Oxford.
von Herzen, R. P. (1979). *EOS* **60**.
Walcott, R. E. (1970). *J. Goephys. Res.* **75,** 3941–3954.
Walcott, R. E. (1972). *Bull. Geol. Soc. Am.* **83,** 1845–1848.
Walstrom, J. E., Mueller, T. D. and McFarlane, R. C. (1967). *J. Petr. Tech.* **19,** 1595–1603.
Waples, D. W. (1980). *AAPG Bull.* **64,** 916–926. [For additional discussion of this paper see Cohen, C. R. and Waples, D. W. (1981). *AAPG Bull.* **65,** 1647–1649.]
Waples, D. and Connan, J. (1976). *AAPG Bull.* **60,** 884–886.
Ward, S. H., Ross, H. P. and Nielson, D. L. (1981). *AAPG Bull.* **65,** 86–102.
Warren, J. E. (1979). *AIME Hydrocarbon Economics and Evaluation Symposium, Dallas*, 11–13 February.
Watts, A. B. and Ryan, W. B. F. (1976). *Tectonophys.* **36.** 25–44.
Weeks, L. G. (1949). *AAPG Bull.* **33,** 1029–1124.
Weeks, L. G. (1952). *AAPG Bull.* **36,** 2071–2124.
Welte, D. H. (1965). *AAPG Bull.* **49**, 2246–2268.
Welte, D. H. (1972). *Petr. Explor.* **1,** 117–136.
Welte, D. H. and Yükler, M. A. (1980). *Org. Geochem.* **2,** 1–8
Welte, D. H. and Yükler, M. A. (1981). *AAPG Bull.* **65,** 1387–1396.
Welte, D. H., Yükler, M. A. Radke, M. and Leythaeuser, D. (1980). *In* "Origin and Chemistry of Petroleum" (Atkinson, G. and Zuckermann, T., Eds.), 67–88. Pergamon Press, Oxford and New York.
White, D. (1915). *J. Wash. Acad. Sci.* **5,** 189–212.
White, D. (1930). *AAPG Bull.* **14,** Pt. 2, 1227–1228.
White, D. A. and Gehman, H. M. (1979). *AAPG Bull.* **69,** 2183–2192.
Whitten, E. H. T. (1964). *Bull. Geol. Soc. Am.* **75,** 455–464.
Wright, N. J. R. (1980). *J. Petr. Geol.* **2,** 411–425.
Yellow, P. C. (1965). *Brit. Coal Utilisation Res. Assoc. Monthly Bull.* **29,** 285–308.
Yükler, M. A. (1976). Ph.D. Thesis, Univ. of Kansas.
Yükler, M. A. and Welte, D. H. (1980). *In* "Fossil Fuels", 267–285. Editions Technip, Paris.
Yükler, M. A., Cornford, C. and Welte, D. H. (1978). *Geol. Rundschau* **67,** 960–979.
Yükler, M. A., Cornford, C. and Welte D. H. (1979). *Init. Rep. Deep Sea Drill. Proj.* **47,** Pt. 1, 761–771.
Zhdanov, M. A. (1962). *Petr. Geol.* **6,** 177–184.

Applications of Biological Markers in Petroleum Geochemistry

Andrew S. Mackenzie*
K. F. A., Jülich, Germany.

I. Introduction

Many papers cite the discovery of porphyrin pigments in geological materials by Treibs (1934) as the beginning of organic geochemistry. It was certainly the first step in that part of organic geochemistry which is now usually referred to as the study of "biological marker" organic com-

* Currently with BP Research Centre, Sunbury-on-Thames, U.K.

ADVANCES IN PETROLEUM GEOCHEMISTRY Vol. 1.
ISBN 0-12-032001-0

pounds. Although the terms "chemical fossil" (Eglinton and Calvin, 1967) and "molecular fossil" (Calvin, 1969) have also been introduced, it is the epithet "biological marker", first coined by Speers and Whitehead (1969) which is most commonly used, and often shortened to "biomarker" (e.g. Seifert and Moldowan, 1981). A biological marker is any organic compound detected in the geosphere whose basic skeleton suggests an unambiguous link with a known, contemporary natural product. This link can easily be seen by comparing the structures of the vanadyl deoxophylloerythroetioporphyrin reported by Treibs (1934, 1936) and its proposed natural product precursor, chlorophyll *a* (Fig. 1). The geochemical reactions which convert the latter to the former are discussed in detail below.

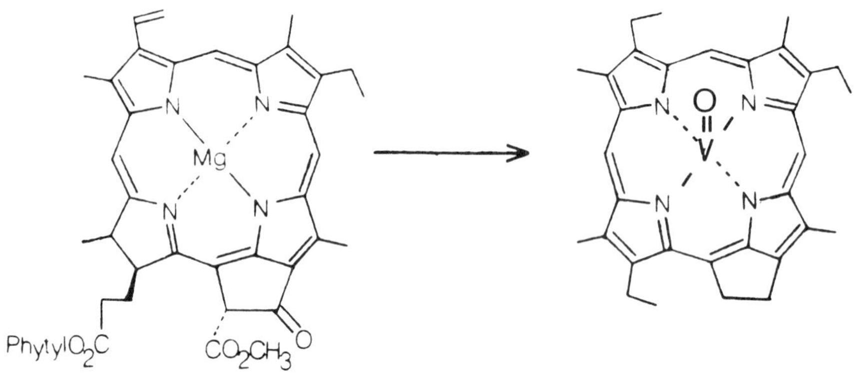

I II

FIG. 1. When Treibs discovered porphyrins of structural type II in sediments and petroleums (Treibs, 1934), he made the obvious link with chlorophyll *a* (I) of present day photosynthetic organisms. A biological marker is any compound, such as the vanadyl porphyrin (II), whose carbon skeleton shows an unambiguous link with a known and common natural product.

Figure 1 adopts the convention for drawing chemical structures normally used by organic chemists: the lines represent carbon–carbon bonds, and carbon atoms lie at the line intersections. All carbon atoms have four bonds, and where less are shown the necessary additional bonds to hydrogen atoms are implied. Double bonds are shown by two parallel lines, and atoms other than carbon and hydrogen are indicated by their standard abbreviation (e.g. N for nitrogen and O for oxygen in Fig. 1). The structures are not planar but are projected onto the plane of the paper. Methods for indicating three-dimensional structure are introduced in the discussion of stereochemistry below.

The obvious link between the biological and geological compounds in Fig. 1 was the first direct evidence that sedimentary organic matter, including petroleum, had a biological origin (Treibs, 1934, 1936).

Major developments have occurred in biological marker geochemistry in the last decade and a half. Although the study and structure elucidation of porphyrin- and chlorin-type molecules in geological samples remains an active field (e.g. Baker and Palmer, 1978; Eglinton *et al.*, 1980; Baker and Louda, 1983), most of these developments have been associated with the geological equivalents of terpenoid and steroid natural products. This stems partly from the rapid improvement of computerized gas chromatography–mass spectrometry, an analytical technique to which the nitrogen-containing porphyrin mixtures are not yet amenable on a routine basis. It is the intention of this review to introduce key papers for all types of biological marker reported to date. However, detailed discussion is restricted to those which have been shown to have an application to solving problems in petroleum geology.

Biological markers exist because their carbon skeletons survive, in a recognizable form, through the processes of diagenesis and thermal maturation which result mainly from the temperature rise associated with the burial of the host sedimentary rocks. Tissot and Welte (1978) introduced a classification of the alteration of organic matter less mature than greenschist facies metamorphism, where three zones were proposed: diagenesis, catagenesis and metagenesis.

Since metagenesis involves the total degradation of organic molecules and the evolution of gas, biological marker compounds are not detected at this stage.

Diagenesis starts in recently deposited sediments and is associated with mild conditions: the temperature and pressure increases are small; one major agent of alteration in such shallow sediments is microbial activity. The chemical transitions with respect to biological markers in this zone are highly complex. As will be discussed, in many cases these transitions lead to a certain simplification of the biological marker compounds: oxygen containing functional groups are lost, carbon–carbon double bonds are either hydrogenated (reduced by the addition of hydrogen which leads to a fully saturated system) or serve as starting points for the aromatization of cyclic systems. The biological marker products of diagenesis are normally recognized, therefore, as saturated hydrocarbons (alkanes) and aromatic hydrocarbons.

The catagenesis stage which follows includes the generation of petroleum. Here, the reactions which affect the biological marker compounds are more straightforward. The reactants and products of specific transitions can be identified with more certainty, and the extent to which the reactions have occurred can be used to assess the extent of thermal maturation.

Forming sediments act as sinks for the natural products from algae, plants and other organisms. The small proportion of natural products which survives the extensive microbial degradation between the deaths of the host organisms and incorporation impart "fingerprints" to the sediment, which, in addition to allowing assessment of maturation, are diagnostic of their original biological origin. These fingerprints are primarily recorded for the organic solvent soluble (and therefore mobile) fraction of a sediment's organic matter.

The petroleums which originate in the catagenesis zone entrain the biological markers. The biological markers, therefore, are useful for the correlation of oils with each other and with their proposed source rocks, and for the assessment of their thermal maturity. They may also play a role in the understanding of migration ranging from the redistribution of minute traces to the accumulation of commercial oil fields and in the monitoring of severe cases of in-reservoir biodegradation.

II. Background

A. Methods

It is not intended to give full references here. Each research group has its own technique and the reader is referred to the many publications introduced throughout the text. The key instrumentation for the biological marker organic geochemist is computerized gas chromatography–mass spectrometry (GC–MS).

Sedimentary organic extracts and crude oils are complex chemical mixtures. Prior to GC–MS analysis, they are separated by liquid chromatography into fractions, each certainly containing at least a thousand single components amenable to partial analysis by gas chromatography. Often the biological markers are merely trace components in these fractions (i.e. $\leqslant 1\%$). In GC–MS analysis the fractions of interest are resolved by the gas chromatography column (different compounds pass through the column at different rates) and the effluent from the column is monitored by coupling it to a mass spectrometer. Clearly it is advantageous to make full use of the high-resolution capillary columns presently available and of a fast scanning mass spectrometer (<2s to scan a mass range of 50–600 a.m.u.). This means the analysis of a single typical fraction of sedimentary organic matter can produce around 3000 complete mass spectra, which require a computer for both their aquisition and subsequent appraisal.

Although full mass spectra are required to determine the chemical structures of the components, often one ion (mass) in the mass spectra of a series of similar compounds is diagnostic of their structural type. Computer-constructed plots of the intensity of this ion against gas chromatography retention time, therefore, allow the isomer and carbon number distribution of the structural type to be conveniently evaluated. The ion chosen is usually the most abundant in the mass spectra (the base peak) and examples for some sedimentary steroid structures to be introduced below are shown in Fig. 2. Furthermore, many such patterns show sufficient similarity between samples, that instead of the mass spectrometer acquiring a full spectrum every two seconds, it need only record the intensity of the ions (masses) of interest. This means a much longer time is spent on these ions within one scan and results in major improvements in sensitivity and reproducibility, and in reduced amounts of data requiring computer storage. The construction of the plots of individual ion intensity against retention time is called "mass fragmentography", whilst the selection of key fragment ions for aquisition by the mass spectrometer has many descriptions, of which "multiple ion detection" (MID) will be used here.

The power of present GC–MS systems is such that pre-separation steps could now be simplified, because much more complex mixtures can be handled. This could provide helpful reductions in total analysis time. In future GC–MS systems as applied to biological markers will further develop using the methods of metastable transition spectra (Gallegos, 1976) and high resolution mass fragmentography (Gallegos, 1975a; Bjorøy *et al.*, 1983). At present most biological marker researchers employ electron-impact ionization techniques, but other "softer" ionization methods may be important in certain circumstances, e.g. chemical ionization (Mackenzie *et al.*, 1981a), field ionization (Payzant *et al.*, 1980). Most biological marker organic geochemistry to date has avoided absolute quantification of the components by GC–MS, and has concentrated on the relative abundances of certain species, usually assuming that the current due to a certain ion represents a similar absolute mass for compounds with similar mass spectra. Seifert and Moldowan (1979) introduced a method for absolute quantification of biological marker compounds in sedimentary organic mixtures, which involved an internal standard. The full potential of this method has neither been published nor realized.

The first biological marker compounds identified, the porphyrins, are just becoming amenable to routine GC–MS. Recent and earlier developments suggest that the problems of involatility can be overcome (e.g. Alexander *et al.*, 1980) and that chemical as opposed to electron-impact ionization may provide more distinctive mass spectra (Shaw *et al.*, 1978). In the meantime, the porphyrins from geological samples are analysed by high-performance liquid chromatography (Hajibrahim *et al.*, 1978) after

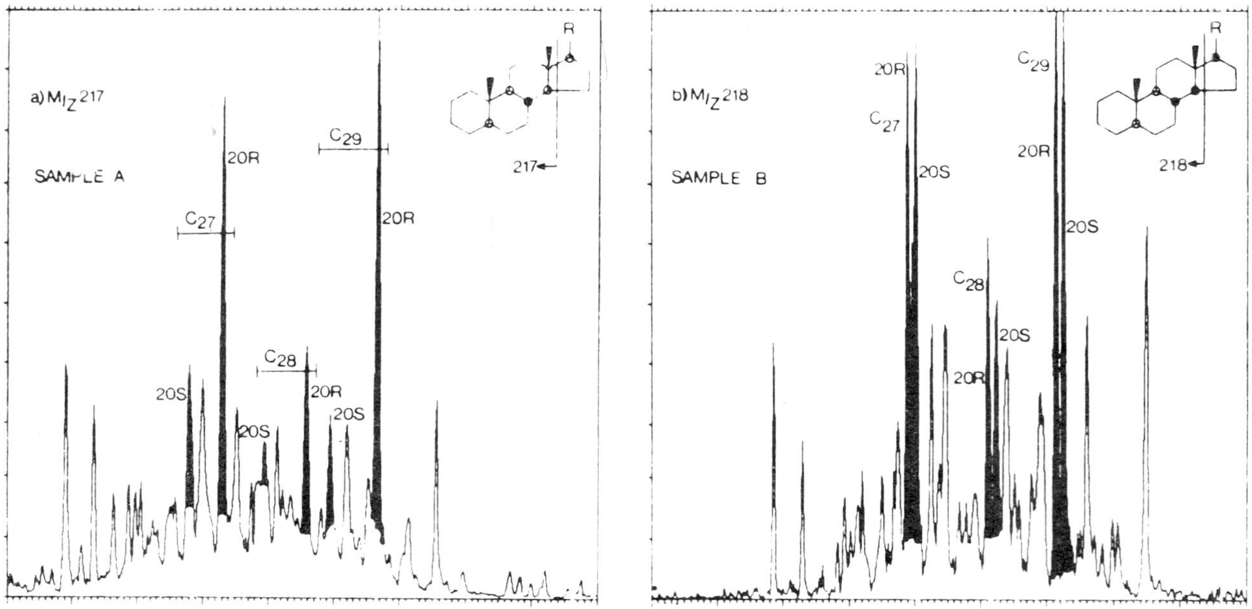

a) M/Z 217
SAMPLE A
C27
20R
20S
C28
20R
20S
C29
20R
R
217
b) M/Z 218
SAMPLE B
20R
C27
20S
C28
20S
20R
C29
20R
20S
R
218

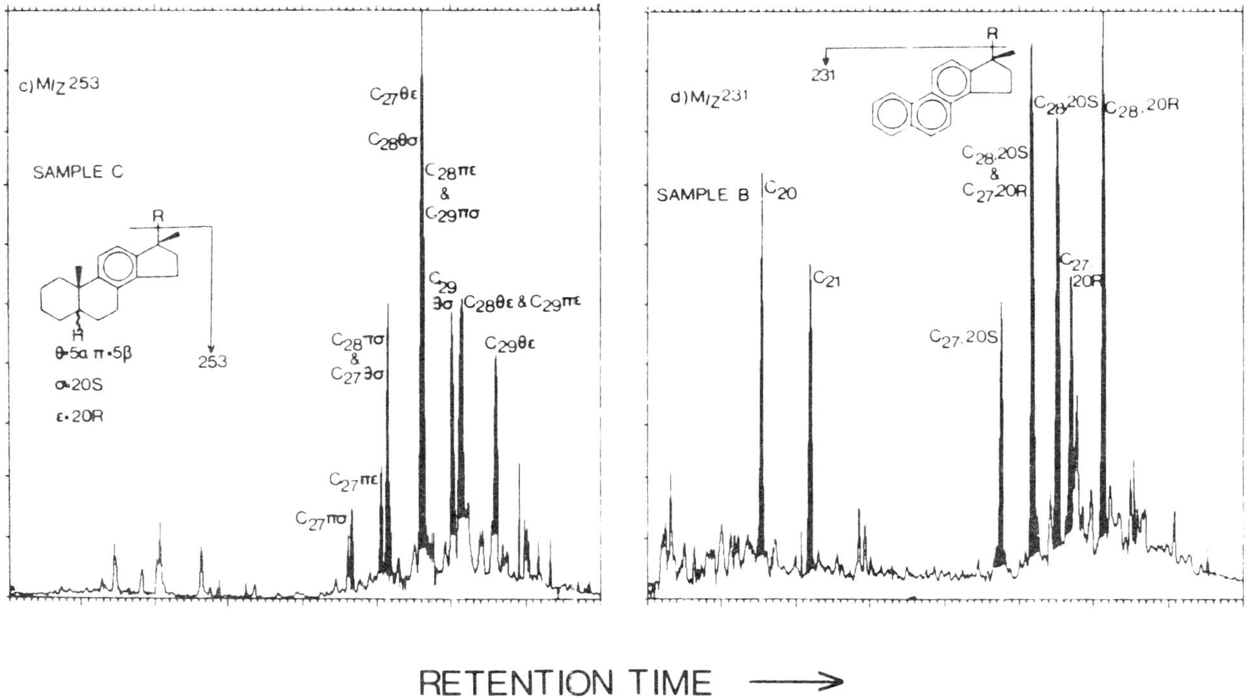

FIG. 2. Series of compounds of similar structural type can share a common mass-spectral fragment ion. Plots of the intensity of this ion against gas chromatography retention time can reveal the distributions of these components from the complex amount of data recorded during the appropriate GC–MS analysis. Such ions include: (a) *m/z* 217 for 14α(H)-steranes; (b) *m/z* 218 for 14β(H)-steranes; (c) *m/z* 253 for C-ring monoaromatic steroid hydrocarbons (an inscribed circle denotes an aromatic ring); and (d) *m/z* 231 for triaromatic steroid hydrocarbons.

a time-consuming series of "wet" chemical pre-separation steps (e.g. Didyk, 1975).

Nearly all biological marker organic geochemists concentrate on components which occur in an "unbound" form; that means, when they are present in sedimentary rocks, they can be extracted, along with many other species, with common organic solvents. Increasingly, the techniques are being applied to biological marker components generated by the pyrolytic breakdown of kerogens (Seifert, 1978; Huc, 1978) and oil asphaltenes (Rubinstein *et al.*, 1979) or by the chemical degradation of kerogens (Michaelis and Albrecht, 1979).

B. Stereochemistry

It is impossible to study biological markers without a basic knowledge of stereochemistry. Since the three-dimensional structures of biological natural products are highly specific, when the full stereochemistry (three-dimensional structure) of a biological marker compound in the geosphere can be determined, its correlation with a proposed natural product precursor can be assisted. It is because of their high structural specificity that biological marker compounds are such useful geochemical tools.

Saturated carbon atoms (Fig. 3a) have their four bonds pointing to the corner of a tetrahedron, and such a carbon atom has two forms if A, B, C and D are all different: one is the mirror image of the other and the carbon atom is referred to as a "chiral centre". If the carbon, as in Fig. 3a, is not part of a ring system, then the two possible configurations are described as R or S, based on an atomic mass priority ordering of the four substituents (see e.g. Gutsche and Pasto, 1975, for details). R refers to a chiral centre where, if the substituent of lowest priority is lined up away from the viewer, the remaining three groups have a decreasing mass priority in a clockwise direction; similarly, anticlockwise becomes S. It is more convenient if the chiral centre is part of a ring system (Fig. 3b) to describe the two configurations as α or β. α implies that the defined bond points into the page when a structure is conventionally drawn, and β that it points out of the page. The chiral centre in Fig. 3b is either β(OH) or α(H).

Each carbon in a molecule is given a number by agreed convention. In this article numbered carbons of interest are either introduced in the Figures or shown in the Appendix, but the interested reader is referred to any comprehensive introductory text to organic chemistry for the principles involved.

Most natural product organic compounds contain many chiral centres, and usually enzymatic biosynthesis produces only one configuration at each centre—that most suited to the function of the molecule in the host organism. Each component can in theory have 2^n stereoisomers, where n is the number of chiral centres, but biological systems produce only one or a very small number out of those theoretically possible. This high specificity is to varying extents preserved in the geosphere. Where two stereoisomers are mirror images of each other (i.e. all chiral centres have the opposite configuration in the other isomer), they are referred to as

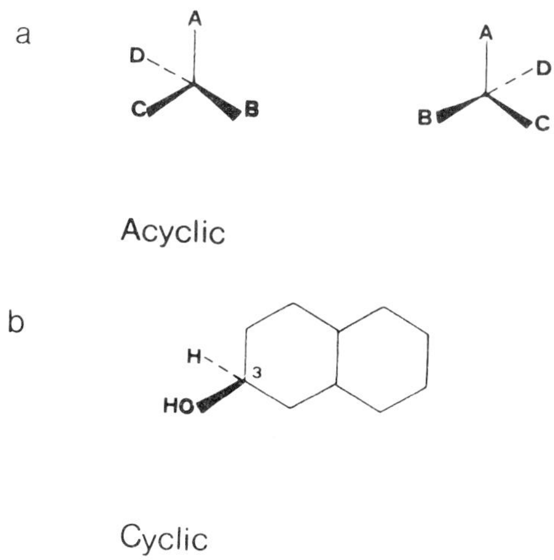

FIG. 3. A saturated carbon atom can be envisaged as sitting at the centre of a tetrahedron, with its four bonds pointing to the corners. When A, B, C and D are all different the molecule can exist in two different forms which are the mirror images of each other. Carbon atoms of this type are known as "chiral centres". For acyclic centres the two forms are called R or S depending on a complicated convention (Gutsche and Pasto, 1975). For centres part of a ring system, the terms α and β are used. α implies a bond pointing into the page (dotted line), β a bond out of the page (solid triangle).

"enantiomers", whilst if at least one of many chiral centres is the same, then the two isomers are not mirror images, and are called "diastereoisomers". Enantiomers have identical chemical properties, whilst diastereoisomers can be distinguished chemically and under the correct conditions resolved by gas chromatography with achiral stationary phases. Different diastereoisomers may, but do not always, have different mass spectra. Chiral centres are drawn in two different ways (see Figs. 5–23): a solid/open dot represents a carbon–hydrogen bond pointing out of/into the plane of the paper, i.e. β(H)/α(H), a solid triangle/dotted line represents a carbon–carbon bond (or carbon bond to a shown heteroatom, e.g. oxygen (O) or nitrogen (N)) above/below the plane of the paper, and a sinuous line means both the latter possibilities exist.

C. Isomerization

Petroleum geochemists can derive much information from the study of biological marker alkane stereochemistry. This is because the chiral centres of structurally specific skeletons of natural products are isomerized with maturation in the geosphere. This configurational isomerization occurs because a hydrogen is removed from the centres at elevated temperatures, probably either as a hydride ion (Ensminger, 1977) or as a hydrogen radical (Seifert and Moldowan, 1980). The positively charged carbonium ion or radical intermediate is often near planar for acyclic carbons and has an equal probability of regaining either a hydride ion or a hydrogen radical, on either side. Where both possible configurations have similar stabilities (as is usually the case for acyclic carbons), then this process converts the single isomeric form of living systems to an equal or near equal mixture of the two possible isomers (all-isomer mixture, e.g. C-20* in steranes, see paragraph (iv) of section IV.B.2., subsection (*a*), below). For chiral centres part of an alkane ring system the two possible configurations usually have different thermodynamic stabilities. Frequently organisms choose to synthesize the less stable form of the precursor natural product. In this case, geological configurational isomerization will tend to convert the skeleton to the more stable form (e.g. C-17 and C-21 in the hopanes, and C-14 and C-17 in the steranes, see paragraphs (i) and (iii) respectively of section IV. B.2., subsection (*b*), below). The exact mechanism of configurational isomerization of alkanes in the geosphere is not known. It will only occur at a chiral centre where one of the four substituents is a hydrogen atom, since it is the exchange of this atom which causes the isomerization.

The only other type of geological isomerization to be discussed in this communication is often also described as "rearrangement". This is restricted here to the intramolecular movement (shift) via a charged carbonium ion of a hydrogen or methyl substituent from one carbon to another, often adjacent, carbon, and is quite a common organic geochemical reaction type. The nomenclature for other significant reactions will be introduced at the appropriate place.

III. Selected Known Biological Marker Compounds

This section intends to introduce most types of biological marker reported to date, with a particular emphasis on those which have an application to petroleum geochemistry. Proposed natural product precursors will be dis-

* C-20 means carbon number 20, C_{20} means a compound with 20 carbon atoms.

cussed together with the geochemical reactions which connect the natural product and its many geological equivalents, although again the emphasis will be on those reactions which are of immediate interest to the petroleum geochemist.

A. Acyclic isoprenoids

These compound types as biological markers have recently been reviewed by Volkman and Maxwell (1984). The heading is normally associated with the ubiquitous pristane and phytane (2,6,10,14-tetramethylpentadecane and 2,6,10,14-tetramethylhexadecane; see Fig. 4). These C_{19} and C_{20} alkanes tend to dominate the branched and cyclic alkane fractions of petroleums and sedimentary rock extracts, although C_{14}, C_{15}, C_{16} and C_{18} members (Fig. 4) are also normally present. Albaiges (1980a) discussed the presence of a large number of different structural types of longer chain acyclic isoprenoid alkanes (up to C_{45}) in petroleums.

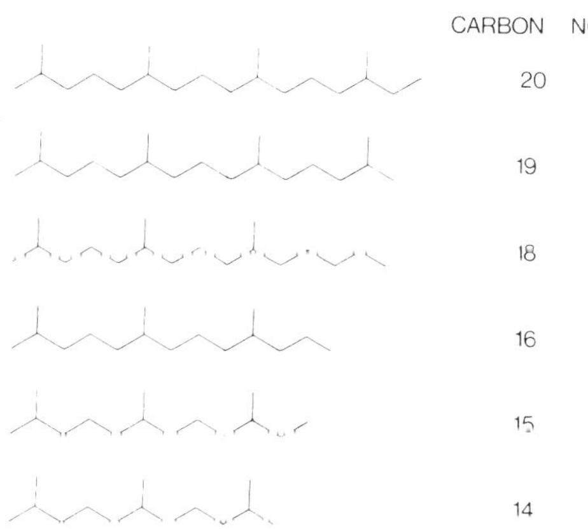

FIG. 4. The common isoprenoid alkanes normally observed in sedimentary rocks and petroleum; C_{20} is phytane and C_{19} is pristane.

Acyclic isoprenoid acids, alcohols and ketones have also been reported from sediments (see Volkman and Maxwell, 1984, and references therein), but they are rarely present in useful quantities in sediments of direct interest to the petroleum geochemist. The pristane of the immature Green River shale has a dominantly 6R, 10S configuration (Maxwell *et al.*, 1972). This is exactly the configuration that would be expected if it had derived from the natural product, phytol (Fig. 5). The regular isoprenoid alkanes are thought to arise from phytol (Maxwell *et al.*, 1972; Ikan *et al.*, 1975). Pristane could arise via oxidation and decarboxylation of phytol, whilst phytane could result from dehydration and reduction (Ikan *et al.*, 1975). Thus pristane/phytane ratios have been proposed as measures of the redox potential of sediments (e.g. Didyk *et al.*, 1978).

With increasing temperature, configurational isomerization of pristane occurs via hydrogen exchange by an unknown mechanism at the two chiral

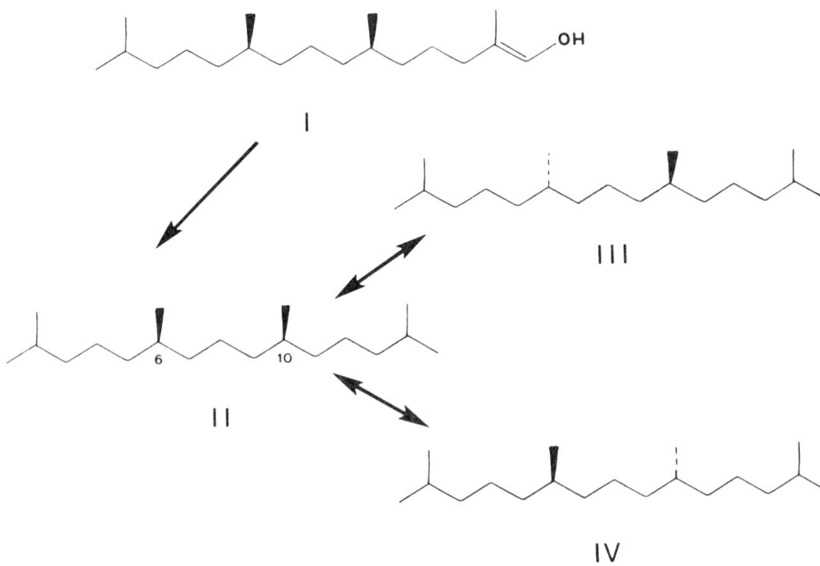

FIG. 5. Pristane in immature sediments comprises only the 6(R), 10(S) or *meso* form (II) which is the stereochemistry inherited from its natural product precursor, phytol (I). With increasing temperature, isomerization produces the other isomers: 6 (S), 10 (S) (III) and 6 (R), 10 (R) (IV). The equilibrium ratio of II: III: IV is 2: 1: 1.

centres (C-6, C-10) to produce an all-isomer mixture of the three possible isomers (Fig. 5) (Patience *et al.*, 1978). A similar process affects phytane (Patience *et al.*, 1980). Just as pristane and phytane are sufficiently abundant components of the branched and cyclic alkane fraction to allow the measurement of their relative concentrations by a gas chromatograph with a flame ionization detector, so the relative stereochemistry of pristane can be assigned routinely by gas chromatography, with only occasional mass-spectrometric checks on the purity of the peaks concerned. The "geological" 6R, 10R and 6S, 10S isomers can under certain conditions be resolved by gas chromatography from the "biological" 6R,10S isomer (Patience *et al.*, 1978). A steady increase in relative proportions of the "geological" isomers is normally observed with increasing burial depth in sedimentary sequences (e.g. Mackenzie *et al.*, 1980a) until an all-isomer mixture is obtained (equilibrium is established).

Although the evidence for an origin from phytol for the six structures observed in the geosphere is good, many other, normally less abundant, structures have been reported, which clearly have a different origin. The regular series (shown in Fig. 4) extends to C_{40} (Albaiges, 1980a) and is joined by the so-called head-to-head (Michaelis and Albrecht, 1979; Moldowan and Seifert, 1979) and tail-to-tail series (Brassell *et al.*, 1981). Examples are shown in Fig. 6. These are thought to arise from the cell wall ethers of archaebacteria (e.g. Fig. 6; de Rosa *et al.*, 1977; Michaelis and Albrecht, 1979). Michaelis and Albrecht (1979) observed the archaebacterial acyclic isoprenoid structures in the products of the chemical degradation of Messel shale kerogen, whilst Moldowan and Seifert (1979) reported them from crude petroleums. A contribution by carbon–carbon bond cracking of the higher molecular weight archaebacterial components or from appropriate ethers (e.g. Kates *et al.*, 1967) to the geological components shown in Fig. 4, thought previously to arise exclusively from phytol, cannot be discounted.

Chappe *et al.* (1982) detected alkane I (Fig. 6) in the alkanes of a North

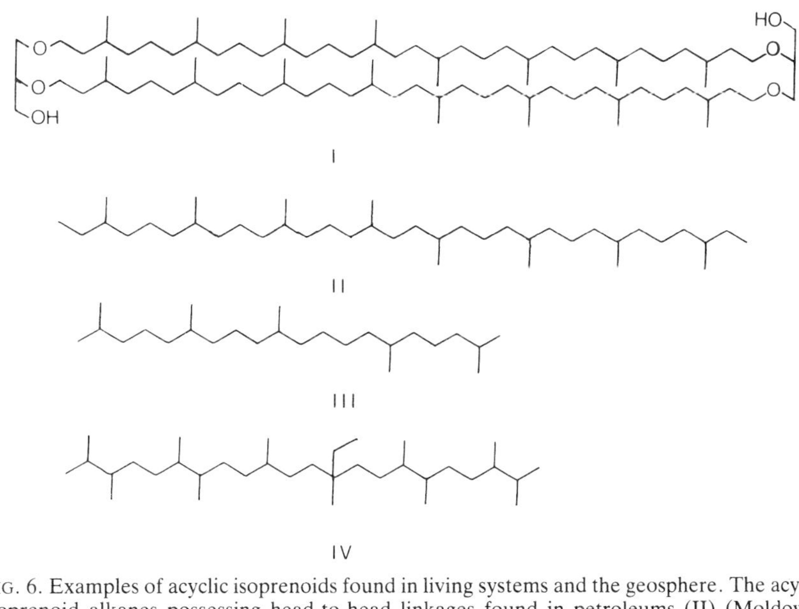

FIG. 6. Examples of acyclic isoprenoids found in living systems and the geosphere. The acyclic isoprenoid alkanes possessing head-to-head linkages found in petroleums (II) (Moldowan and Seifert, 1979) are thought to derive from the dibiphytanyl glyceryl ethers of archaebacteria (I) (de Rosa *et al.*, 1979). As well as regular head-to-tail compounds (e.g. Fig. 4) $> C_{20}$ tail-to-tail isoprenoid alkanes have been reported from sediments (e.g. III) (Brassell *et al.*, 1981). IV is botryococcane.

Sea oil sample, and the same compound and similar structures, also found in archaebacteria, which contain one or two five-membered carbon rings, in the products of chemical degradation of the polar lipids of sediments and petroleums. Thus an important contribution from the membranes of micro-organisms to the organic matter of sediments and petroleum can be postulated.

Also shown in Fig. 6 is botryococcane, which is an unusual C_{34} isoprenoid alkane. The alkene natural product precursor botryococcoene has only been reported from one algae *Botryococcus braunii* (Maxwell *et al.*, 1968). The alkane has been found in an Indonesian crude of algal origin (Moldowan and Seifert, 1980).

B. Hopanoids

1. *Basic skeleton*

The finding and structural elucidation of hopanoids as hydrocarbons (e.g. Fig. 7) in sediments and crude oils (Ensminger *et al.*, 1974) gave much of

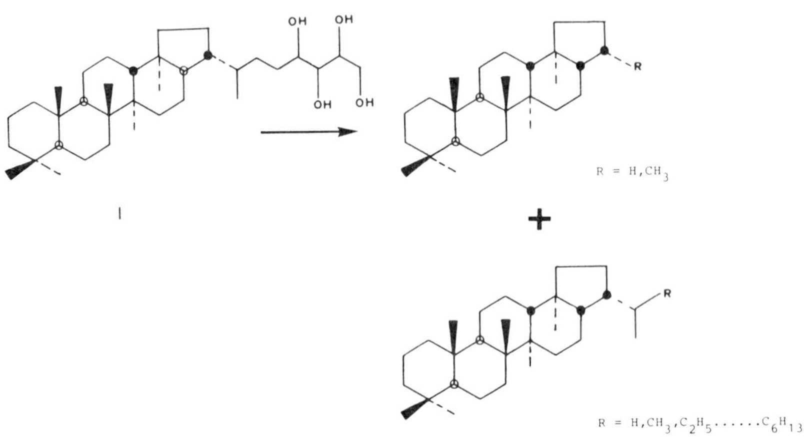

FIG. 7. Polyhydroxybacteriohopanes (e.g. I) are the presumed precursors of the ubiquitous extended hopanes found in recent and ancient sediments and in petroleums (Ourisson *et al.*, 1979), of which the least stable 17β(H), 21β(H) series is shown (but see Fig. 9).

the impetus to the present applications of biological markers within petroleum geochemistry. It was their finding as hydrocarbons, ketones, alcohols and acids in geological samples that spawned a more detailed study of biological samples to locate appropriate natural product precursors and possible functions for this molecular type (see Ourisson *et al.*, 1979, for a recent review, and references therein). Ourisson *et al.* (1979) believe the hopanoids serve as "rigidifiers" for cell wall membranes and form a link in the pathway taken by chemical evolution of such rigidifiers which leads ultimately to steroid structures. The hopanoids are pentacyclic triterpenoids and are found in almost every geological sample, including petroleums, which is sufficiently immature to have allowed preservation. All of their applications to petroleum geochemistry use the alkanes shown in Figs. 7 and 8 (e.g. Seifert and Moldowan, 1978, 1979). These alkanes range from C_{27} to C_{35}, but do not include a regular C_{28} member (Fig. 7). Ourisson *et al.* (1979) proposed an origin from the tetrol shown in Fig. 7, which has been isolated from several bacteria (e.g. Rohmer and Ourisson, 1976). The recent report of extended hopanes up to C_{40} in a bitumen (Rullkötter and Philp, 1981) could require other biological precursors.

In addition to the basic skeleton shown in Fig. 7, a number of slightly modified carbon skeletons have been observed in petroleums and source rocks. These are shown in Fig. 8. Compound I (25,28,30-trisnormoretane) has 27 carbons and, in comparison with the basic hopanoid skeleton, has "lost" carbons number 25 and 28 and has a C_2 side chain (Bjorøy and Rullkötter, 1980). This structure is now considered to have a 17α(H),21β(H)-hopane stereochemistry (Volkman *et al.*, 1983). Compound II (28,30-bisnorhopane) lacks carbon 28, but it also has a different configuration at C-17 and C-21 from that of compound I (Seifert *et al.*, 1978). Compounds of type III (25-norhopanes) (Rullkötter and Wendisch, 1982) have only been seen to date in highly biodegraded crude oils and asphalts. In contrast to I–III, where methyl groups have been "lost" relative to the basic hopane skeleton, the C_{27} structure IV (18α(H)-trisnorneohopane) has the C-28 methyl attached to C-17 instead of C-18 (Whitehead, 1973). Structures I–IV have been determined using GC–MS, X-ray crystallography and proton magnetic resonance spectroscopy (^{1}H NMR); structural type V has only been inferred from GC–MS analysis (Seifert and Moldowan, 1978). Schmitter *et al.* (1982) have recognized hopanes and possibly other polycyclic triterpanes with Ring C opened in a Nigerian oil sample. Many other hopanoid structures (which are not alkanes) have been reported from geological samples, but they are normally restricted to immature sediments (see Ourisson *et al.*, 1979, for a review). Of particular note are the more persistent species: the hopenes, where evidence has been obtained for the migration of a side chain double bond into the ring system (Ensminger, 1977; Brassell *et al.*, 1980); the hopanoic acids (Ensminger, 1977; Schmitter *et al.*, 1978); and the hopanoid aromatic hydrocarbons (Spyckerelle, 1975; Greiner *et al.*; 1976, Spyckerelle *et al.*, 1977).

The hopanoid story (Ourisson *et al.*, 1979) provides a good example of how an organic geochemist can be likened to a molecular palaeontologist, whose fossils are the molecular structures in the geosphere called biological markers. Just as palaeontology can reveal areas where our knowledge of biology is lacking, so can the organic geochemist assist the biochemist. The principles adopted by Ourisson *et al.* (1979) to predict the biochemical function of hopanoids were identical to those followed by the palaeontologist studying functional morphology. The occurrence of hopanoids in the

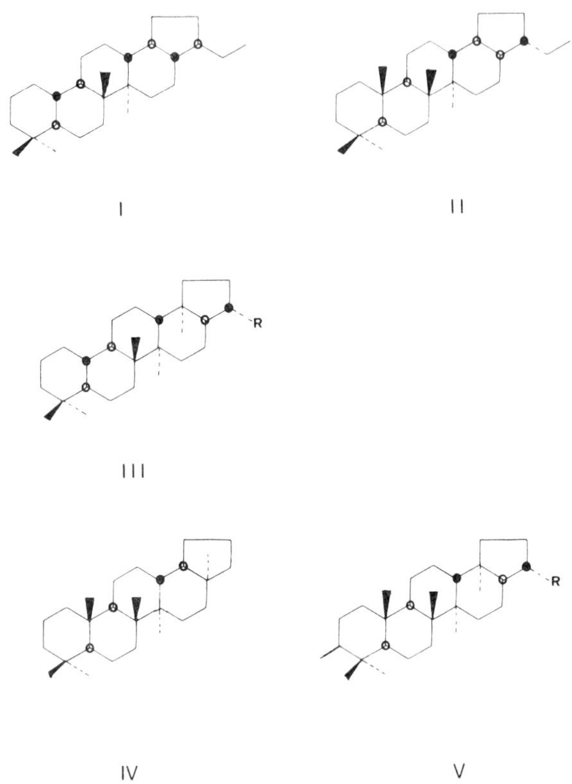

FIG. 8. The structures of some hopanes observed in sediments, whose substituents show changes from the major skeletal type (Figs. 7 and 9).

geosphere led to more understanding of the connected biochemistry. Already the work of Rullkötter (Bjorøy and Rullkötter, 1980; Rullkötter and Wendisch, 1982) suggests that C-10 desmethylation of hopanoids is an important microbial process in the geosphere, and this has not yet been observed by biochemists. There are and will be other examples of how the cooperation of organic geochemists and natural product chemists (biochemists) has won and will win many rewards.

2. *Stereochemistry*

It was for the hopanes that it was first suggested that the temperature-dependent conversion of one molecule to another could be applied to the assessment of the thermal maturity of the host sediment (Ensminger *et al.*, 1974). The preferred "biological" 17β(H),21β(H) configuration inherited by the alkanes of immature sediments is lost with increasing maturity in favour of the more stable 17α(H),21β(H) and 17β(H),21α(H) configurations, but mainly the former (Ensminger *et al.*, 1977; Allan *et al.*, 1977) (see Fig. 9). At higher levels of maturity, the 17β(H),21α(H)-hopanes also convert to the 17α(H),21β(H) form (Seifert and Moldowan, 1980). Although the former reaction can be envisaged as a simple isomerization via hydrogen exchange at one chiral centre (see above), the later conversion is less straightforward (see Seifert and Moldowan, 1980, for details). The situation is slightly more complicated than is often imagined. Whilst the 17β(H),21β(H) stereochemistry dominates the hopanoids found in organisms to date (Ourisson *et al.*, 1979), the 17α(H),21β(H) and

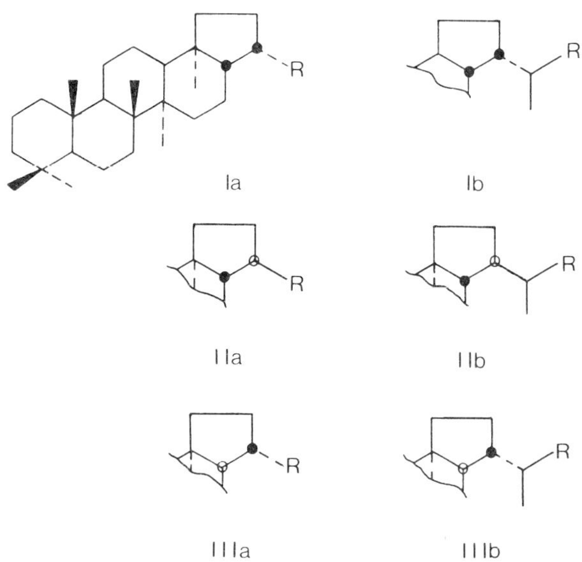

FIG. 9. Stereoisomers represented by the hopane structures I–III are present during early diagenesis. Stability increases in the order I (17β(H), 21β(H)) < II (17β(H), 21α(H)) < III (17α(H), 21β(H)). The major biological input is I, but geological isomerization of the alkanes means that this is converted by isomerization at C-17 to mainly III with increasing maturity. At still higher temperatures structural type II appears to be converted to III by a more complicated isomerization (see text). Ib, IIb and IIIb have a chiral centre at C-22. In immature sediments only the 22R configuration (C–H bond, not shown, into the page) derived from the natural product precursors (see Fig. 7) exists. Isomerization converts this to a 3:2 mixture of 22S and 22R components.

17β(H),21α(H) components have been reported from biological material (Quirk *et al.*, 1980, and Galbraith *et al.*, 1965, respectively). The nomenclature in the literature of the three main types observed can be confusing—all 21α(H)-hopanoids are often called "moretanoids" and the C-21 stereochemistry not defined. Where the C-21 stereochemistry is not defined along with the name hopanoid then a 21β(H) stereochemistry is implied.

When R in Fig. 9b is greater than -CH_3 ($\geq C_{31}$) then a chiral centre exists at C-22 (R refers to alkyl group as well as a stereochemistry). Only the 22R configuration has been seen in hopanoid natural products, and dominates in immature sediments. With increasing maturity, this preference is lost in favour of a 60:40 mixture of the S and R isomers (Ensminger *et*

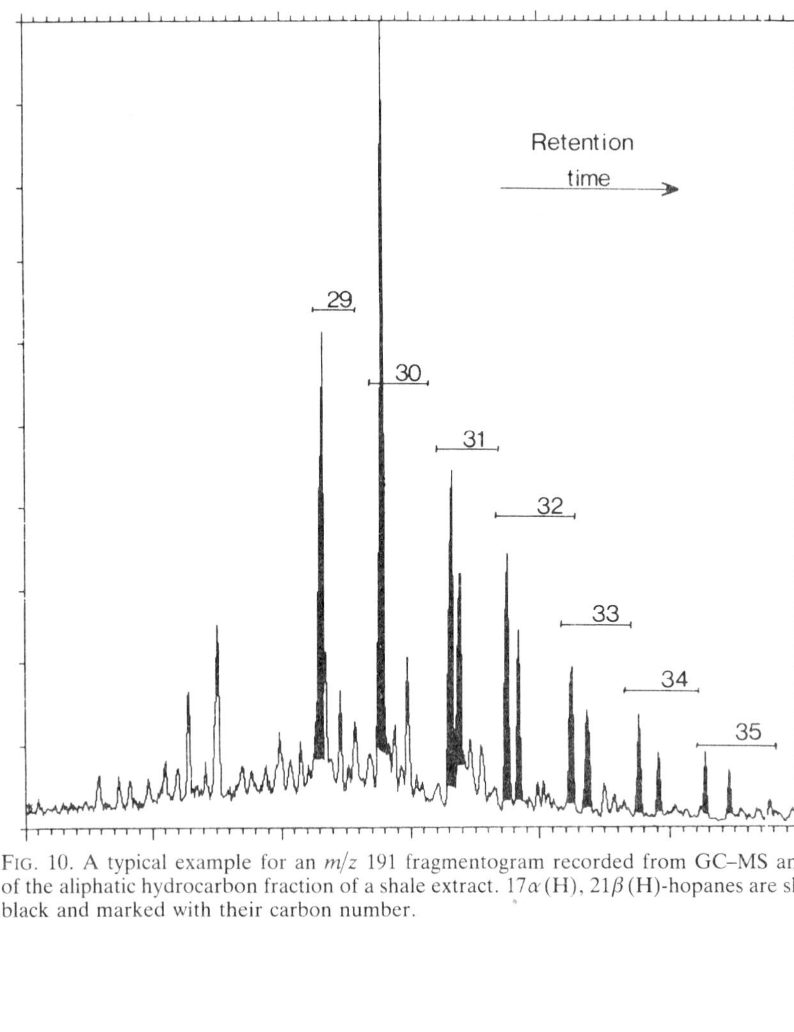

FIG. 10. A typical example for an *m/z* 191 fragmentogram recorded from GC–MS analysis of the aliphatic hydrocarbon fraction of a shale extract. 17α(H), 21β(H)-hopanes are shaded black and marked with their carbon number.

al., 1974; Seifert and Moldowan, 1980). So, the products of all the hopane isomerizations are distributions in mature sediments and petroleums characterized by a high relative abundance of hopanes with the stable 17α(H), 21β(H) configuration with mixtures of the C-22 isomers in members $\geq C_{31}$ (e.g. van Dorsselaer *et al.*, 1974). These distributions are revealed by mass fragmentograms for *m/z* 191, the most abundant peak (base peak) in the spectra of 17α(H),21β(H)-hopanes (van Dorsselaer *et al.*, 1977) (see Fig. 10).

C. Other polycyclic triterpenoids

Because the regular hopanoids are so widespread, it is often the study of the non-hopanoids and the modified hopanoids (see above) which gives the triterpenoid distributions some specificity. Although a number of structures have been proved, in most cases possible precursors are not known with any certainty.

1. *Pentacyclic triterpenoids other than hopanoids* (Fig. 11)

(*a*) *Gammacerane* (Fig. 11 I)
Gammacerane was first identified by Hills *et al.* (1966) in Green River shale by X-ray crystallography. It has subsequently been observed in Indonesian and Chinese crude oils (Seifert and Moldowan, 1981; Shi Jiyang *et al.*, 1982). It could derive from tetrahymanol (Fig. 11 II) which is a constituent of a protozoan.

(*b*) 18α(H)-*oleanane* (Fig. 11 III)
X-ray crystallography of a purified fraction showed 18α(H)-oleanane to be present in Nigerian Delta crude oils and sediments (Smith *et al.*, 1970). This has also been observed by GC–MS (Pym *et al.*, 1975; Ekweozor *et al.*, 1979a). It has also been reported in crude oils and sediments from the Mahakam Delta, Indonesia (Mackenzie *et al.*, 1982a). The higher plant-derived oleanenes (Fig. 11 IV, V) found in the immature sediments of the deltas probably gave rise to the alkane (Ekweozor *et al.*, 1979b).

(*c*) *Ring A degraded and altered pentacyclic triterpenoids* (The ring nomenclature is included in the Appendix)
A number of these structures (e.g. Fig. 12) have been postulated for both crude oils and sediments. Smith (1970) showed, using X-ray crystallographic studies, that a spirotriterpane of structure I (Fig. 12) was present in Nigerian crudes. Corbet *et al.* (1980) showed a whole range of such structures (e.g. Fig. 12 II–IV) to be present as acids and hydrocarbons in recent and

immature sediments from the Mahakam Delta, and II–IV have subsequently been found with a similar species (V) in a Nigerian crude oil (Schmitter *et al.*, 1981). Ring A degradation of triterpenoids is a common geochemical process, which may result from microbial (Spyckerelle *et al.*, 1977) or photochemical alteration of higher plant triterpenoid alcohols and ketones (Corbet *et al.*, 1980). Other Ring A degraded triterpenoids have been suggested (Ekweozor *et al.*, 1981), but no structures have been proved.

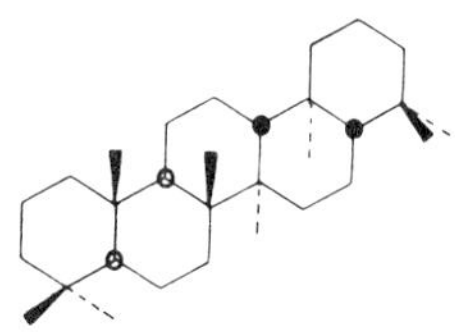

I

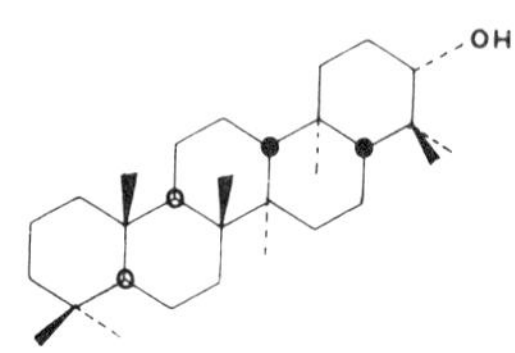

II

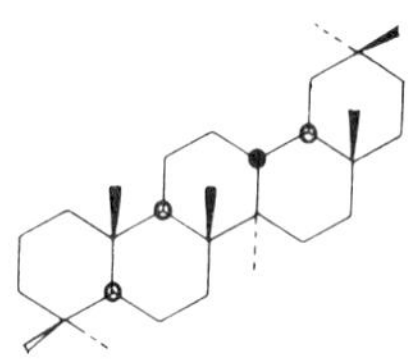

III

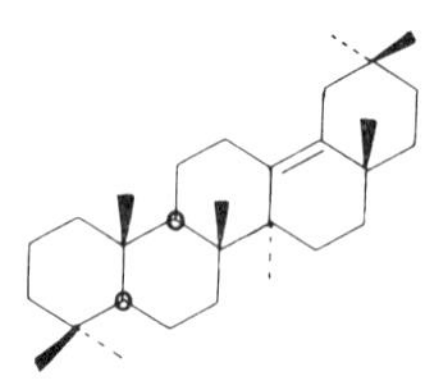

IV

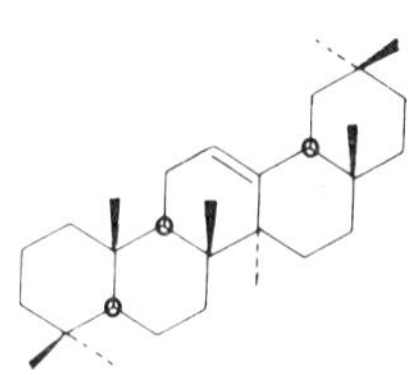

V

FIG. 11. The structures of some pentacyclic triterpanes and triterpenes other than hopanes which have been reported from sediments and crude oils. I is gammacerane which perhaps derives from the natural product, tetrahymanol (II). III is 18α(H)-oleanane, the equivalent in mature sediments and crude oils of the olean-13(18)-ene (IV) and olean-12-ene (V) found in immature sediments.

(*d*) *Other lupanoids*

Other lupanoid structures are always sought by geochemists, but conclusive proof exists in only one case: Rullkötter *et al.* (1982) reported the finding of two isomers of 28,30-bisnorlupanes (Fig. 12 VI, VII) in high abundance in immature sediments from West Greenland and Suez.

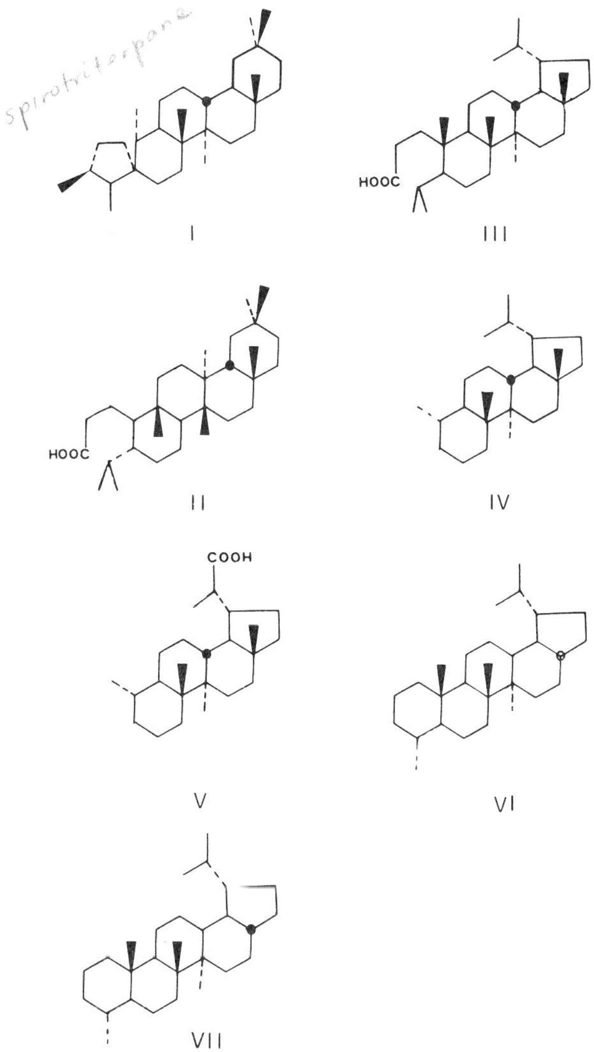

FIG. 12. I–V are examples of some of the ring-A degraded triterpenoids found in sediments and crude oils. III–VII include some of the lupanoid structures reported to date.

2. *Tricyclic and tetracyclic triterpenoid derivatives* (Fig. 13)

Anders and Robinson (1971) and Gallegos (1971) reported a series of C_{20}–C_{26} tricyclic alkanes from the Green River shale. These are known to be near ubiquitous components of crude oils and organic rich ancient sediments (e.g. Reed, 1977; Seifert and Moldowan, 1978) and to extend to C_{45} (Ekweozor and Strausz, 1982; Moldowan *et al.*, 1983). The work of Aquino Neto *et al.* (1982) has shown them to have the structure drawn in Fig. 13. There is an acyclic chiral centre at C-22, which means that the C_{25}–C_{45} members are normally present as a mixture of isomers at this centre. Possible natural product precursors for all members of the series are not known, but may derive from the incomplete biosynthetic cyclization of a regular hexaprenol. Trendel *et al.* (1982) reported a series of tetracyclic alkanes of structural type II as ubiquitous components of crude oils.

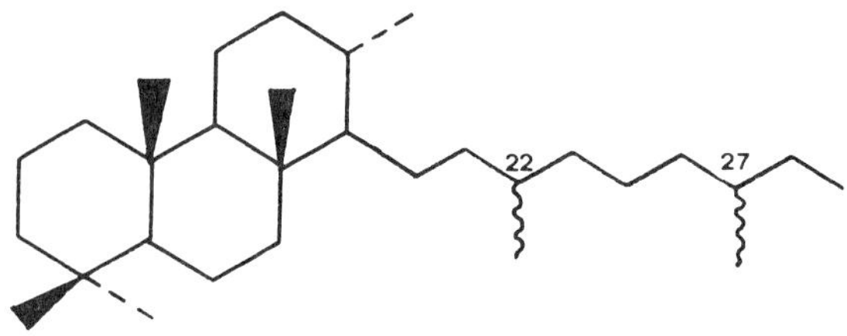

I

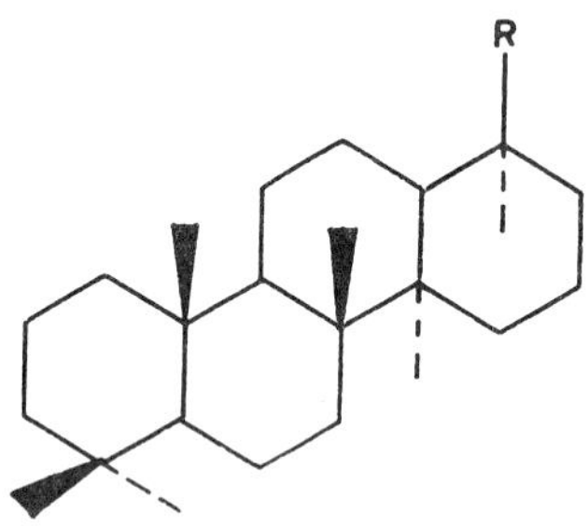

II

FIG. 13. I is the basic skeleton of tricyclic terpenes found in sediments and crude oils. The molecule drawn has 30 carbon atoms. A series with 19 to 45 carbon atoms has been reported. The lower carbon numbers can be derived from cleavage of the side chain. Mixtures of the two possible isomers are to be expected when C-22 is chiral (C_{25}–C_{29}) and of the four isomers when C-22 and C-27 are chiral (C_{30}). II is the basic skeleton of tetracyclic terpanes found in sediments and crude oils. R = H, CH_3, C_2H_5, C_3H_7; when R = H it has an α stereochemistry (Trendel *et al.*, 1982).

D. Steroids

This group of compounds is without doubt the most understood in terms of both biological origin and geological fate. Consequently they represent a powerful tool for the petroleum geochemist. A very detailed review of the organic geochemistry of steroids appeared recently (Mackenzie *et al.*, 1982a), so only an outline is provided here. Figure 14 reproduces a plate from that review, in order to convey the complexity now known.

The compounds of interest to the petroleum geochemist are the partially stabilized hydrocarbons which result from the diagenesis of natural product sterols (e.g. cholesterol, R = H in Fig. 15 I). Diagenesis can lead from cholesterol, for example, via chemical dehydration and microbial reduction of the remaining double bonds (in either order and often one process occurring in the water column (Wakeham *et al.*, 1980)) to an alkane-cholestane (R – H in Fig. 15 IV) (Rhcad *et al.*, 1971; Gaskell and Eglinton, 1975; Dastillung and Albrecht, 1977). Alternatively at the sterene stage (V) the A ring may be aromatized, possibly by microbial activity (Hussler *et al.*, 1981) to give VIII and IX, or the Δ2 double bond in III (this means C-2 and C-3 are connected by a double bond) may move to the more stable Δ4 and Δ5 positions. This often leads to the complete clay-catalyzed rearrangement of the steroid skeleton to give rearranged sterenes (VI) (Rubinstein *et al.*, 1975) called diasterenes. This rearrangement has been simulated in the laboratory (Sieskind *et al.*, 1979). Because the double bond is now in position 13(17) the biologically inherited sterol configuration at C-20—20R—is rapidly isomerized to an equal mixture of 20R and 20S. When this double bond is reduced (by hydrogen addition to give a single (saturated) bond), the abundant diasteranes seen in ancient sediments and petroleums are formed (VII) (Ensminger *et al.*, 1978; Seifert and Moldowan, 1978).

Ring C monoaromatic steroid hydrocarbons (Fig. 15 X) occur in sediments and petroleums, and appear to form just before the steroidal alkanes (Fig. 15 IV) as a mixture of isomers at C-20 (Ludwig *et al.*, 1981; Mackenzie *et al.*, 1982b; Scifert *et al.*, 1983). Their origin is not clear, although Mackenzie *et al.* (1982a, b) suggested, from laboratory simulation experiments, that they may be the products of the rearrangement of the stera-3,5-dienes, which can be formed by diagenetic dehydration of sterols (see above and Brassell *et al.*, 1980). All of the processes *summarized* in Fig. 15 are relatively low temperature effects, and have been documented by the analysis of cores from the International Phase of Ocean Drilling of the Deep Sea Drilling Project (DSDP) (e.g. Brassell *et al.*, 1980; Rullkötter *et al.*, 1980). At the onset of the later stages of diagenesis and catagenesis of interest to the petroleum geochemist, the lipid extracts of sediments contain mainly alkanes and monoaromatics as their major steroidal components.

Three major maturity related processes have been observed in steroid hydrocarbons at this stage—configurational isomerization, aromatization and the enrichment of lower molecular weight components.

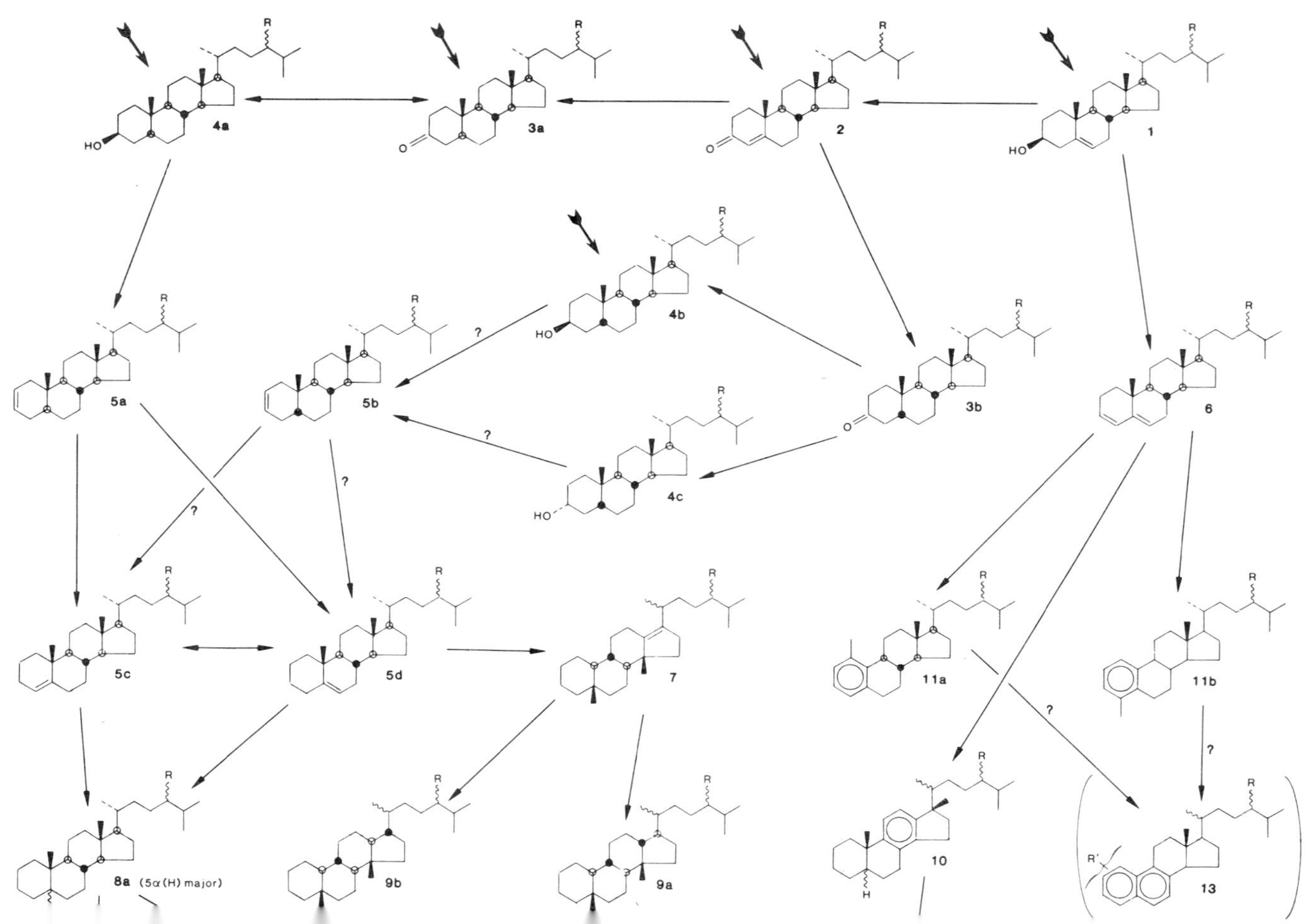
1
2
3a
4a
3b
4b
4c
5a
5b
5c
5d
6
7
8a (5α(H) major)
9a
9b
10
11a
11b
13

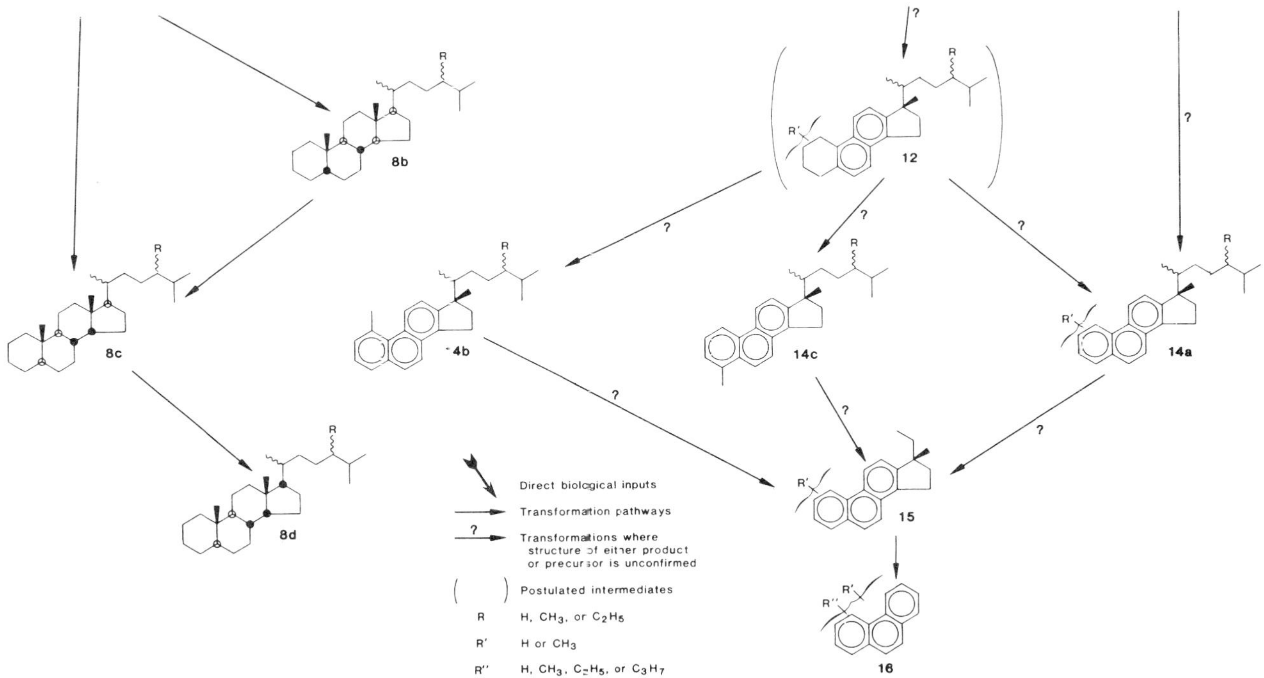

FIG. 14. Simplification of the proposed pathways of sterol diagenesis and catagenesis. The various processes involved include defunctionalization, reduction, rearrangement, isomerization and aromatization. This is not discussed in detail in the text, but further simplified in Figs. 15–17. A hexagon with an inscribed circle is an aromatic ring. Taken from Mackenzie *et al.* (1982a).

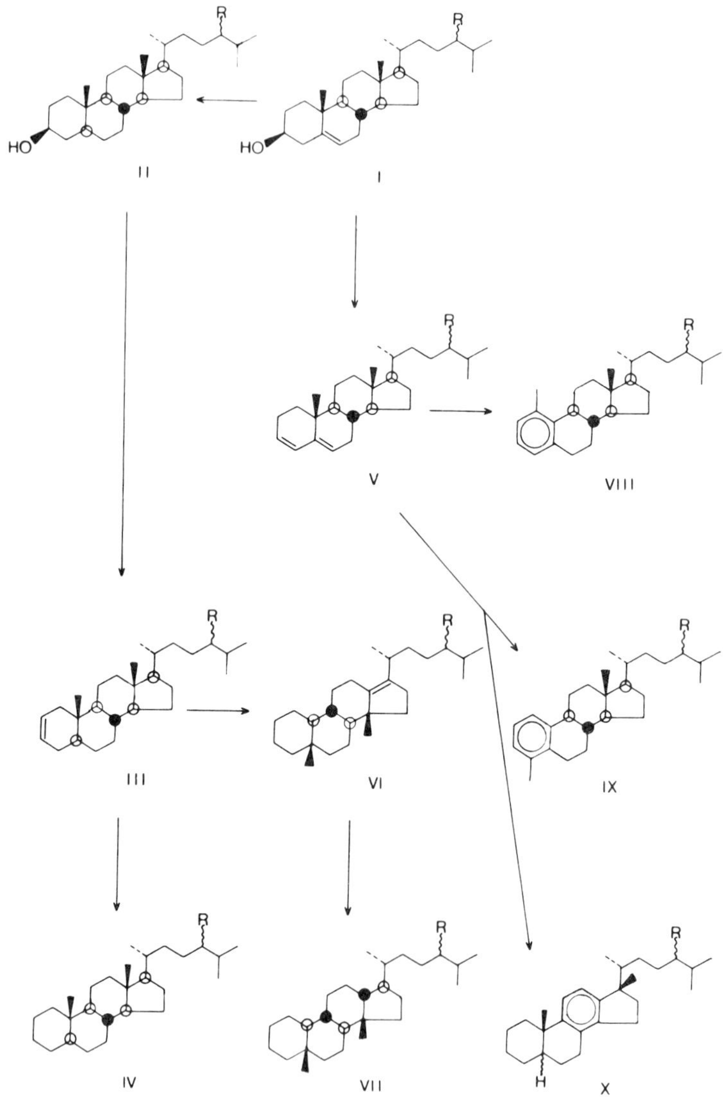

FIG. 15. Highly simplified schematic summary of proposed pathways of steroid diagenesis. Only direct biological inputs of sterols (stenols, I, and stanols, II) are considered. For most measurements of isomerization and aromatization reported in the text, geological derivatives of the C_{29}, 24-ethylcholesterols ($R = C_2H_5$) are used. R is mainly H, CH_3 or C_2H_5. When R = H, structure I is cholesterol. A hexagon with an inscribed circle is an aromatic ring.

Most sterols without a methyl group at C-4 reported to date range from C_{27} to C_{29} (e.g. Huang and Meinschein, 1979) and indeed this carbon number range is seen to dominate in petroleum and sedimentary steroid alkanes and monoaromatic steroid hydrocarbons (Ensminger, 1977; Seifert and Moldowan, 1978; Connan *et al.*, 1980; Mackenzie *et al.*, 1981a). Variable amounts of C_{21} and C_{22} members are also often present (Connan *et al.*, 1980; Mackenzie *et al.*, 1981a). C_{27}–C_{29} implies C_8 to C_{10} side chains and the extra alkylation normally occurs at C-24 by the addition of an extra methyl or ethyl group (see Fig. 15 IV). C_{27} steranes have eight chiral centres whilst C_{28} and C_{29} species have nine, the extra one occurring at C-24 (Fig. 15 IV). The complex isomeric geological mixtures of the steranes and monoaromatic steroid hydrocarbons are revealed by mass fragmentography (see above and Fig. 2).

Both the biosphere and the geosphere have large amounts of 4-methyl sterols. These therefore have three methyl substituents on the cyclic systems (three nuclear methyls) (e.g. Boon *et al.*, 1979). Their geological fate is not considered here, for simplicity, but the reader is referred to Mackenzie *et al.* (1982a) and references therein for details. All the steranes and aromatic steroid hydrocarbons considered in detail here derive from sterols with two nuclear methyl groups (e.g. Fig. 15). Ring A-norsteranes (five carbon membered A ring) have recently been reported from black shales in the Massif Central France (van Graas *et al.*, 1982a).

1. *Configurational isomerization*

Since sterols are proposed to function as rigid inserts in cell membranes (Nes, 1974), a certain degree of "molecular flatness" is required, which is achieved with a 8β(H), 9α(H), $10\beta(CH_3)$, $13\beta(CH_3)$, 14α(H), 17α(H) configuration (e.g. Fig. 15 I). At certain carbons (C-8, C-9) the most stable configuration already exists, while that at C-10 and C-13 cannot be changed by hydrogen exchange. These four centres will not be discussed further. The diagenetic processes described above do not affect the biological stereochemistry, so that the steranes of immature sediments preserve the original configurations (e.g. Fig. 15 IV). Only one sterol with a 20S configuration has been reported (Tsuda *et al.*, 1958), and only the 20R configuration is seen in immature non-rearranged steranes (Mulheirn and Ryback, 1975). The chiral centres at C-5 and C-24 are less stereospecific. That at C-5 comprises a mixture of isomers (α(H) and β(H)) which result indirectly from the reduction of sterols and sterenes during early diagenesis (mainly to the more stable 5α(H)) and directly from the precursor stanols (e.g. Nishimura, 1977). The configuration at C-24 is also a mixture (Mulheirn and Ryback, 1975) which depends on the origin of the precursor sterols. In a very general sense green and blue-green algae synthesize sterols which give rise to 24S-steranes, whilst higher plant sterols would chiefly convert to 24R steranes (Patterson, 1971).

Isomerization has been observed with increasing burial depth at C-14 and C-17 (Mackenzie *et al.*, 1980a; Seifert and Moldowan, 1979), C-20 (Mackenzie *et al.*, 1980a) and C-24 (Maxwell *et al.*, 1980). The biologically derived 20R isomers are converted to a near-equal mixture of 20R and 20S, which implies similar stabilities (Fig. 16). The "molecular flatness" (14α(H), 17α(H)) is lost in favour of the thermodynamically more stable 14β(H), 17β(H) form (Fig. 16). Typically, the major non-rearranged steranes (up to 75% of mature sediments and petroleums) have, however, a

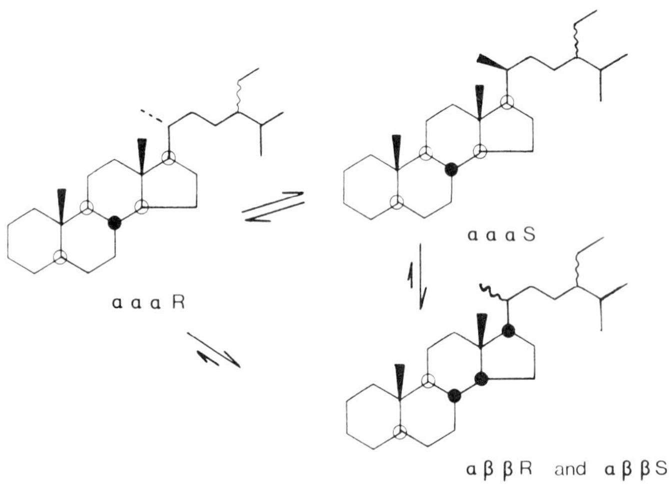

FIG. 16. The principle isomerizations of 5α(H), 14α(H), 17α(H), 2OR–C_{29} steranes which occur with increasing temperature in the sedimentary column. αααR means 5α(H), 14α(H), 17α(H), 2OR etc. The αααR form is inherited from the natural product steroid precursors and consequently dominates the sterane distributions of immature sediments. With increasing maturity, isomerization results in increasing amounts of the other isomers until the equilibrium ratio of αααR: αααS: αββR: αββS is 1:1:3:3.

5α(H), 14β(H), 17β(H) configuration, with the sterol-derived configuration still preserved at all the other chiral centres that are part of the ring system (Mackenzie *et al.*, 1980a; Seifert and Moldowan, 1979, 1981). The other non-rearranged steranes are mainly of the "flat" sterol derived configuration—5α(H), 14α(H), 17α(H), although other isomers are present in minor amounts (Seifert, 1981, Mackenzie, 1980).

The mature geological distributions have been simulated in the laboratory by the isomerization of 5α(H), 14α(H), 17α(H), 20R-steranes on platinized charcoal in the presence of H_2 (Petrov *et al.*, 1976; Seifert and Moldowan, 1979). Many of the structures were proved using such mixtures (Seifert and Moldowan, 1979). The mature geological distributions also agree with the thermodynamic stabilities of several cholestane isomers predicted by theoretical calculations (van Graas *et al.*, 1982b).

Under normal conditions of GC–MS analysis, the C-24 isomers are not resolved (e.g. Mackenzie *et al.*, 1980a). Using specialized conditions, separation is possible (Maxwell *et al.*, 1980). Such analyses have suggested any stereopreference is sufficiently rapidly lost, so that for most medium-mature to mature sediments and crude oils, all steroids can be considered as equal mixtures of the two C-24 isomers (Maxwell *et al.*, 1980; Mackenzie *et al.*, 1980a).

2. *Aromatization*

The aromatization of C-ring monoaromatic steroids to ABC-ring triaromatics (Fig. 17 IV) (Ludwig *et al.*, 1981) has been well studied (Mackenzie *et al.*, 1981a, 1982c; Shi Jiyang *et al.*, 1982) and has shown that the dominant process involves the loss of the nuclear methyl group (C-19) on the A/B ring junction (Mackenzie *et al.*, 1981a). This means that the carbon number range of the major triaromatics arising from C-ring monoaromatics of type II in Fig. 17 is shifted to one carbon number lower. There is one other major triaromatic series in ancient sediments and petroleums which still has two nuclear methyl groups (Fig. 17 VII, VIII) (Ludwig *et al.*, 1981). This probably derives (i) partly from aromatization of the A-ring monoaromatics (Fig. 17 V, VI) (Hussler *et al.*, 1981) reported from immature sediments; (ii) partly from the aromatization of C-ring monoaromatics of Type II (Fig. 17) where the C-19 methyl group is retained by rearrangement; and (iii) partly from the aromatization of C-ring monoaromatics with three nuclear methyl groups (Mackenzie *et al.*, 1981a). The latter must arise from the 4-methyl steroids present in the biosphere and geosphere and not discussed in detail here (see above). Since A-ring aromatic steroids with only one nuclear methyl have not been observed in significant amounts (Hussler *et al.*, 1981), the dominant process changing the relative concentrations of C-ring monoaromatic steroids of Type II in Fig. 17 to ABC-ring triaromatics of Type IV in Fig. 17 is the conversion of II to IV by geological aromatization. Type IV ABC-ring triaromatic steroids are not seen in immature sediments and have been seen to increase in concentration as II decrease in over ten sedimentary sequences between 80 and 110°C (Mackenzie *et al.*, 1982b). This is a thermally driven aromatization process. Only small concentrations of BC-ring diaromatics (Fig. 17 III) have been observed (Schaeflé, 1979; Mackenzie *et al.*, 1981a), so aromatization of ring B in C-ring monoaromatics must be followed rapidly by the aroma-

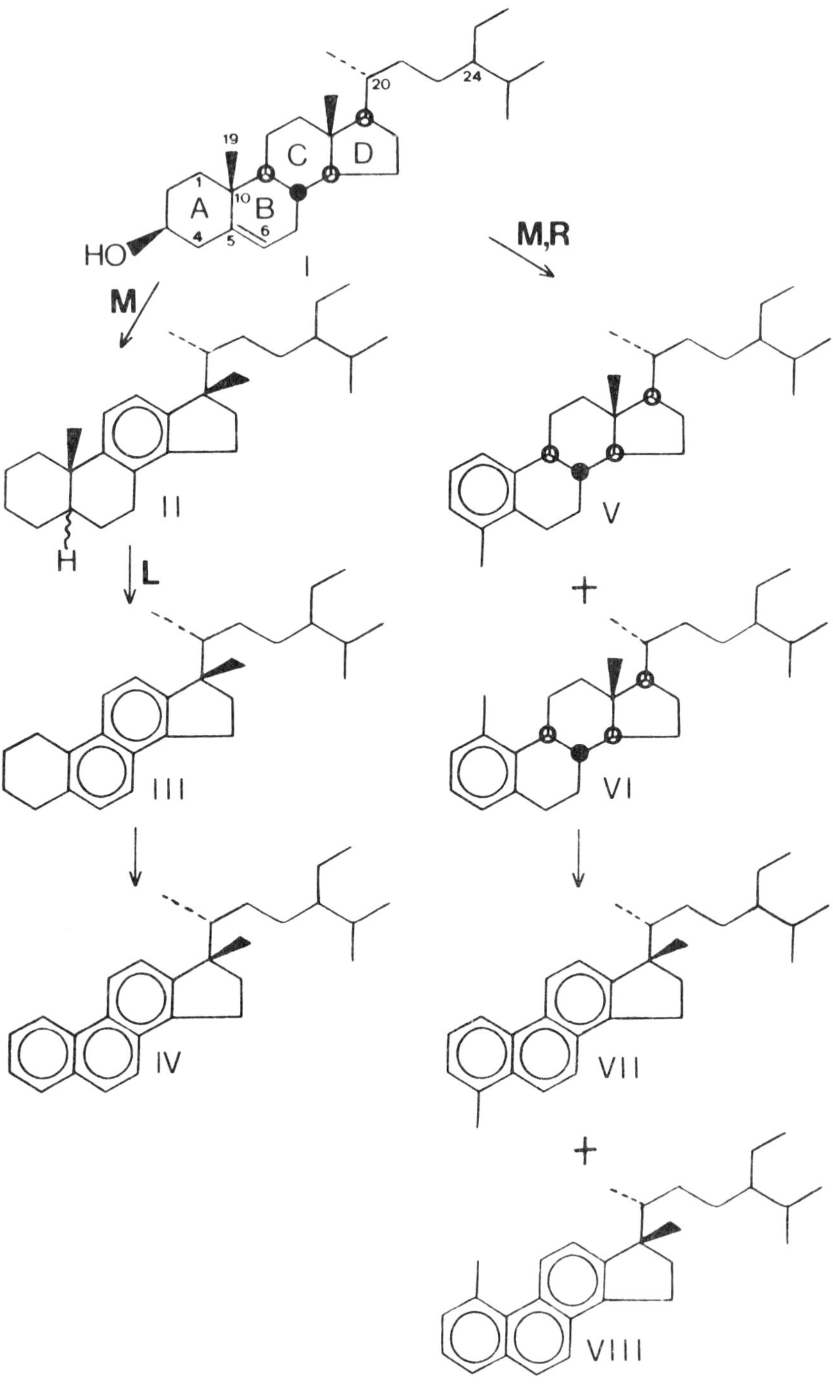
19
1
A
10
B
4
5
6
C
D
20
24
HO
I
M
M,R
II
H
L
III
IV
V
+
VI
VII
+
VIII

tization of ring A. Therefore monitoring the aromatization of C-ring monoaromatic steroids to ABC-ring triaromatics involves the study of the rate at which the conversion of C-ring monoaromatic steroids to BC-diaromatic steroids occurs (the rate-determining step).

It is not possible to discuss here all the possible stereoisomers of every compound type observed. Instead, the two types of most immediate interest to the applied petroleum geochemist—II and IV in Fig. 17—are concentrated on. II has two chiral centres which have a hydrogen substituent (C-5 and C-20) not including C-24 (see above). Both forms exist at each centre, giving four peaks per carbon number (C_{27}–C_{29}) although this is often simplified to two mature samples (Mackenzie *et al.*, 1983a), where the 5α(H) forms are absent. IV has only the centre at C-20, and again both isomers are present, giving two peaks per carbon number (C_{26}–C_{28}). (See Fig. 2).

3. *Enrichment of lower molecular weight components*

The lower molecular weight steroids (C_{20}–C_{22}) invariably show an increase in their abundance relative to the higher molecular weight components (C_{26}–C_{29}), particularly so for the triaromatic steroid hydrocarbons, as sediments move through the oil window (Mackenzie *et al.*, 1981a; Sajgò *et al.*, 1984). The extent to which this reflects a higher stability to thermal degradation for the lower molecular weight components or a direct conversion by carbon–carbon single bond cleavage in the side chain (e.g. Fig. 18) is not known. Indeed, the required cleavages would not be predicted by first principles. Results for the triterpanes would support the former, since normally a parallel increase can be seen in the ratio of the tricyclic triterpanes (Fig. 13) to pentacyclic terpanes (Figs. 7–12) and it is highly unlikely that the pentacyclics could be cracked to provide the tricyclic structures.

FIG. 17. Simplified reaction scheme to show the conversion of a sterol, I, to various aromatic hydrocarbons. An arrow labelled **M** implies that the reaction occurs in more than one step. One labelled **R** implies aromatization of the A ring, but with retention of the C-19 methyl group which is rearranged from C-10 to either C-1, C-4 or other positions. One labelled **L** implies aromatization of the B ring with the loss of the C-19 methyl group. The monoaromatic hydrocarbons II, V and VI appear during early diagenesis (Ludwig *et al.*, 1982; Hussler *et al.*, 1981), but are both further aromatized to triaromatic steroids (IV, VII and VIII). The C-19 methyl group is chiefly lost when the order in which the aromatization of the rings occurs is C → B → A, but is retained if the order is A → B → C. Hence neither V nor VI are precursors of IV, and the reaction rate can be obtained from the ratio of the concentrations of II and IV. The structure of III is not proven, and is proposed on the basis of the GC–MS results. The conversion of II and III is the basis of the GC–MS results. The conversion of II to III is the rate-determining step in the two reactions which convert II into IV. A hexagon with an inscribed circle is an aromatic ring.

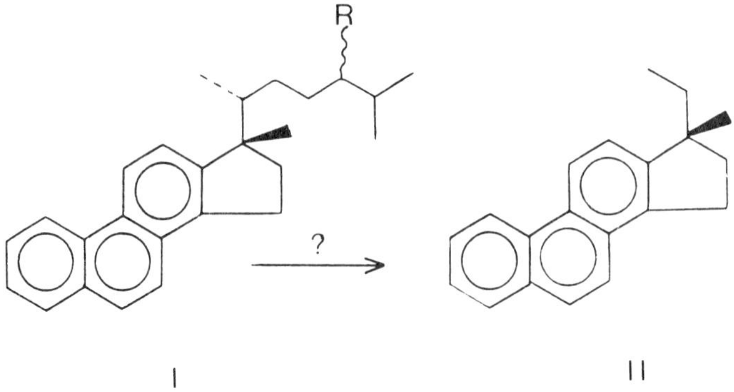

FIG. 18. A possible cleavage in C_{26}–C_{28} triaromatic steroid hydrocarbons (I, R = H, CH_3, C_2H_5) which could explain the enrichment of C_{20} species (II) relative to C_{26}–C_{28} components with increasing maturity. A hexagon with an inscribed circle is an aromatic ring.

E. Porphyrins

The recent study of these components can be divided into two halves: the detailed investigation of the geochemical steps and stabilized intermediates of the conversion of the biological chlorophylls to the metallated alkyl porphyrin pigments of sedimentary rocks and petroleum (e.g. Fig. 1), and the detailed structural elucidation of the porphyrins. The whole subject was reviewed by Baker and Palmer (1978), and Eglinton *et al.*, (1980) described briefly the progress on the structural elucidation in the intervening years.

Alkyl porphyrins occur in varying concentrations in most immature sedimentary rocks as complex mixtures of two major series: the deoxophylloerythroetio (DPEP) and aetio types (I and II respectively in Fig. 19), together with three minor series, the rhodo-DPEP, rhodo-aetio and di-DPEP types (III–V) (Treibs, 1934; Baker *et al.*, 1967; Shaw *et al.*, 1978; Barwise and Whitehead, 1980). Porphyrins ranging from C_{26} to C_{63} have been reported (Baker and Palmer, 1978; Barwise and Whitehead, 1980), but most distributions are dominated by C_{27}–C_{33} species (Hajibrahim *et al.*, 1981).

Most of these porphyrins occur as vanadyl or nickel (II) complexes (e.g. Baker, 1969), although both copper (Palmer and Baker, 1978) and gallium (III) porphyrins (Bonnett and Czechowski, 1980) have been reported.

Four major effects have been proposed to occur with increasing extent of maturation. These are: (1) a decrease in DPEP/aetio ratio by cracking of the isocyclic ring arrowed in Fig. 19 (Corwin, 1959; Yen and Silverman, 1969; Didyk *et al.*, 1975); (2) a decrease in average molecular weight resulting from carbon–carbon bond cleavage in the alkyl substituents (Didyk *et al.*, 1975); (3) an increase in average carbon number, proposed to result from an alkyl exchange reaction (Baker, 1966; Baker *et al.*, 1967; Bonnett *et al.*, 1978); and (4) a decrease in the ratio of nickel to vanadyl porphyrins (Baker *et al.*, 1978). All effects have been observed in Miocene sediments adjacent to a sill (Baker *et al.*, 1978) and with increasing burial

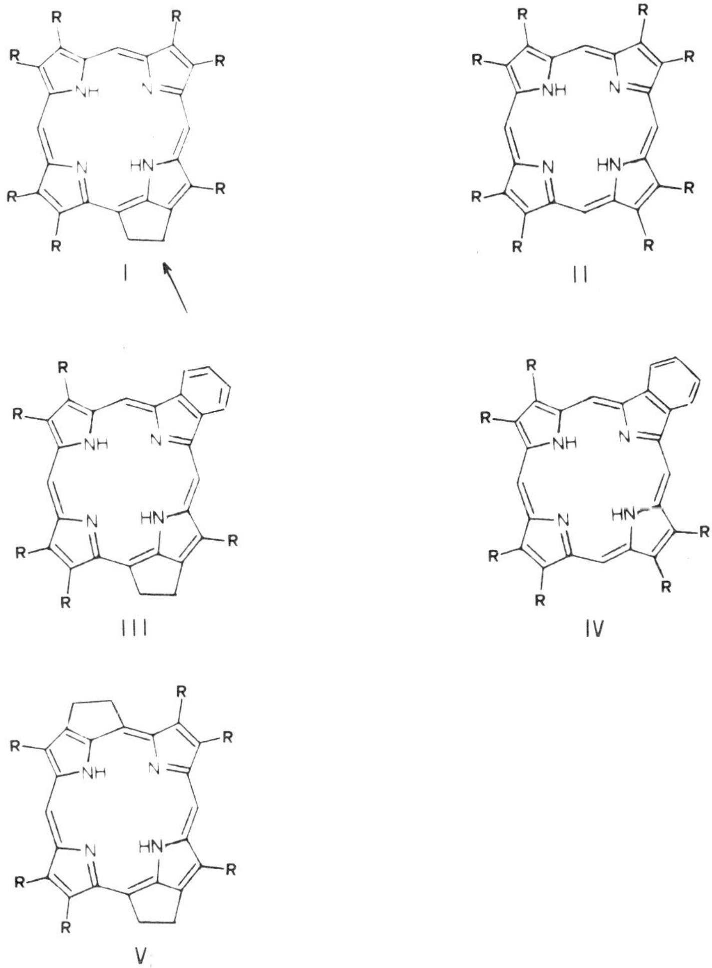

FIG. 19. The basic porphyrin structures encountered to date in sediments and crude oils. R = H or alkyl chain. I, deoxophylloerythroetioporphyrin (DPEP) and II, aetio are the most common, usually complexed to vanadyl or nickel ions. III is rhodo-DPEP, IV is rhodo-aetio and V is di-DPEP. With increasing maturity, bond cleavage in the isocyclic ring of structural type I (arrowed) converts it to structural type II.

depth in the Toarcian shales of the Paris Basin (Mackenzie *et al.*, 1980b). Mackenzie *et al.* (1980b) saw them occur quicker in the nickel than in the vanadyl porphyrins, and proposed that the generation of higher molecular weight species (C_{35}–C_{40}) arose from thermal cracking of kerogen (see also Quirke *et al.*, 1980).

Most of the more detailed structural work on the prophyrins has primarily involved ^{1}H NMR studies of individual components isolated from a bitumen (Quirke *et al.*, 1979). These structures were shown to be present in typical crude oils and source rocks, and three pseudohomologous series of aetio-porphyrins proposed (Hajibrahim *et al.*, 1981).

Tracing the porphyrins of oils and sediments back to the chlorophylls has been possible from the detailed study of Deep Sea Drilling Project cores (e.g. Baker and Louda, 1980, 1983). A complex series of reactions and their reactants and products has been determined, although the sequence is often different from that originally proposed by Triebs (1934, 1936) and summarized by Baker (1969). In simple terms it can be envisaged as the demetallation (removal of magnesium) from chlorophyll, followed by a variable sequence of defunctionalization-type reactions. After the chlorin intermediates (dihydroporphyrins with one less double bond in the nucleus) are "aromatized" (they lose two hydrogens from the nucleus) to porphyrins, these so-called "free base" porphyrins chelate with nickel or vanadyl and the above maturation reactions follow (Baker and Palmer, 1978). There are often deviations from this simplified scheme. For example, in highly oxic sediments the isocyclic ring, normally cleaved at depth by thermal cracking, is opened by oxidative cleavage (Baker and Louda, 1980). The distributions of chlorins and alkyl porphyrins can be revealed by a combination of high performance liquid chromatography (HPLC) and direct insertion probe mass spectrometry (e.g. Baker, 1966; Hajibrahim *et al.*, 1978; Baker and Louda, 1980; Mackenzie *et al.*, 1980b). The present structural knowledge neither suggests nor permits the stereochemical approach applied to other compound classes.

F. Other compounds

1. *2,6,10-trimethyl-7-(-3-methylbutyl)-dodecane*

This unusual three-pronged C_{20} alkane compound (Fig. 20 I) has been reported from a crude oil and several recent and ancient sediments (Yon *et al.*, 1982). It coelutes with pristane and could interfere with pristane/phytane ratios etc. Its origin, i.e. precursor natural product, is not known, although a reasonable biosynthetic route has been suggested (Yon *et al.*, 1982). Its only reported occurrence in an oil was at Rozel Pt., Utah. This is quite an unusual oil, containing large amounts of steranes and no triterpanes, and a hypersaline source rock has been postulated (Wardroper, 1979).

2. *Diterpenoids and sesquiterpenoids*

Although these components, often associated with resins, should be expected in crude oils, appropriate documentation is sparse. Sesquiterpenoids of the cedrene- and cuprene-type (e.g. II–IV in Fig. 20), have been observed in the Oligocene Bovey Tracy lignite beds of south England

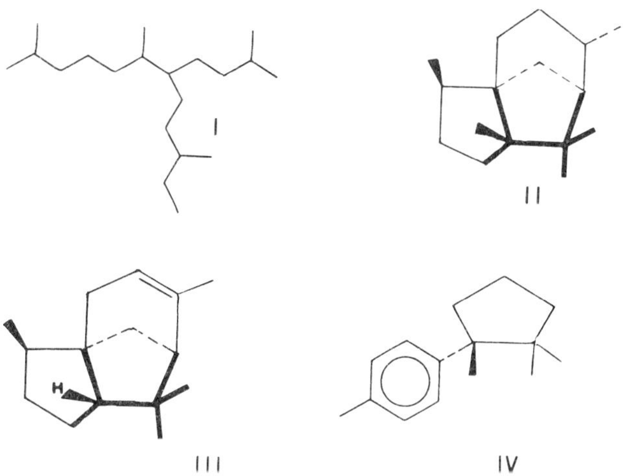

FIG. 20. A novel acyclic diterpenoid (I) and some of the sesquiterpenoids (II–IV) recently reported from sediments and petroleums. II is cedrane, III is α-cedrene and IV is cuparene.

(Grantham and Douglas, 1980), but not from oils or source rocks. Alexander *et al.* (1983a) have identified the bicyclic sesquiterpenes drimane and endesmane in oils. Simoneit (1977) reviewed the occurrence of diterpenoids in ocean sediments, particularly those associated with abietic acid (Fig. 21 I) which degrades via dehydroabietic acid (II) and further aromatization to give simonellite (III) and retene (IV) (Laflamme and Hites, 1978). These (III and IV) have been reported from a number of sediments (e.g. Brassell *et al.*, 1980), but presumably also occur in crude oils, and are swamped by the complex mixture of aromatic hydrocarbons generated by kerogen breakdown. Shaw *et al.* (1980) concentrated on the pimerane skeletal type (e.g. the pimaric acids V), and its diagenesis. Alkanes of this type occur in oils from north-west Canada, which are designated as immature and are thought to derive mainly from resinite-rich source rocks at low levels of overall maturity (Snowdon, 1980). The third type of diterpenoid acid, communic acid, polymerizes rapidly, so its detection in the geological sample is less favoured (Carman *et al.*, 1970).

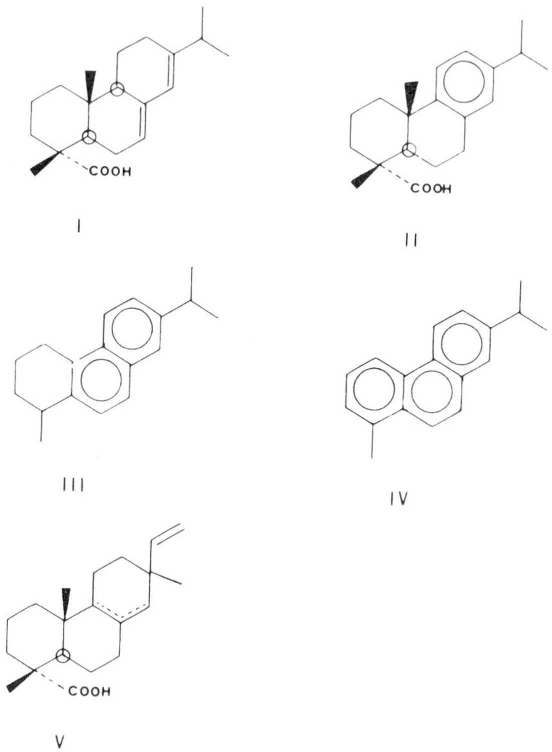

FIG. 21. Some examples of the tricyclic (resin) diterpenoids observed in sediments. I is abietic acid, II dehydroabietic acid, III simonellite, IV retene and V pimaric acid. A hexagon with an inscribed circle is an aromatic ring.

3. *Carotenoids* (Fig. 22)

β-carotane was reported by Murphy *et al.* (1967) from Green River shale. It derives from carotenoid pigments. It has also been found in an oil and Oligocene sediments probably dominated by algal material from the Shengli Basin, China (Shi Jiyang *et al.*, 1982). The carotenoids are excellent organic geochemical indicators of sedimentary palaeoenvironments (Watts *et al.*, 1977; Tibbetts, 1980). Unfortunately, most of them do not appear at present to survive even mild diagenesis in a recognizable form. Notable exceptions are β-carotane (Fig. 22 I) and three aromatic hydrocarbons from a Paris Basin Toarcian shale, including isorenieratane (Fig. 22 II), (Schaeflé *et al.*, 1977).

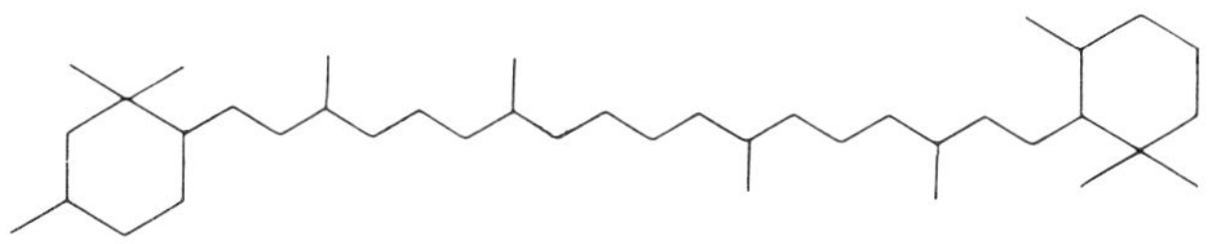

I

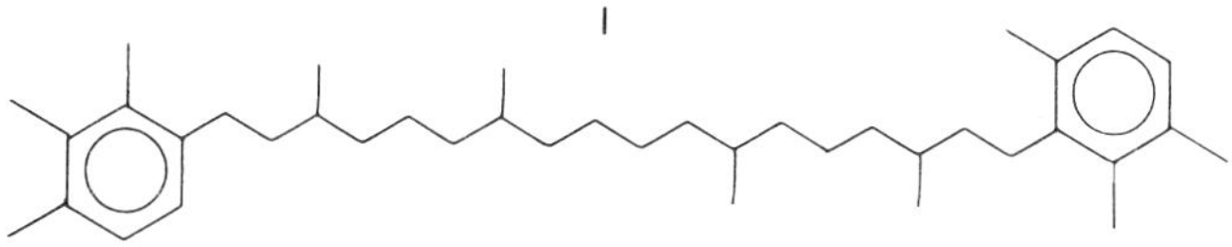

II

FIG. 22. Examples of carotenoid-derived hydrocarbons found in sediments and petroleums. I is β-carotane and II is isorenieratane. A hexagon with an inscribed circle is an aromatic ring.

4. *Bicyclic alkanes* (Fig. 23)

These were observed in Texan and Australian crude oils (Bendoraitis, 1974; Seifert and Moldowan, 1979; Philp *et al.*, 1981), although they are presumably more widespread. An obvious natural product precursor has as yet not been proposed. It is unlikely that they are the products of severe in-reservoir biodegradation (Philp *et al.*, 1981) as suggested by Seifert and Moldowan (1979).

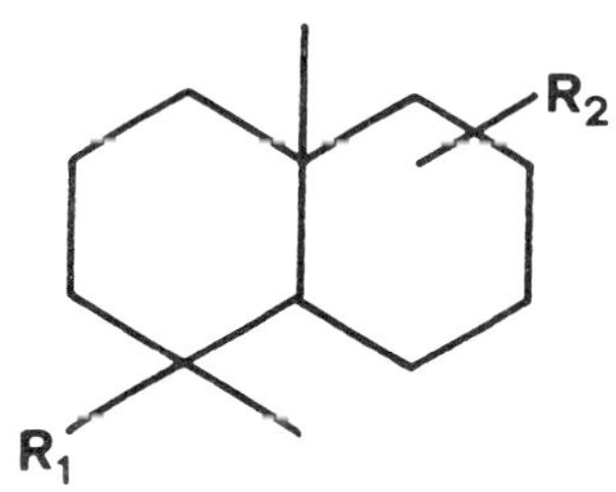

FIG. 23. Basic bicyclic skeleton reported from crude oils and sediments. R_1 = H or CH_3, R_2 = C_2H_5 or C_3H_7.

5. *Bile acids*

The findings and detailed stereochemical studies of C_{22} and C_{24} steroid acids (Fig. 24 I–IV) of a Californian petroleum suggested an animal contribution to the petroleum (Seifert *et al.*, 1972). This oil contains no alkanes (Seifert, personal communication), and it must be considered that these acids are the intermediates of a microbial oxidation of steranes in the reservoir. In addition, Boon *et al.* (1978) suggest, from the results of Mandava *et al.* (1974), that the distribution of these components could be explained by a higher plant contribution to the oil.

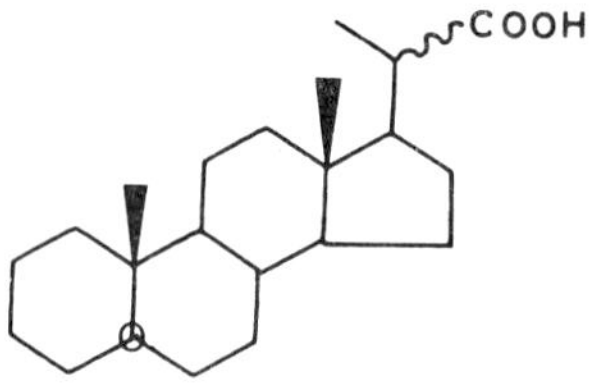

I

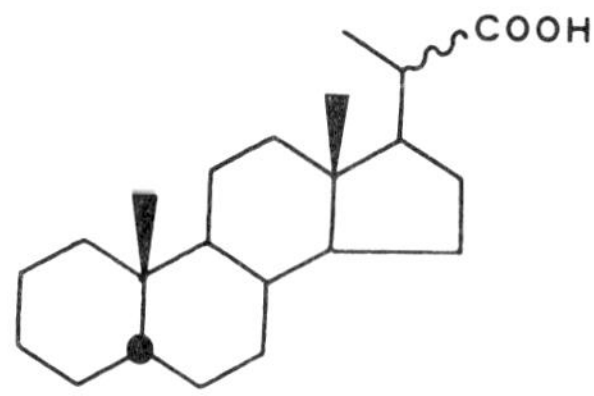

II

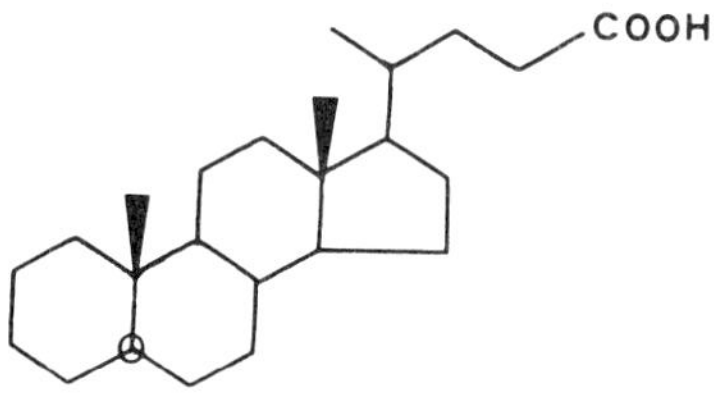

III

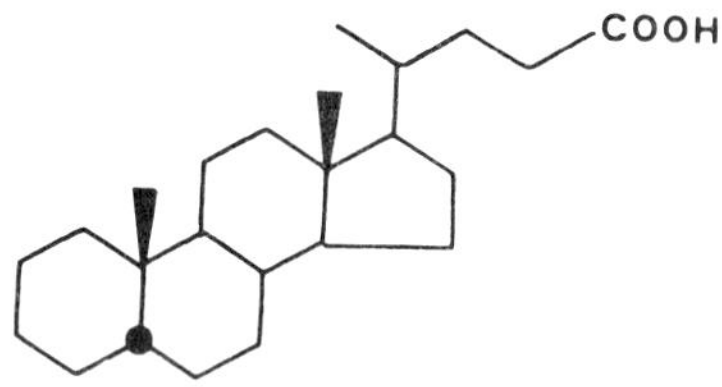

IV

FIG. 24. Steroid acids reported from a petroleum. I and II are 5α(H) and 5β(H)-pregnane acids respectively. III and IV are 5α(H) and 5β(H)-cholanic acids.

6. *Additional*

Other compounds which clearly have a biological origin have been ignored, where the actual evidence is less obvious (e.g. phenanthrenes, *n*-alkanes, alkyl benzenes and cyclohexanes). The work of Schmitter *et al.* 1980) tells that the justified devotion of biological marker petroleum geochemists in the past to hydrocarbons (with the exception of the porphyrins and the acids, see above) has ignored the biological information which will soon be found in the heterofractions of crude oils and sedimentary extracts.

IV. Applications for the Petroleum Geochemist

The potential of biological markers to assist the petroleum geochemist and geologist is great. Most of their contribution comes after drilling has commenced and often when oil shows have already been recorded. At an early stage of drilling, ideas on maturity and sedimentary heating rates and thermal history can be formed, which will assist in the determination of the timing of oil generation within a certain horizon. Later, oil occurrences can be grouped into sets of similar type and each type tentatively assigned to a possible source area in a basin. Such an exercise can determine preferred avenues of migration and help to predict favourable areas for further exploration. When the biological marker data are combined with other organic geochemical, geological ("geology" including sedimentology and palaeontology) and geophysical data, it can play a decisive role in the reconstruction of the processes which lead to the accumulation of oil in a sedimentary basin. This reconstruction is vital because a good understanding of migration and its mechanisms is still lacking (Roberts and Cordell, 1980) and because earlier ideas on petroleum generation (e.g. Tissot and Welte, 1978) may be oversimplified (e.g. Price *et al.*, 1981). It is also vital because exploration teams are increasingly turning to mathematical models (e.g. Welte and Yükler, 1981; McKenzie, 1981) to predict the processes of oil formation. These models can serve as the mouthpiece of organic geochemists who are necessarily steeped in much highly technical data and jargon, and allow the experience gained from drilled basins to be applied to unknown areas, often before drilling or seismic surveys have begun.

Despite this great potential (to be demonstrated below), far from fully realized as yet, there has been a tendency for some biological marker geochemists, including the present author, to overestimate the power of their technology and to dangerously oversimplify the geological systems and processes involved. This is particularly true in the area of oil/source-rock correlation (see below). Seifert and Moldowan (1981) divided discussion of the role of biological markers in petroleum exploration into four areas: biodegradation, maturation, correlation and migration. A similar subdivision is adopted here, and under biodegradation the effects of water washing and evaporation are also considered. It is hoped to honestly evaluate the contribution that biological markers can make in each division and the extent to which the data can be relied on in the face of conflicting interpretations arising from other areas of petroleum geology and geochemistry.

A. Biodegradation

1. *In sediments and the water column*

It is beyond the scope of this review to fully consider this topic. Clearly much of the alteration of organic material in the water column and shallow sediments is microbial. Coleman *et al.* (1979) have shown, from stable isotopic studies of diagenetic carbonates, that this microbial influence extends to great depths (hundreds of metres) and includes several distinctly different stages. The nature of this microbial alteration, and the lipids of the microbes themselves, will influence the nature of the biological marker fingerprint to be retained by the sediments at a more mature stage and ultimately inherited by any crude oil derived from the sediment. A more detailed preamble on the imprint of the biosphere on contemporary bottom sediments is included in part C below.

2. *In the oil reservoir*

Active in-reservoir biodegradation can destroy an accumulation's commercial value and has done so many times in the past (Demaison, 1977). Many authors have observed that a stepwise uptake of hydrocarbon types occurs with increasing extent of biodegradation: *n*-alkanes are utilized first by the microbes, followed by the branched alkanes, such as the acyclic isoprenoid alkanes, and alkyl cyclohexanes, which in turn are followed in very extreme cases by the most used (see below) biological marker alkanes, the steranes and triterpanes (Winters and Williams, 1969; Bailey *et al.*, 1973a; Rubinstein *et al.*, 1977; Seifert and Moldowan, 1979; Goodwin *et al.*, 1983).

At the time the *n*-alkanes have been lost, aromatic hydrocarbons are also attacked, and a sequence of uptake can also be seen here (Walker *et al.*, 1975; Rubinstein *et al.*, 1977; Aldridge *et al.*, 1977; Connan, 1981). Aromatic steroid hydrocarbons are rarely utilized in the reservoir. Most of the degradation is thought to result from the combined activity of aerobic and anaerobic bacteria (Winters and Williams, 1969; Bailey *et al.*, 1973a; Jobson *et al.*, 1979; Mackenzie *et al.*, 1983b) transported by and supplied with molecular oxygen and nutrients from downward-moving surface waters. Most of the detection of biodegradation relies on the absence or low concentration of the easily degraded components, e.g. *n*-alkanes. However, there are cases where *n*-alkanes have not been degraded, but certain aromatic hydrocarbons, e.g. alkyl naphthalenes, have (Connan, 1981). The generalized sequence of uptake presented above, however, usually holds. Positive evidence for the presence of bacteria on the oil can be obtained by studying the monoacid fraction. Mackenzie *et al.* (1983b) observed distributions of monoacids in biodegraded crudes, which included unsaturated straight-chain C_{18} acids (fatty acids) and 3(R), 7(R), 11(R), 15-tetramethylhexadecanoic acid (phytanic acid which still had the stereospecificity known for the acyclic isoprenoid ether lipids of certain bacteria) and which could suggest a recent bacterial origin. The finding of bacterial lipids does not, however, mean the bacteria were active on the oil (e.g. they had used it as a carbon source to a great extent). Biodegradation is not detected for reservoirs which have been at temperatures >80°C since filled with oil (Phillipi, 1977).

The consequence of the stepwise uptake is that only in very extreme cases are most of the biological marker distributions of interest altered. In less extreme situations the acyclic isoprenoids can be removed and pristane/phytane ratios altered (e.g. Connan *et al.*, 1980), but no change in the distributions of the higher molecular weight (up to C_{45}) acyclic isoprenoids has been reported to date. The general resistance to in-reservoir biodegradation of the optically active polycyclic hydrocarbon biological markers means that as the biodegradation (removal of the optically inactive *n*-alkanes) proceeds an increase in optical activity is observed (e.g. Winters and Williams, 1969). Initial results of Mackenzie *et al.* (1983b) for the pristane of biodegraded oils suggest that the increase in optical activity is not associated with the preferential consumption of the optical isomers of certain compounds (Phillipi, 1977), but rather results from the concentration of the source of the optical activity in oils, which is less easily degraded. This resistance to microbial alteration is an advantage for correlation studies, both in the geosphere (see below) and in environmental studies of oil spillages (e.g. Albaiges, 1980b). The biological markers are also unaffected by the other two main processes which will affect the composition of an oil after its spillage: photo-oxidation and evaporation of volatile components.

(*a*) *Biodegradation of polycyclic biological marker hydrocarbons*

(i) *Steranes.* Reed (1977) reported the curious absence of steranes in a weathered petroleum seep. Seifert and Moldowan (1979) reinterpreted this as extreme biodegradation, because crude oils from shallow Californian reservoirs (ca. 300 m), which had no *n*-alkanes or branched alkanes, showed evidence that the steranes had also been utilized by the microbes. The non-rearranged steranes, known to be present at the time of the oil's migration and accumulation since they were detected in more deeply pooled oils from common sources, had disappeared, and the rearranged steranes had been partially removed. Both skeletal types are removed by the microbes, but the non-rearranged steranes are either attacked first or faster. Mackenzie *et al.* (1983b) saw a similar effect in the Athabasca tar sand, where the non-rearranged steranes had been removed, but the rearranged steranes remained. In severely degraded oils from Pre-Cambrian basins in Australia, McKirdy *et al.* (1983) suggested that the 14α(H),17α(H)-steranes (biological configuration) were removed in advance of the 14β(H), 17β(H)-steranes (products of geological isomerization). Most of these effects were simulated in a laboratory culture by Goodwin *et al.* (1983), who also saw the C_{27} non-rearranged steranes utilized in advance of the C_{28} and C_{29} members.

(ii) *Triterpanes.* Reed (1977), again for a weathered seep with no regular hopanes, postulated that series of desmethylhopanes were the products of triterpane biodegradation. These were also found in the highly microbially degraded Californian petroleums (Seifert and Moldowan, 1979). Rullkötter and Wendisch (1982) have shown that they have been demethylated at C-10 (the C-25 methyl was removed) (see Fig. 8 III). They cover the carbon numbers C_{26}, C_{28}–C_{34}; the C_{30}–C_{34} members occur as mixtures of stereoisomers at C-22. Rullkötter and Wendisch (1982) separated the C_{29} component from a Madagascar asphalt by preparative gas chromatography and proposed its structure from ^1H NMR experiments. Since they have lost one carbon, the full distributions can be revealed by mass fragmentograms of *m/z* 177 (cf. *m/z* 191 in Fig. 10). In the other oils and laboratory cultures with reported sterane degradation the triterpenes were unaffected (Mackenzie *et al.*, 1983b; McKirdy *et al.*, 1983; Goodwin *et al.*, 1983).

Rullkötter and Wendisch (1982) also suggested that ring C opening may occur with the biodegradation of triterpanes.

(iii) *Aromatic steroid hydrocarbons and porphyrins.* Even in the highly degraded oils from California where both steranes and triterpanes were attacked, reported on by Seifert and Moldowan (1979), all types of aromatic steroid hydrocarbon and porphyrin which occur in crude oils were unaltered (Wade *et al.*, 1982). The porphyrins of Athabasca tar sand show no evidence of microbial alteration when compared with the undegraded Bellshill oil from the same source (Hajibrahim, 1978).

Since in nearly all oils produced to date the biodegradation of polycyclic alkanes has not occurred, and since the destruction of the commercial value of extreme cases is more than apparent without the biological markers, the detailed study of the sequence of uptake is to a large extent of academic value. More important is that the biological distributions are mainly unaffected by biodegradation and are still valid for the maturity and correlation applications discussed below.

B. Thermal maturation

1. *Framework*

The assessment of the extent of thermal maturation experienced by an organic-rich sedimentary rock in the temperature range ca. 50–150°C can be determined most accurately by so-called organic molecular parameters. Thermal maturation here encompasses the temperature-dependent chemical and physicochemical reactions which are experienced in the diagenesis and catagenesis zones of Tissot and Welte (1978). The best and most reliable examples of this approach involve the monitoring of the extent of certain reactions by measuring the relative concentrations of the reactants (A) and products (B):

$$[A] \underset{k_2}{\overset{k_1}{\rightleftharpoons}} [B]. \tag{1}$$

Ideally, the concentration of B should be zero at the start of the maturation process. The best measure of the reaction extent is

$$\frac{[B]}{[A] + [B]}. \tag{2}$$

This can be expressed as a percentage. This is preferable to the often-used ratio of products to reactants ([B]/[A]), since here very minor fluctuations (perhaps due to errors in the measurements) in the relative concentrations of A and B, when [B] is very much greater than or very much smaller than [A], are unnecessarily exaggerated. The relative concentrations of A and B are normally determined by GC–MS, and occasionally by gas chromatography, high-performance liquid chromatography and probe mass spectrometry (see below).

When $k_1 \gg k_2$ in (1) then the value of (2) will, in an ideal case, rise from 0 to 1 (or 0 to 100%) with increasing maturity. For most of the configurational isomerization reactions to be discussed below, there is a noticeable back reaction (k_2) and the value of (2) never reaches 1 (100%). Say for example $k_1 = k_2$, i.e. the equilibrium constant for the reaction is unity ($K_{eq} = k_1/k_2 = 1$), then with increasing maturity (2) will rise from 0 to 0.5 (50%). At 0.5 equilibrium is established and the reaction has reached its "end value". When a reaction has reached completion it can no longer be used to assess the extent of thermal maturation. Maturity assessment from all of these types of reaction, including those which are not reversible ($k_2 = 0$; (2) varies from 0–1) involves monitoring the approach to equilibrium at rates which are significant with respect to geological time. These rates are temperature dependent and the reactions are governed by the activation energies of the rates and not the difference between the free energies of reactants and products. There is no reported use in organic geochemistry of reversible reactions which occur rapidly on a geological time scale, and which because of a temperature dependence of the equilibrium constant could be used to evaluate palaeotemperatures.

Although very few molecular parameters of thermal maturation based on biological markers fit the ideal requirement that [B] = 0 when the reaction starts, only such reactions are evaluated in detail here. Discussion of measures of thermal maturation, which are based purely on the changes in the relative concentrations of a number of components observed with depth in a sedimentary sequence, but not obviously linked to a single reaction, is also limited. In both of the above non-ideal cases there is a real danger that the measures will show a dependence on organic matter type and/or the conditions of deposition and early diagenesis. This severely curtails such parameters' usefulness to the petroleum geochemist. Where [B] starts as zero (or at least always starts at a constant value), then the measurement should be independent of these effects in a direct sense. The only possible indirect problem is that the depositional palaeoenvironment will partly determine the catalytic activity of the sediment with respect to the reaction. The importance of this is discussed below. The above rules and concepts will be further explained with actual examples and summarized by recommending the most valuable measurements. For each molecular measurement of maturity discussed a detailed description of how the measurement is made is included. Practical advice based on the author's own experience is also provided.

2. *The reactions and measurements*

Figure 25 includes the four reactions, where the initial value of (2) when the reaction starts is nearly always zero. Table I includes the starting and end values of all reactions. The reactions are initially introduced as applied to the organic extracts of sedimentary rocks; details of applications to petroleums and the products of kerogen pyrolysis will be discussed later. Where it is mentioned that a measurement changes within the zone of hydrocarbon generation, then it is also capable of application to maturity assessment of crude oils.

(*a*) *Configurational isomerization at chiral centres not part of a ring system* (*acyclic*)

(i) C-6 *and* C-10 *in pristane* (Figs. 5 and 25). This isomerization at the two chiral centres of pristane is followed by gas chromatography of the alkane fraction with 100 m glass capillary columns (ca. 0.25 mm i.d.) coated with diethylglycolsuccinate (e.g. Patience *et al.*, 1978; Mackenzie *et al.*, 1980a). The *meso*isomer ("biological" 6R, 10S stereochemistry) is separated from the two enantiomers formed by geological isomerization (e.g. Fig. 25). In Expression (2) [A] is the area of the *meso* peak whilst [B] is the area of the peak containing the other two isomers. Values of (2) range from 0 to 0.5 (50%) with increasing maturity. This isomerization is apparently complete at a relatively early stage of maturity (Mackenzie *et al.*, 1980a). In order to determine the composition of the first eluting peak (the relative amounts of the two geological isomers), the pristane must be purified by

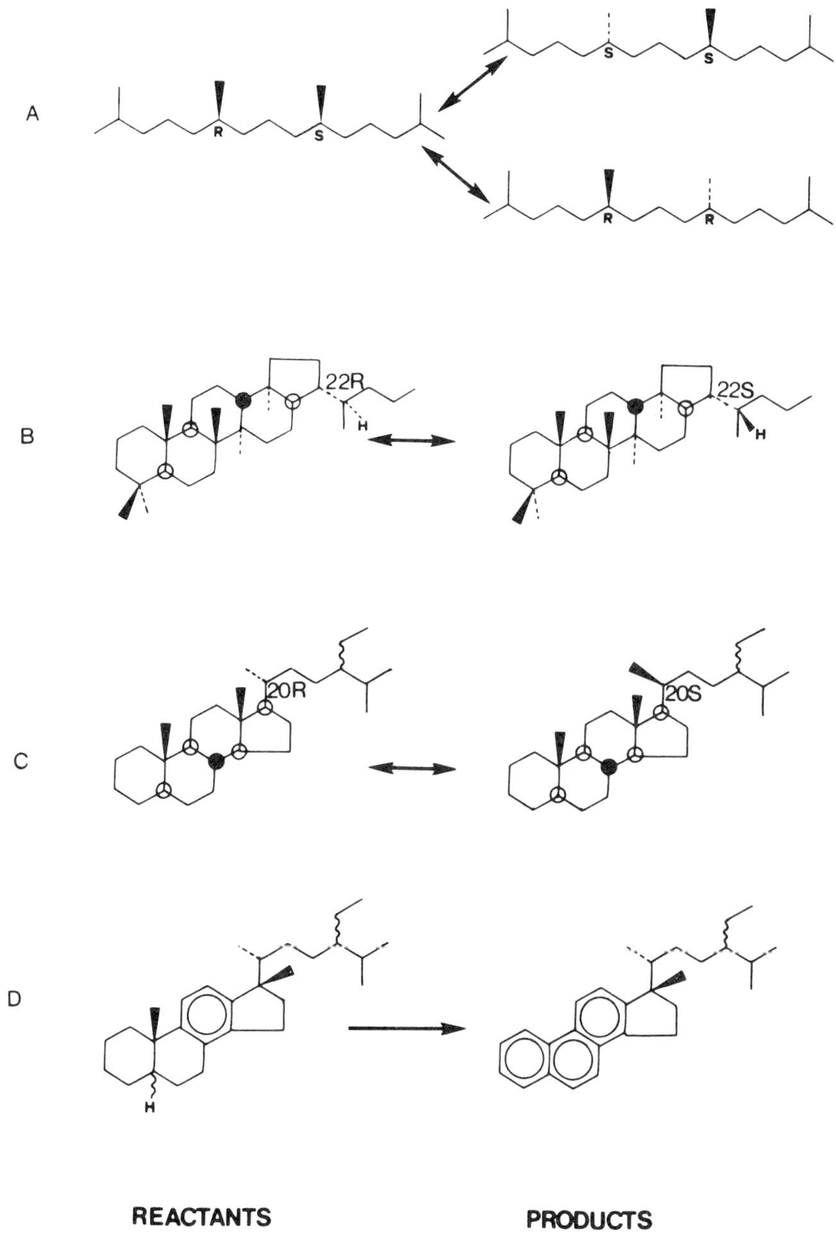

FIG. 25. Four temperature-dependent geochemical reactions which can be used for the assessment of the extent of thermal maturation in sediments and crude oils.

TABLE I. *Reactions for determination of thermal maturity.*

Reaction	Ratio* (2)	Starting Value of (2)	End Value of (2)
Configurational isomerization	$\frac{\text{RR} + \text{SS}}{\text{total pristane}}$	0	0.5
	$\frac{\text{22R}}{\text{(22R + 22S)} - 17\alpha\text{(H)-hopanes}}$	0	0.6
	$\frac{\text{24R}}{\text{(24R + 24S)-steranes}}$	variable	0.5
	$\frac{\text{20S}}{\text{(20R + 20S)-steranes}}$	0	0.5–0.55
	$\frac{17\alpha\text{(H)},21\beta\text{(H)-hopanes}}{\text{total hopanes}}$	variable	0.9
	$\frac{5\alpha\text{(H)},14\beta\text{(H)},17\beta\text{(H)-steranes}}{\text{total steranes}}$	variable	0.8
Aromatization	$\frac{\text{triaromatic steroid hc.}}{\text{mono + triaromatic steroid hc.}}$	0	1.0
Apparent C–C bond cleavage	(e.g.) $\frac{C_{20}\text{ triaromatic steroid hc.}}{C_{20} + C_{28}\text{ triaromatic steroid hc.}}$	variable	1.0
	$\frac{\text{Aetio}}{\text{DPEP + Aetioporphyrins}}$	variable	1.0

* see text for details

preparative gas chromatography and then oxidized and the derivitized acids that arise examined by gas chromatography (e.g. Patience *et al.*, 1978). Such a process, when applied to phytane (Fig. 4), which has three chiral centres and eight stereoisomers, suggests the isomerization may be more complicated than first imagined, since part of it may occur when the acyclic isoprenoid moiety is in a "bound" form, perhaps to kerogen (Patience *et al.*, 1980)

(ii) C-22 *in* 17α(H),21β(H)-*hopanes* (Figs. 9 and 25). This isomerization can be followed in the C_{31}–C_{35} hopanes, as revealed by the *m/z* 191 fragmentograms for the GC–MS analysis of a separated alkane fraction (Fig. 10). The 22S isomer elutes before the 22R isomer in each case. Either an average of the isomer ratios for all five carbon numbers can be measured or one carbon number consistently chosen. C_{31} should be avoided, but occasionally it is the only one of the five present in sufficient quantity. The results for C_{31} are prone to error, because often the two isomers are not resolved to baseline, and gammacerane, another pentacyclic triterpane, coelutes with the 22R isomer under certain conditions (e.g. Shi Jiyang *et al.*, 1982). The C_{32} hopane is probably the best to use for making the measurement. The value of Expression (2), where A is 22R and B is 22S, varies from 0 to ca. 0.6 (60%) with increasing maturity. (Ensminger *et al.*, 1974; Seifert and Moldowan, 1980), although the actual end point (equilibrium value) varies slightly between carbon numbers. The isomerization requires a higher level of maturity than at C-6/C-10 in pristane to reach completion, but it is normally complete before the onset of intense hydrocarbon generation (Mackenzie and Maxwell, 1981). Under rare conditions, very immature oils, presumably derived from the beginning of the zone of hydrocarbon generation, can have values of (2) which suggest this isomerization is incomplete (Seifert and Moldowan, 1980; Shi Jiyang *et al.*, 1982). In general, however, this measurement has no application to the measurement of the extent of thermal maturation in crude oils.

(iii) C-24 *in steranes*. Isomerization at this centre was first postulated by Mulheirn and Ryback (1975) using ^{1}H NMR, but has since been primarily studied by the direct GC with the column employed for pristane (see above) of the thiourea adduct of a separated branched and cyclic alkane fraction, and has mainly been monitored for (20R)-24-methyl-5α(H), 14α(H),17α(H)-cholestane (R = CH_3 in Fig. 15 IV) (Maxwell *et al.*, 1980). The starting value of Expression (2) is variable, because both configurations are biosynthesized as sterols. In the three sedimentary sequences studied (Green River shale (Mulheirn and Ryback, 1975), Toarcian shales, Paris Basin (Mackenzie *et al.*, 1980a), Cretaceous shales, Wyoming Overthrust Belt (Mackenzie *et al.*, 1983a)), the immature shales have shown a preference for 24S, which is lost at a similar maturity to that level at which the isomerization of pristane at C-6/C-10 is complete. Isomers at C-24 are *not* resolved under the GC–MS conditions used for all the other steroid transformations.

(iv) C-20 *in steranes* (Figs. 15, 16 and 25). The elution patterns for steranes, seen by way of *m/z* 217 and 218 fragmentograms (Fig. 2) are highly complex (i.e. many isomers), and because of overlap of rearranged and non-rearranged steranes, it is only possible to follow the isomerization of C_{29} components ($R = C_2H_5$ in Figs. 15 and 17). Mass fragmentograms of the sterane molecular ions can help when the steranes are major components of the alkane fraction being analysed by GC–MS. The extent of C-20 isomerization is measured for 5α(H),14α(H),17α(H)-steranes using the peak areas of the two isomers in the *m/z* 217 mass fragmentograms. In Expression (2) A is the 20R isomer and B is the 20S isomer. Values of (2) rise from 0 to 0.5–0.6 (50–60%). Because of the complexity of the sterane distributions, it is very difficult to determine the equilibrium value accurately, and the possibility of a temperature dependence for the equilibrium constant must be considered. This isomerization is the most stubborn of all achiral centre isomerizations studied. It can extend well into the zone of hydrocarbon generation (Mackenzie and Maxwell, 1981) and appears to have an application to the assessment of thermal maturity of both sedimentary rocks and crude petroleums (Mackenzie *et al.*, 1980a; Seifert, 1981; Seifert and Moldowan, 1981).

(*b*) *Configurational isomerization at chiral centres part of a ring system (cyclic).*

In general, the picture is much more complex for these centres, and the reader is advised to use them as supplementary measures of maturity to confirm and extend trends seen in acyclic centres and the aromatization of steroids (see below).

(i) C-17 *and* C-21 *in hopanes* $\geqslant C_{29}$ (Fig. 9). Most hopanoids are biosynthesized with the 17β(H),21β(H) stereochemistry (Ourisson *et al.*, 1979), but 17α(H),21β(H) and 17β(H),21α(H) forms have also been reported in organisms (Quirk *et al.*, 1980, and Galbraith *et al.*, 1965, respectively). The stability increases in the order 17β(H),21β(H) < 17β(H),21α(H) < 17α(H),21β(H). 17β(H),21β(H) is usually the dominant stereochemistry of very immature sediments and it disappears fairly rapidly with increasing maturity (Ensminger *et al.*, 1974, 1977). This disappearance is thought to be due to the isomerization of 17β(H),21β(H)-hopanes to their more stable forms, but mainly to the most stable 17α(H),21β(H) form. Thus Expression

(2) can be simplified as

$$\frac{[17\alpha(H),21\beta(H)]}{[17\beta(H),21\beta(H)] + [17\alpha(H),21\beta(H)]} \qquad (3)$$

Values of (3) rise from 0.2–0.3 (20–30%) to 1 (100%). 100% 17α(H),21β(H) is normally reached again at a similar maturity level to that at which pristane isomerization is complete (Ensminger *et al.*, 1977; Mackenzie *et al.*, 1980a) and also to the total disappearance (saturation?) of hopenes and rearranged sterenes. At this stage the less stable 17β(H),21α(H) stereochemistry is present at up to half the concentration of the more stable 17α(H),21β(H) stereochemistry (Seifert and Moldowan, 1980). The conversion of the former to the latter is more complicated (see above) and is delayed (the value of

$$\frac{[17\alpha(H),21\beta(H)]}{[17\beta(H),21\alpha(H)] + [17\alpha(H),21\beta(H)]} \qquad (4)$$

remains unchanged for a stage) until sufficient thermal energy is available (Seifert and Moldowan, 1980).

This normally occurs in the early stages of oil generation and the values of (4) rise to ca. 0.9–1.0 (90–100%) (Seifert and Moldowan, 1980).

The measurement of (3) and (4) is best confined to the C_{30} hopanes. Only with this carbon number can the peak areas in *m/z* 191 fragmentograms (Fig. 10) alone be used, since for the other species a given current from *m/z* 191 will represent a much greater difference in absolute mass for the three isomeric types (cf. van Dorsselaer *et al.*, 1977).

(ii) C-17 *in* C_{27} *hopanes* (Fig. 9). Since C_{27} hopanes have no side chain (R = H in Fig. 9 IIa and IIIa) and therefore no chiral centre at C-21, the conversion of the less stable 17β(H) compounds to 17α(H) forms is a slightly different reaction than that discussed for hopanes $\geqslant C_{29}$. It reaches completion (100% 17α(H)) at a maturity level between the disappearance of 17β(H),21β(H)-hopanes and the start of significant conversion of 17β(H),21α(H)-hopanes to 17α(H),21β(H)-hopanes—i.e. before the onset of intense hydrocarbon generation (Seifert and Moldowan, 1980).

(iii) C-14 *and* C-17 *in steranes* (Fig. 16). Again only C_{29} components are considered. Immature sterane distributions include mainly 5β(H), 14α(H),17α(H),20R- and 5α(H),14α(H),17α(H),20R-steranes. The latter are normally two to three times more abundant than the former (Gallegos, 1971). This means six peaks in the *m/z* = 217 fragmentograms (two per carbon number) acquired under normal conditions. Increasing maturity sees the appearance of the 20S isomers and the 14β(H),17β(H) species (Seifert and Moldowan, 1979; Mackenzie *et al.*, 1980a; Seifert, 1981).

Unfortunately, the major 5α(H),14β(H),17β(H) components coelute with the 5β(H),14α(H),17α(H)-steranes under most conditions (Seifert and Moldowan, 1979; 1981). Seifert and Moldowan (1979) and Mackenzie *et al.* (1981b) suggested elaborate calculations to overcome the problem and to allow the measurement of

$$\frac{[5\alpha(\mathrm{H}),14\beta(\mathrm{H}),17\beta(\mathrm{H})\ 20\mathrm{R} + 20\mathrm{S}]}{[5\alpha(\mathrm{H}),14\alpha(\mathrm{H}),17\alpha(\mathrm{H})\ 20\mathrm{R} + 20\mathrm{S}] + [5\alpha(\mathrm{H}),14\beta(\mathrm{H}),17\beta(\mathrm{H})\ 20\mathrm{R} + 20\mathrm{S}]} \qquad (5)$$

for C_{29} steranes.

These exploited the facts that the major peak in the mass spectra of 14β(H)-steranes is *m/z* 218, whilst it is *m/z* 217 for 14α(H)-steranes (Mulheirn and Ryback, 1977), and that *m/z* 151 is a significant fragment of 5β(H)-steranes but not of 5α(H)-steranes (Tökes and Amos, 1972). The measurements required mass fragmentograms for *m/z* 151, 217 and 218. In fact for mature samples the errors in these measurements are probably as great as those caused by simply calculating the ratio of the absolute area of the peaks in the *m/z* 218 fragmentograms corresponding to the C_{29} 5α(H),14β(H),17β(H)-steranes, but also including a small contribution from C_{29} 5β(H),14α(H),17α(H)-steranes, to the absolute areas of the two peaks in the *m/z* 217 fragmentograms equivalent to the C_{29} 5α(H),14α(H),17α(H)-steranes (Fig. 2). With the column coatings (SE-52, OV-1) used by this author the C_{29} 5β(H),14α(H),17α(H) 20R isomer elutes between the 20R and 20S isomers for C_{29} 5α(H),14β(H),17β(H)-steranes. Seifert and Moldowan (1981) suggest that with Dexsil phases a more accurate value for (5) can be obtained by just using the 20R isomers, since GC–MS with Dexsil columns causes the 5β(H),14α(H),17α(H) 20R isomer to elute entirely with the 20S isomer of 5α(H),14β(H),17β(H)-steranes.

There remain further real problems with this measurement. Values of zero for (5) are rarely observed, even in highly immature sediments with little or no 5α(H),14α(H),17α(H), 20S-steranes. Usually the value for (5) climbs suddenly over the first half of the oil-generation zone from anything between 0 and 0.5 (50%) to as high as 0.8 (80%, Mackenzie *et al.*, 1980a; Seifert and Moldowan, 1981). This increase occurs over a similar maturity range as the conversion of 17β(H),21α(H)-hopanes to 17α(H),21β(H)-hopanes (see above). The variable values of (5) in immature sediments normally show no trends with increasing maturity. In the past the high amounts of the "mature" 5α(H),14β(H),17β(H)-steranes in relatively immature sediments have been ascribed to a reworked contribution. The consistency of this discrepancy in several sedimentary sequences (listed below) suggests that another explanation is required. Perhaps functionalized steroids (sterenes, sterols) can be "rearranged" to the 14β(H),17β(H) stereochemistry at an early diagenetic stage (cf. Caspi *et al.*, 1975) and the extent of this "rearrangement" is dependent on the mineral matrix and/or other factors. Whatever the cause, clearly the use of this reaction to assess the thermal maturity of both sediments and crude oils requires extreme caution, especially when the value of (5) is less than 0.5.

(*c*) *Configurational isomerization in monocarboxylic acids*
The analysis of these species requires lengthy work-up procedures, and the necessary components are often not present in sufficient concentrations. The isomerization of hopanoic acids at C-17, C-21 and C-22 (Ensminger, 1977) and of C_{14}–C_{21} acyclic isoprenoid acids (Mackenzie *et al.*, 1982c, 1983a) has been studied. In both cases the overall isomerization of the acids is slower than that of the corresponding alkanes. For example, 17β(H),21β(H)-hopanoic acids have been reported from oils and 17β(H),21β(H)-hopanes have not (Schmitter *et al.*, 1978). Both the inductive effect and increased steric hindrance of the carboxylic acid group slows the isomerization. However, the observation that in acyclic isoprenoid acids chiral centres nearest the carboxyl group isomerize fastest probably means that these centres are isomerized by a different mechanism from the more extreme acid centres and the alkane centres (see Mackenzie *et al.*, 1982c, for details).

(*d*) *Aromatization of C-ring aromatic steroid hydrocarbons* (Fig. 17)
This transformation is the only aromatization-type reaction where quantification has been attempted (Mackenzie *et al.*, 1981a). The conversion of C-ring aromatic steroid hydrocarbons with two nuclear methyl substituents to ABC-ring triaromatic steroid hydrocarbons with one nuclear methyl substituent via diaromatics (Fig. 17 II, IV) is thought to be the dominant process changing the relative concentrations of proposed reactants (II) and final products (IV). The measurement of Expression (2) concentrates only an aromatic steroid hydrocarbons with a 20R stereochemistry and C_{10} side chains (C_{29} monoaromatics and C_{28} triaromatics, Fig. 25) even though 20S isomers and components with C_2, C_3, C_8 and C_9 side chains exist. In the proposed reaction (Fig. 25) four monoaromatic steroid hydrocarbons isomeric at C-5 and C-24 are converted to two triaromatic steroid hydrocarbons, because the chiral centre at C-5 is lost. The C-24 isomers are not resolved under the GC–MS conditions normally used, so the calculation of (2) for the reaction requires the relative concentrations represented by three GC–MS peaks. It is assumed that the currents of the fragments from the cleavage of the 17(20) bond in the ion source of the mass spectrometer (see Fig. 2; *m/z* 253 for monoaromatics, *m/z* 231 for triaromatics) represent a similar absolute mass for both com-

pound types. In both cases these fragments are by far the major ones in the mass spectra (Mackenzie *et al.*, 1981a). Therefore the ratios of reactants to products can be calculated with *m/z* 231 and 253 fragmentograms acquired from the GC–MS analysis of the appropriate aromatic fraction (see Mackenzie *et al.*, 1981a, for details). Because the C_{29} $5\beta(H)$ 20R-monoaromatic steroid hydrocarbon, coelutes with the C_{28} $5\alpha(H)$ 20R-monoaromatic, it is necessary to correct for this by assuming the C_{27}/C_{28} ratio in the triaromatics is the same as the C_{28}/C_{29} ratio in the monoaromatics. This correction is not necessary for mature samples, where the $5\alpha(H)$monoaromatics are absent.

If

a = area (total ion count) of the C_{28} 20R triaromatic peak in the *m/z* 231 fragmentogram,

b = area of the C_{27} 20R-triaromatic peak in the *m/z* 231 fragmentogram,

c = area of the C_{29} $5\alpha(H)$ 20R-monoaromatic peak in the *m/z* 253 fragmentogram,

d = area of the C_{29} $5\beta(H)$ 20R- and C_{28} $5\alpha(H)$ 20R-monoaromatic peak in the *m/z* 253 fragmentogram,

then (2) for the aromatization reaction is

$$\frac{a}{d - c(b/a - 1) + a} \tag{6}$$

It is often difficult to identify the relevant peaks in the *m/z* 253 fragmentograms, because molecular ions can only be recorded with chemical ionization, and because in certain sequences new monoaromatic steroid hydrocarbons, not explicable in terms of the structural possibilities shown in Figs. 2 and 17, appear in the *m/z* 253 fragmentograms. Special care is therefore required (see section 4 below).

This measurement (6) increases from 0 to 1 (100%) with increasing maturity over a range which can extend from about the beginning of the acyclic isomerization reactions to the peak of oil generation (but see section 9 below), and which is similar to that covered by isomerization at C-20 in steranes. It has an application to crude oils, but unfortunately it is altered by fractionation effects resulting from migration (see below).

(*e*) *Apparent carbon–carbon single-bond cleavage.*
The adjective "apparent" is used because in most cases there is no concrete evidence that this reaction is converting one species to the other measured component.

(i) *Increase in relative amounts of lower molecular weight steroid hydrocarbons.* This can be measured from the appropriate mass fragmentograms (e.g. Fig. 2; *m/z* 217 for steranes, *m/z* 231 for ABC-ring triaromatic steroid hydrocarbons, *m/z* 253 for C-ring monoaromatic steroid hydrocarbons). An idealized version of Expression (2) based on Reaction (1) is not possible. However, if in (2) A is a C_{28} 20R-triaromatic steroid and B is a C_{20} species (Fig. 18) then an increase from a low value to 1 (100%) is always seen from the onset of intense oil generation to the end of the oil window (vitrinite reflectance ca. 0.6–1.5%) (Mackenzie *et al.*, 1981a, Mackenzie and Maxwell, 1981). In sediments less mature than this range, the value often shows variations which are not explicable as maturity trends. Short side-chain sterols do occur in recently deposited sediments, and these could be natural products (Brassell, 1980). Diagenetic oxidative cleavage of the Δ22 double bond present in many steroids could also be a source for the short side-chain steroid skeletons in geological samples. Similar effects to those in the triaromatics can be seen in the monoaromatic steroid hydrocarbons and the steranes, and all have an application to crude oils, because the major increase in the relative amounts of the short side-chain species straddles the oil-generation zone (Seifert and Moldowan, 1978; Mackenzie *et al.*, 1981a; Sajgó *et al.*, 1984).

(ii) *Conversion of DPEP to aetioporphyrins.*
This is thought to arise from the cleavage of the isocyclic ring in the DPEP porphyrins (Corwin, 1959; Didyk *et al.*, 1975). It can be measured from the total averaged probe mass spectra of demetallated porphyrin concentrates from sedimentary organic extracts and oils (e.g. Mackenzie *et al.*, 1980b). Appropriate values of Expression (2) therefore include all the major carbon numbers detected and can vary from 0 to 1 (100%) with increasing maturity. Unfortunately, under oxidizing depositional conditions the isocyclic ring in the precursor chlorins is cleaved (see above) so that the value of the appropriate form of (2) is not always zero, when the thermal cleavage of the isocyclic ring starts (Baker and Louda, 1980).

(*f*) *Other reactions*
There exist many other proposed organic molecular measurements of thermal maturity. Most of them relate to the highly complicated area of early diagenesis, where the simple concepts outlined in section 1 above are difficult to apply. They include the conversion of sterenes to rearranged sterenes, aromatics and steranes (e.g. Rubinstein *et al.*, 1975; Giger and Schaffner, 1981; Rullkötter and Welte, 1983); the movement and stabilization of hopene double bonds from the side chain to the ring system (Ensminger, 1977; Brassell *et al.*, 1980) and of sterene nuclear double bonds (Giger and Schaffner, 1981); and the reduction of rearranged sterenes (Ensminger *et al.*, 1978). Seifert and Moldowan (1978) have suggested two additional hopane ratios for crude oil maturity assessment: an increase in

the ratio of 18α(H)-trisnorneohopane to 17α(H)-trisnorhopane (Figs. 7–9) and in the amounts of the C_{27} and C_{29} regular hopanes relative to the C_{30} regular hopane. The driving force for these ratio changes is not clear.

3. *Possible uses for the petroleum geochemist*

Many of these molecular parameters, especially those discussed in detail, have been successfully applied to the following sedimentary sequences: (1) Paris Basin, Lower Toarcian shales with variable maximum burial depths (Ensminger *et al.*, 1977; Mackenzie *et al.*, 1980a, b, 1981a); (2) Cretaceous shales of the Wyoming Overthrust Belt (Seifert and Moldowan, 1980; Mackenzie *et al.*, 1983a); (3) Pliensbachian shales of north-west Germany around a proposed igneous intrusion (Mackenzie *et al.*, 1981a, c); (4) Oligocene shales from the Zhanhua Basin, China (Shi Jiyang *et al.*, 1982); (5) Jurassic shales from the North Sea (Mackenzie *et al.*, 1984); (6) Miocene coals and shales from the Mahakam Delta, Indonesia (Hoffmann *et al.*, 1984; Schoell *et al.*, 1983); (7) Pliocene shales and siltstones of the Pannonian Basin, Hungary (Sagjó *et al.*, 1984). In addition, this author has learned of other helpful applications, as yet confidential, from other researchers in the field. Most lacking are published studies on carbonate sequences.

Within each reaction type (e.g. isomerization of acyclic alkane centres) a sequence of changes is always observed, and Fig. 26 attempts a summary of the ranges covered by certain measurements in relation to vitrinite reflectance and oil generation for the Mesozoic samples studied. It is only a rough guide, since the relative rates of the reactions are governed by the kinetic constants of the reaction rates and vary with thermal history (see Section 9 below). Those reactions whose ranges extend into the oil-generation zone have an application to the measurement of crude oil maturity as well as sedimentary rocks. However, the greatest contribution to maturity assessment occurs in the area above the generation zone. Some parameters cover a wide range (e.g. C-20 isomerization and aromatization of steranes; see Fig. 26) whilst others change rapidly over narrower ranges (e.g. C-22 isomerization of hopanes). The latter situation can provide a sensitive maturation measurement, when the sediments to be assessed fall within the maturation stage over which the change occurs. That most of the reactions involved are different from those associated with petroleum generation means the biological markers should be regarded as being built-in measures of thermal stress and not as mimicking the oil generation processes. In this way their total concentration in an extract or an oil is unimportant.

The sequence of changes in the isomerization of acyclic chiral centres, whereby the apparent isomerization rates decrease in the order C-6/C-10 in pristane, ≃C-24 in non-rearranged steranes, >C-22 in 17α(H), 21β(H)-hopanes, >C-20 in non-rearranged steranes, probably reflects the increasing steric hindrance at the centres concerned. The more hindered a centre, the lower the isomerization rate (Mackenzie *et al.*, 1980a). The steric hindrance probably influences the isomerization by controlling the access of a centre to a catalyst.

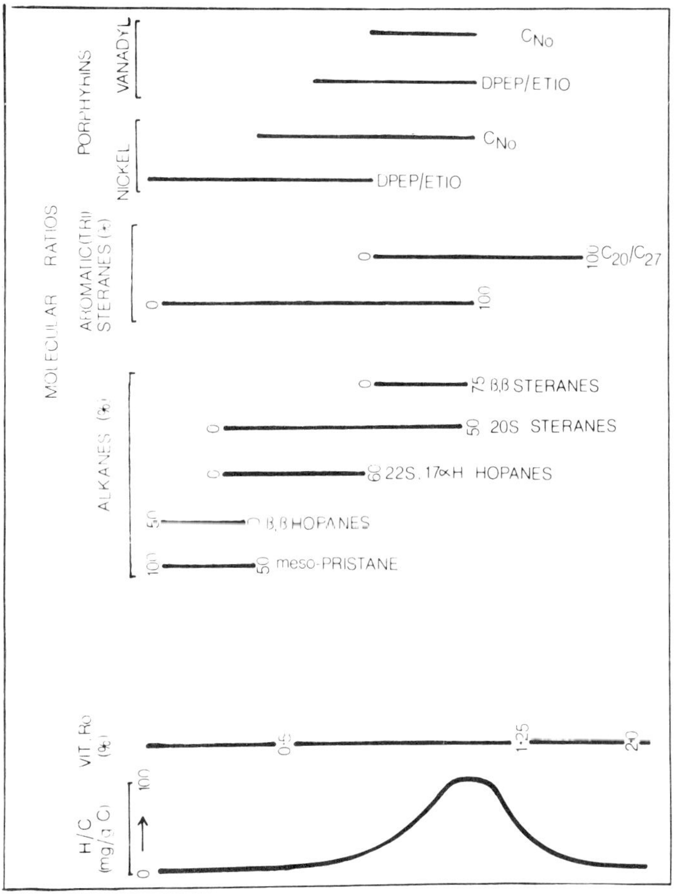

FIG. 26. Ranges of individual molecular measurements of thermal maturation, including those for steroids, plotted against the downhole hydrocarbon generation curve and vitrinite reflectance values. This was compiled from results for the Toarcian shales of the Paris Basin and for the Pliensbachian shales of N.W. Germany.

4. *Reproducibility*

Such a topic is of great importance, and all biological marker measurements should be accompanied by replicate analyses of standardized samples and of selected samples from the sample suite being investigated. It would be of considerable help if an internationally agreed set of standard samples were available in all laboratories using GC–MS in petroleum geochemistry. Preliminary studies of reproducibility (Mackenzie, 1980 and unpublished results) suggest that typically the worst errors in Expression (2) for all the hydrocarbon measurements discussed above in detail are about ±0.05. It is not always possible to record the maturity parameters for all samples. Sometimes the components are simply not present, and sometimes coelution with a component of the ratio with an unknown species prevents accurate measurement of its relative concentration. For example, occasionally such a component coelutes with the 20S isomer of 5α(H), 14α(H), 17α(H) C_{29}-steranes. This component has *m/z* 217 in the mass spectrum but no *m/z* 218 or *m/z* 400 (C_{29} sterane molecular ion) so the isomerization of steranes at C-20 can be studied using *m/z* 218 or *m/z* 400 fragmentogram instead of *m/z* 217. Such solutions are not always possible, and at all times attention should be paid to peak shapes in a number of diagnostic fragmentograms and to examination of full mass spectra.

Any study based primarily on cuttings material should include some analyses of cores to check the effects of mud contamination.

5. *Calibration*

There is always pressure for biological marker organic geochemists to conform and to calibrate their measures of organic maturation with more accepted indicators such as vitrinite reflectance. Such a policy is *not* advised by this author. The molecular reactions used to measure thermal maturation are nearly all complete by a reflectance in oil (R_o) of ca. 0.6–0.7%. This means they are much more sensitive and accurate assessors of maturity in the approximate range $R_o = 0.3$–0.7%, where the reflectance measurements are particularly erroneous (Dow, 1977). Secondly, studies on the relative rates of certain reactions suggest that such a calibration over many sedimentary basins is invalid (Mackenzie *et al.*, 1982b, and see section 9 below). It is preferable to calibrate the molecular parameters with more direct measurements of thermal history, such as maximum burial depth, and ultimately with mathematical expressions of a shale's thermal history (see section 9 below). Figures 27 and 28 show examples of plots of maximum burial depth (in both cases equivalent to the present burial depth) against two molecular measures of thermal maturity. Figure 27 shows percentage aromatization of steroid hydrocarbons in the Jurassic of the East Shetland Basin (North Sea) and Fig. 28 shows percentage isomerization of C-20 in 5α(H), 14α(H), 17α(H)-steranes for the Pliocene of the Pannonian Basin. Calibration with maximum burial depth is more difficult when uplift has occurred, since it relies on the geological interpretation of burial history.

Indeed, the biological marker studies on Paris Basin Toarcian shales suggested that earlier estimates of maximum burial depth in the south-east of the basin (Deroo, 1967) were too low (Mackenzie *et al.*, 1980a, b, 1981a). This agreed with the maximum burial depths calculated from sonic logs (Goy, 1979) and from mathematical models for the evolution of the basin (Brunet, 1981).

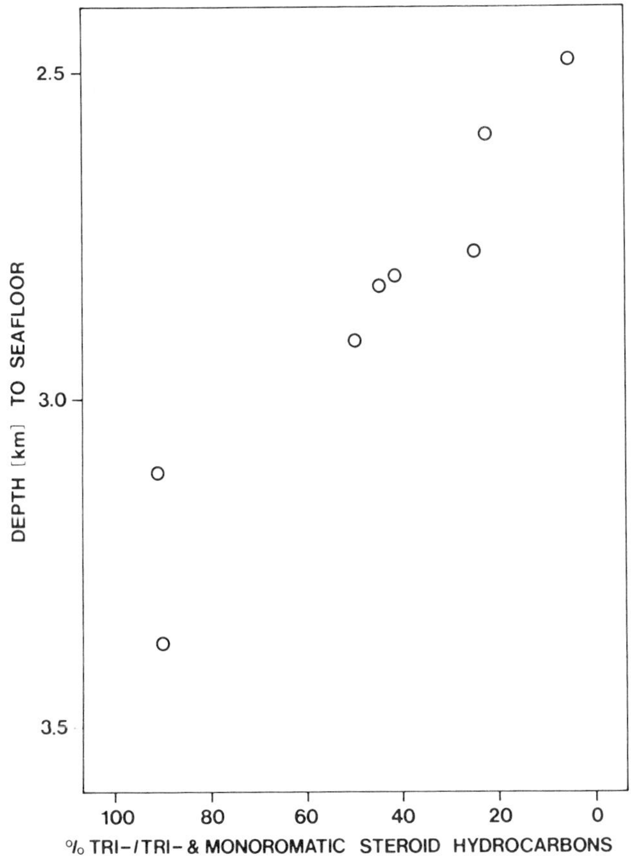

FIG. 27. Increase in the extent of geochemical aromatization of C-ring monoaromatic steroid hydrocarbons to triaromatic steroid hydrocarbons with present-day burial depth in the Jurassic shales of the East Shetland Basin, North Sea.

6. *Problems of migration and reworked material*

Since the soluble organic portion of sedimentary rocks is mobile, i.e. it can migrate (Vandenbroucke, 1972), certain molecular parameters of thermal maturation based on this portion are equally applicable to crude petroleums. However, this mobility also means that care should be taken to ensure that an organic extract is entirely indigenous before a maturity ranking is attempted. Errors can result in two ways: fractionation associated with redistribution can cause changes in the molecular measures (this is apparently not the case for the isomerization reactions at chiral centres,

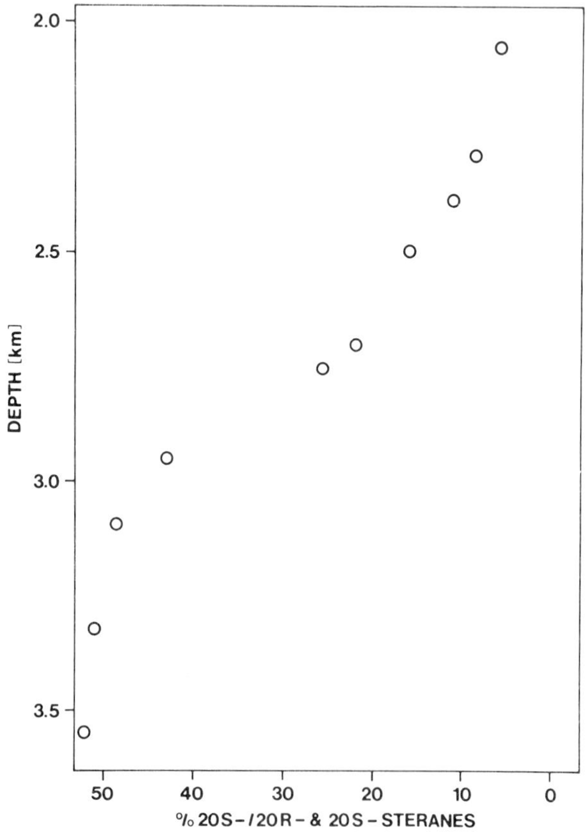

FIG. 28. Increase in the extent of geochemical isomerization at C-20 of 5α(H), 14α(H), 17α(H), 20R-C_{29} steranes with present-day burial depth in the Pliocene shales of the Pannonian Basin, S.E. Hungary.

but may influence many of the other measurements), and a sediment can be impregnated with a more mature crude oil that has migrated from depth. There are many organic geochemical techniques which allow detection of these effects (e.g. Leythaeuser *et al.*, 1983a). More detailed discussion of these problems can be found in part D below.

In an analogous manner, significant amounts of rederived or reworked biological markers from the erosion of, for example, mature shales, and associated with either reworked mineral or organic matter, could cause problems in the interpretation of data. Mackenzie *et al.* (1980a) postulated such a process to explain the high amounts of 5α(H), 14β(H), 17β(H)-steranes in immature samples. The apparently random variation in the relative amounts of these components occurs so frequently with immature samples that other reasons (see above) may apply.

7. *Pyrolysis*

An attractive way of avoiding the migration problems is to generate the biological marker hydrocarbons by pyrolysis of the extracted sediment (e.g. Gallegos, 1975b; Seifert, 1978; Huc, 1978) and the rapid removal of the pyrolysate from the reaction site. In such a way acyclic isoprenoid alkanes, steranes and triterpanes have been generated, but no aromatic steroid hydrocarbons, acids or porphyrins. Such techniques lengthen considerably the analysis time per sample. The increased steric hindrance connected with the binding of these components to kerogen slows the isomerization rates (Seifert and Moldowan, 1980; Mackenzie *et al.*, 1983a).

8. *Recommendations*

Prior to discussing the latest developments with the kinetics of the biological marker reactions, some advice on the use of the basic approach is provided. For routine and quick assessment of maturity, those hydrocarbon parameters where values of zero for (2) are normally recorded in immature samples and which can be measured by a typical GC–MS analysis are recommended. These are the isomerization of 17α(H),21β(H) hopanes at C-22 and of 5α(H),14α(H),17α(H)-steranes at C-20 and the aromatization of C-ring monoaromatic steroids. The hopane reaction has little application to oils, but provides a more sensitive measure of maturity at the less mature stage, where the other two reactions can occur much more slowly. The immature region can be supplemented by the conversion of 17β(H), 21β(H)-hopanes to their 17α(H),21β(H) form and the more mature end by the isomerization of 17β(H),21α(H)-hopanes to the 17α(H),21β(H) stereochemistry and the transformation of steranes at C-14 and C-17.

All of the above measurements can be recorded by two GC–MS runs, each lasting an hour, of the total aliphatic and aromatic fractions simply and quickly separated from a total oil or extract by one chromatographic step (e.g. Radke *et al.*, 1978, 1980). Future developments of high-resolution mass spectrometry and mass fragmentography could allow all the measurements to be made by a single GC–MS analysis of either a total oil or extract, or a total hydrocarbon fraction (e.g. Bjorøy *et al.*, 1983). Pilot studies of such a technique show great promise (Mackenzie *et al.*, 1983c).

It is the experience of this author that in the approximate vitrinite reflectance range 0.3–0.7% these parameters are much more sensitive and reliable. However, only one set of measurements extend beyond this range—the enrichment of lower molecular weight species (see above). Indeed, beyond ca. 0.6%, vitrinite reflectance can be more confidently measured, partly because the increase is more rapid with respect to depth or temperature (Dow, 1977). This means that for most of the lower part of the oil window, assuming the currently proposed reflectance limits apply (e.g. Tissot and Welte, 1978, but see also Price *et al.*, 1981), vitrinite reflectance still offers the best assessment of thermal maturity. It also suggests that the application of biological markers to the maturity of a large number of the world's more mature crude oils is restricted to the somewhat problematic enrichment of lower molecular weight species (see above). The need for more satisfactory molecular parameters covering the range $R_o = 0.7\text{–}2.0\%$, based on the solvent extractable portion of a sediment's organic carbon and therefore applicable also to crude oils, is great. It is perhaps too ambitious to seek reactions of similar specificity to those involved in most of the satisfactory biological marker parameters developed to date, because this maturity range involves a particularly complex set of reactions which includes the cleavage of many carbon–carbon covalent bonds. Probably the approach of the Methylphenanthrene Index of Radke *et al.* (1982) is more realistic, although the effects of variation in organic matter type on this value need to be further evaluated.

9. *Reaction kinetics*

Mackenzie *et al.* (1981c, 1982b) proposed that the expected variation in kinetic parameters between different reaction types was significant within the range of geological conditions encountered to date. Plots of the isomerization of C-20 in 5α(H), 14α(H), 17α(H)-steranes against the aromatization of C-ring monoaromatic steroid hydrocarbons (e.g. Fig. 29) suggested that sequences of sediments whose temperature was increased at high average rates (e.g. for the Pannonian Basin in Fig. 29, ca. 15°C per M.Y.) the rate of aromatization was accelerated to a greater extent than the isomerization rate, when compared with lower average heating rates (e.g., for the East Shetland Basin in Fig. 29, <1°C per M.Y.). This was proposed to be partly due to the aromatization reaction having a higher temperature dependence than the isomerization reaction.

Individual reaction rates (k) should vary with the absolute temperature (T) according to Arrhenius law

$$k = A \exp(-E_a/RT) \qquad (7)$$

where A is the frequency factor, E_a the activation energy and R the gas constant. A varies in most cases as T^n, where n is generally $\frac{1}{2}$ or 1, and since the temperature range to be discussed below varies between 350 K and 530 K, the small variation of A with temperature can be neglected.

In order to estimate the values of A and E_a for the two reactions, plots of $1/T$ against the natural log of the calculated rate constant are required,

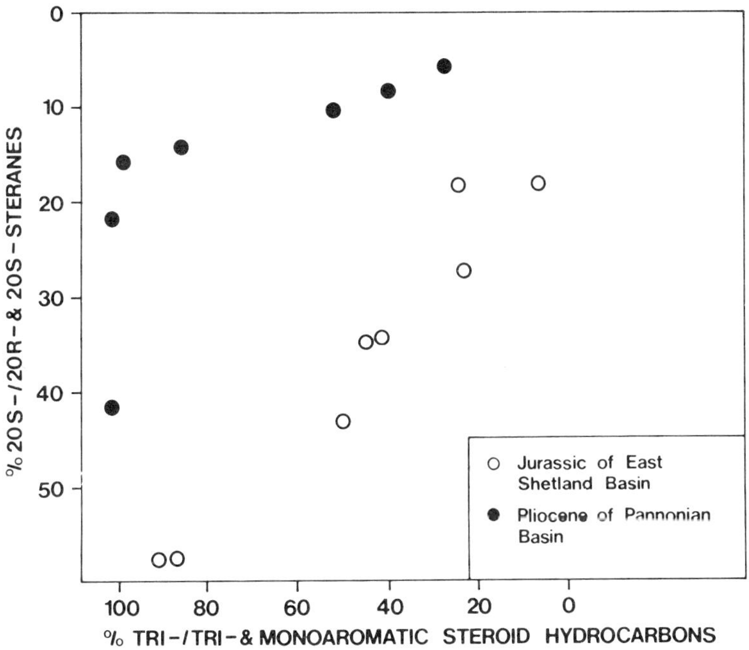

FIG. 29. Comparison of the variation of the extent of aromatization of C-ring monoaromatic steroid hydrocarbons relative to the extent of isomerization at C-20 of $5\alpha(H)$, $14\alpha(H)$, $17\alpha(H)$, 20R-C_{29} steranes, between Jurassic shales of the East Shetland Basin, North Sea and the Pliocene shales of the Pannonian Basin, S.E. Hungary.

since the slope is $(-E_a/R)$ and the intercept $\log_e A$. For the North Sea samples and the Pannonian samples the present temperature can be regarded as the maximum temperature (McKenzie, 1981; Mackenzie and McKenzie, 1983). In this way estimates for the kinetic constants have been obtained, using expressions which calculated the reaction rates from the appropriate values of Expression (2) and an effective heating time of 40% of the time that had elapsed since the crustal stretching, which is proposed to have initiated the formation of the two basins, i.e. 40 M.Y. for the North Sea and 6 M.Y. for the Pannonian (see Mackenzie and McKenzie, 1982, for details). The results of laboratory heating experiments (Mackenzie *et al.*, 1981c) were also included in the aromatization plot (Fig. 30). The estimates obtained were:

(i) for aromatization $E_a = 200\ \text{kJ mol}^{-1}$

$A = 1{\cdot}8 \times 10^{14}\ \text{s}^{-1}$

(ii) for isomerization $E_a = 91\ \text{kJ mol}^{-1}$

$A = 6 \times 10^{-3}\ \text{s}^{-1}$

That Mackenzie and McKenzie (1983) were able to fit the aromatization data from laboratory and field with one value of A and E_a suggests that, despite the major variation in reaction time and temperature, the mechanism could be the same in both cases.

In order to further test these estimates, plots of the type shown in Fig. 29 were generated by computer models which used the estimates and current thinking on the evolution of the basins, with particular respect to changes in the heat flow with time (see Mackenzie and McKenzie, 1983, for details). With very minor adjustments to A, good matches were obtained with the observed trends for the North Sea, Pannonian (e.g. Figs. 31(*a*) and (*b*)) and Mahakam Delta sediments seen previously on diagrams like Fig. 29 (Mackenzie and McKenzie, 1983). Note that the orientation in Fig. 31 is changed from that of Fig. 29. In Fig. 29 maturity increases to the bottom left and in Fig. 31 increases to the top right. In Fig. 29 appropriate forms of (2) are expressed as percentages, but they are not in Fig. 31. It was demonstrated that the extents of organic geochemical reactions can be used to assess the amount and timing of uplift in the Paris and Lower Saxony Basins (Mackenzie and McKenzie, 1983). The plots produced by the computer include depths and temperatures (e.g. Figs. 31(*a*) and (*b*)); these help to further test the fit between observed and predicted trends.

Since the back reaction (k_2 in (1)) is as fast as the forward reaction (k_1 in (1)) for isomerization, but non-existent for aromatization, it is necessary to compare the aromatization rate with twice the isomerization rate (see Mackenzie and McKenzie, 1983, for details). Figure 30 shows a plot of the values against $1/T$. The two trends cross at 90–100°C—above this temperature aromatization is faster and below isomerization proceeds at a faster rate. This allows the reader to grasp why the kinetics and variations in heating rate can conspire to produce the large differences in behaviour seen in Fig. 29. The North Sea samples have spent a longer time in the temperature interval where isomerization is faster.

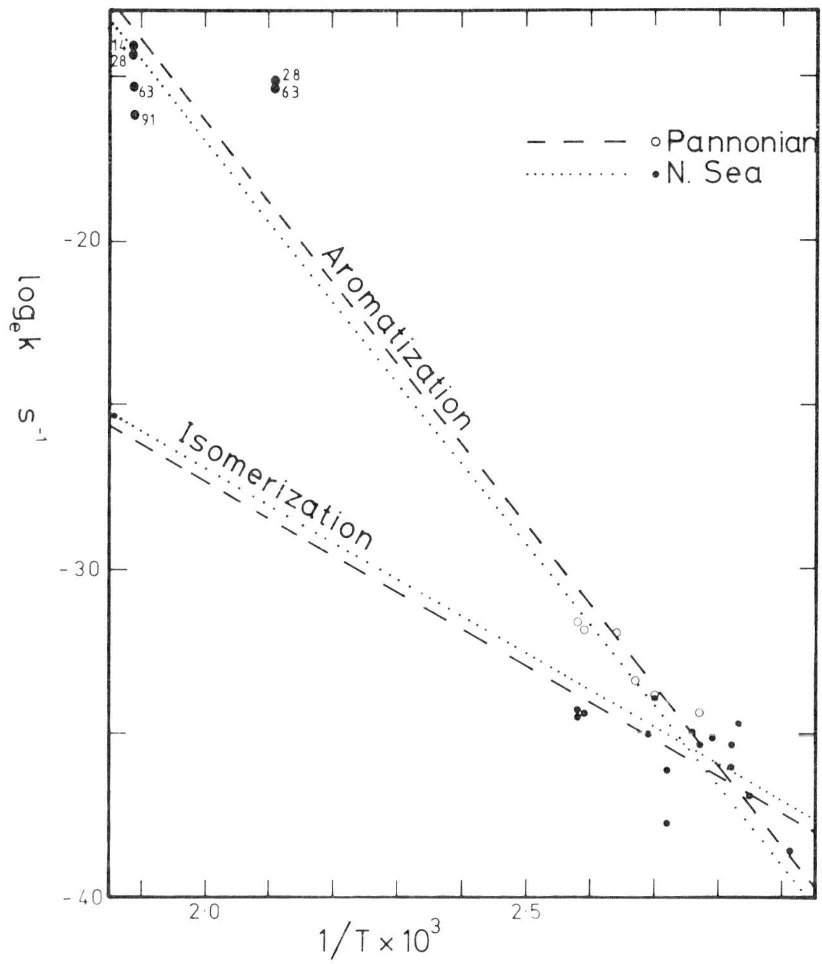

FIG. 30. The rates of the aromatization and isomerization of steroids plotted against an estimate of the reciprocal of the present temperature of the rocks. Estimates of the rates (k) of isomerization at C-20 in the steranes and of aromatization of C-ring monoaromatic steroids were obtained respectively from $(1/2t)\log_e(1-2y)$, where y is 20S/20R + 20S and from $-(1/t)\log_e(1-x)$, where x is tri-/tri-+monoaromatic steroids. t is taken as 6 M.Y. for the Pannonian samples and as 40 M.Y. for the North Sea. The straight lines correspond to isomerization reaction rates with an activation energy of 91 kJ mol^{-1} and frequency factors of 10^{-2} s^{-1} and 7×10^{-3} s^{-1} for the North Sea and Pannonian Basin lines respectively and to aromatization reaction rates with an activation energy of 200 kJ mol^{-1} and frequency factors of 10^{14} s^{-1} and 1.8×10^{14} s^{-1} for the North Sea and Pannonian Basin respectively. Only the aromatization measurements are plotted. The unlabelled dots are the geological samples, whilst those marked with heating time in days are laboratory observations of Mackenzie *et al.* (1981c).

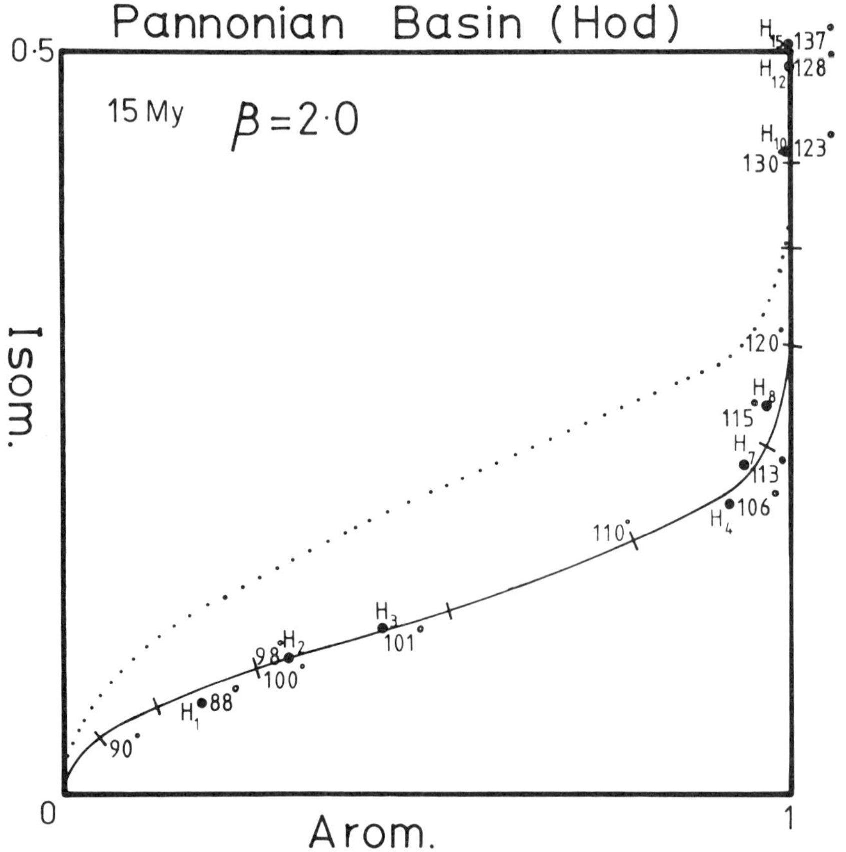

FIG. 31(*a*). The extent of isomerization of steranes at C-20 as a function of aromatization of C-ring monoaromatic steroid hydrocarbons for a number of samples from the Pannonian Basin shown as dots, together with an estimate of the present temperature in °C of the rocks from which they were obtained. The solid line corresponds to a basin produced by extension, by a factor of 2, 15 M.Y. ago, with the present temperature marked at 5°C intervals, and with reaction rates of $7 \times 10^{-3} \exp(-91 \text{ kJ mol}^{-1}/RT)$ and $1.8 \times 10^{14} \exp(-200 \text{ kJ mol}^{-1}/RT)$ for isomerization and aromatization respectively.

10. *Catalysis*

All of the above interpretations assume that the dominant feature controlling the variation in reaction rates is temperature, and that the rates are relatively independent of pressure. Effective pressure independence is generally assumed in organic geochemistry, even though it has only been

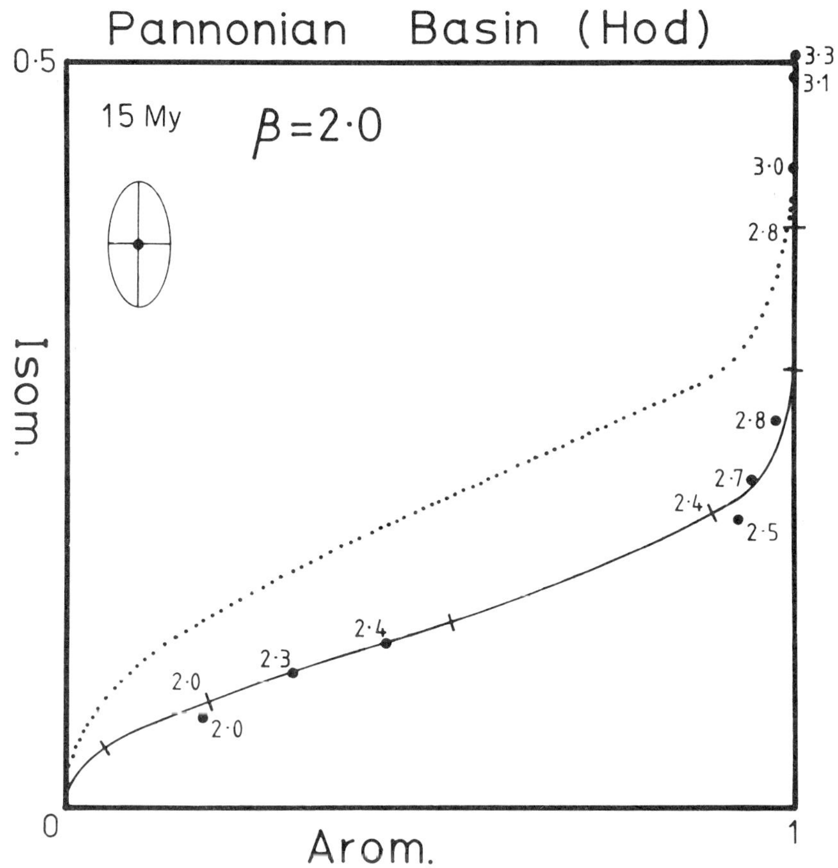

FIG. 31(*b*). As for (*a*) but marked with depths in km instead of temperatures, and with the maximum error in both measurements shown as an ellipse.

partially proved for coalification effects (Tissot and Welte, 1978) and not for oil generation itself. In the case of the isomerization and aromatization reactions mentioned here, such an assumption is, however, probably valid.

Most of the reactions in the sediments could be catalysed by the mineral matter. Important for the above considerations is that the variation in such activity with respect to those reactions is insignificant for geological samples.

Studies of coal/shale pairs from similar burial depths in the Mahakam Delta and of adjacent shales of different mineralogy in the Paris Basin have suggested that this could be the case (Mackenzie *et al.*, 1980a; Hoffmann *et al.*, 1984) but much further work, particularly in carbonate basins, is required. Fuller discussions of this problem are to be found elsewhere (Connan, 1974; Hunt, 1979; Sieskind *et al.*, 1979; Alexander *et al.*, 1983b; Mackenzie and McKenzie, 1983).

The rearrangement of sterenes (Rubinstein *et al.*, 1975) however seems very dependent on the nature of the mineral matrix (Sieskind *et al.*, 1979). Thus the ratio of rearranged steranes to non-rearranged steranes may be a useful marker of the lithology of an oil's source beds.

The calculation of kinetic reaction parameters was based on expressions for the rate constants of unimolecular reactions (see section 9 above). This assumes that, if catalysis occurs, the concentration of the active sites must far exceed the biological marker concentrations.

C. Correlation

The lipid composition of contemporary bottom sediments depends not only on the source organisms, but on the many processes which change this biological input prior to and shortly after its incorporation. All the palaeo-environmental factors which determine the nature of a biological population will also affect the distribution of biological markers ultimately to be inherited by mature sedimentary rocks and crude oils. It must be a distant aim of biological marker organic geochemistry to be able to predict the detailed lipid composition of a sediment at the time of its deposition, so as to comment in detail on the environment of deposition and the agents of early diagenesis. As far as mature sediments and crude oils are concerned this is largely unrealized, partly because so much biological information is lost during the defunctionalization, isomerization and other reactions that occur during diagenesis (see above). More hopeful are the studies of the well preserved but immature ocean sediments that are often sampled by DSDP (e.g. Brassell, 1980).

Nonetheless, the variation in the biological marker distributions of mature sediments and crude oils which results is sufficient to allow the correlation of sets of oils with each other and with possible source-rock organic extracts. Poor correlations can often be explained in terms of differences in maturity (with the knowledge reviewed in part B above) and as variation in the nature of the biological precursors (depositional environment), which normally cannot be defined but simply distinguished from the effects of maturation. For oils the recognition of severe biodegradation (see part A above) and of any changes which result from migration is important.

1. *Steroids*

The initial approach involves simply comparing the biological marker distributions as revealed by the appropriate mass fragmentograms. For many geochemists, correlation involving biological markers concerns steranes and triterpanes. The sterane distributions are apparent from mass fragmentograms of *m/z* 217 or 218 (see Fig. 2). Figure 32 is taken from Shi Jiyang *et al.* (1982). The sterane distribution of oil Yi-18 correlates well with that of shale Yi-21, a slightly less mature equivalent of its suspected source rock. It is apparent, however, that maturity effects interfere with the correlation of oil Cheng 15 with the shale Yi-21. The steranes of Cheng 15 appear more mature, since the products of geological isomerization (i.e. 5α(H), 14β(H), 17β(H)-steranes and (20S)-5α(H), 14α(H), 17α(H)-steranes) are present in higher relative proportions. These increases in the geological isomers are presumed, certainly in the case of isomerization at C-20, to be independent of organic matter type. Counterparts of such depositional- environment- (often called source-) independent maturity measurements—maturity-independent source parameters—are more difficult to find. Huang and Meinschein (1979) believe the relative amounts (C_{27}:C_{28}:C_{29}, the dominant carbon numbers of sterols and petroleum steranes) of biosynthetic sterols in bottom sediments vary with input organisms and with different depositional environments. For ancient sediments triangular diagrams for steranes with the same stereochemistry (e.g. 5α(H), 14α(H), 17α(H), 20*R*) but different carbon numbers, can indicate relationships between samples (Shi Jiyang *et al.*, 1982). Thus, in Fig. 32, oil Cheng 15 has a similar carbon number distribution to shale Yi-21, reflecting perhaps a similar depositional environment; it differs, however, in both maturity and depositional environment from shale Lo-14. Shale Lo-14 is broadly similar to oil Yi-18 in its extent of sterane isomerization, but seems to represent a different environment of deposition.

There are several published accounts of using sterane distributions for correlation of both oils and sedimentary extracts of proposed source rocks (e.g. Leythaeuser *et al.*, 1977; Seifert, 1977; Seifert and Moldowan, 1978; Seifert *et al.*, 1980; Hufnagel *et al.*, 1980). In all cases, the same approach is followed: the isomer distributions are considered for maturity, and ratios involving different carbon numbers in the C_{27}–C_{29} range of the same stereochemistry are used to detect source differences. Seifert and Moldowan (1978), Seifert *et al.* (1980, 1983) and Shi Jiyang *et al.* (1982) have all also used the aromatic steroid hydrocarbon distributions (e.g. Fig. 2) to group oil and source rock samples into series and explain the differences between the series in terms of maturity (see part B above) and source. From consideration of the distributions of monoaromatic steroid hydrocarbons and carbon isotopes, Seifert *et al.* (1980) proposed that the oils in

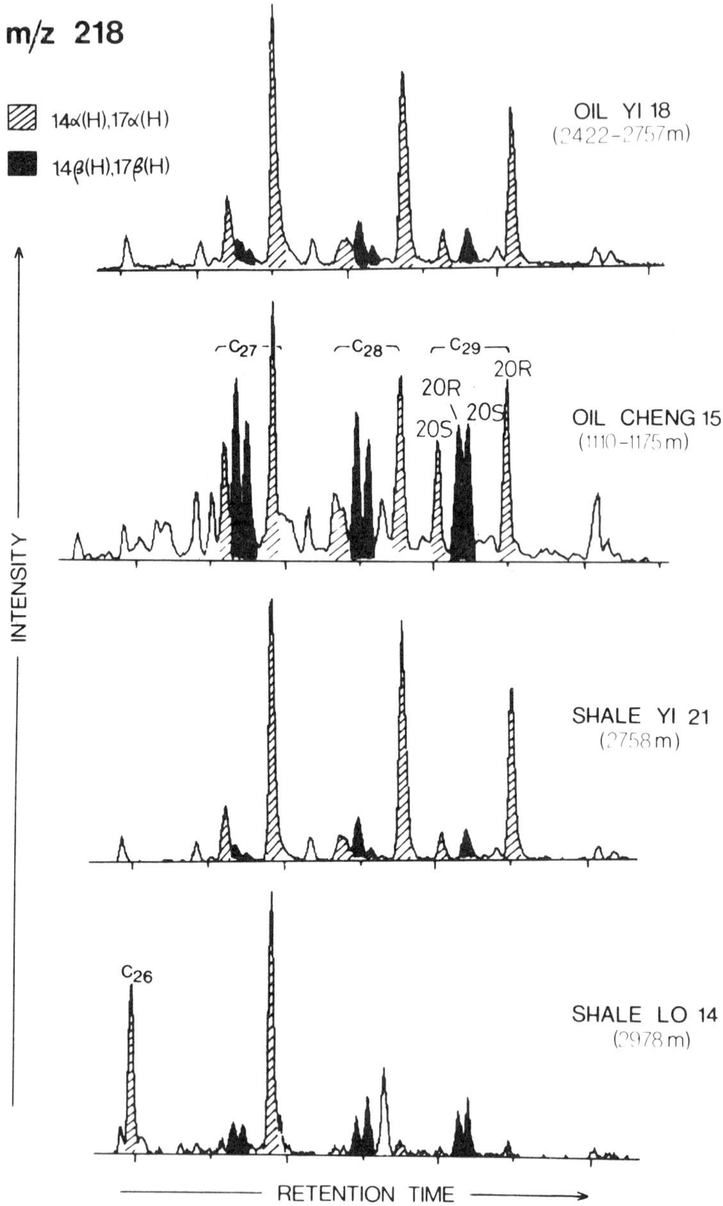

m/z 218
14α(H),17α(H)
14β(H),17β(H)
OIL YI 18
(2422-2757m)
C27
C28
C29
20R
20S
20R
20S
OIL CHENG 15
(1110-1175m)
SHALE YI 21
(2758m)
C26
SHALE LO 14
(2978m)
INTENSITY
RETENTION TIME

certain reservoirs in Prudhoe Bay, Alaska were derived from more than one source formation.

Normally the 4-methylsteranes (Rubinstein and Albrecht, 1975) present in most mature sediments and crude oils are not used for correlation purposes, partly because so little is known of their structures in mature samples. However, the impression is that the ratio of methylsteranes to the more common desmethyl steranes (e.g. Fig. 17) is relatively independent of maturity (Shi Jiyang *et al.*, 1982) and may serve as an additional maturity-independent source (depositional-environment) parameter, of as yet unknown significance.

2. *Triterpanes*

The distributions of nearly all the triterpanes discussed in detail above are best studied with the m/z 191 fragmentograms derived from the GC–MS analysis of the aliphatic hydrocarbon fractions (Brooks *et al.*, 1977). This way all the hopanes and most of the non-hopanes which occur, including the C_{21}–C_{30} tricyclic terpanes, can be considered. It is rare for the hopanes not to be present, but attention to the carbon number distribution can still reveal differences, which allow distinctions of geological importance to be made (e.g. Claret *et al.*, 1977; Seifert and Moldowan, 1978).

The effect of maturity on the triterpane distributions of crude oils is not so well known. Most of the isomerizations are complete before the oil-generation zone. This creates problems in the interpretation of oil/source-rock correlations. High amounts of non-hopanoid triterpanes can help distinguish an oil or a sedimentary rock extract. This has been done with 18α(H)-oleanane (III in Fig. 11) for Niger Delta samples (Pym *et al.*, 1975; Ekweozor *et al.*, 1979a) and for Mahakam Delta samples where a link with deltaic sedimentation was proposed (Hoffmann *et al.*, 1984), and with gammacerane (I in Fig. 11) for Chinese (Shi Jiyang *et al.*, 1982) and Indonesian samples (Seifert and Moldowan, 1981). However, Shi Jiyang *et al.* (1982) were able to correlate an immature oil with its source rock, because both contained gammacerane, but were unable to rule out this source rock as the origin for the other more mature oils which did not contain gammacerane, because the stability of gammacerane to thermal

FIG. 32. Comparison of the steranes of two oils and two shales from the Zhanhua Basin, China, as illustrated by their m/z 218 fragmentograms (see Fig. 2): (i) the fragmentogram of shale Yi-21 correlates well with that of oil Yi-18, but not with that of oil Cheng 15, which is more mature, possessing increased relative amounts of 5α(H), 14α(H), 17α(H), 20S- and 5α(H), 14β(H), 17β(H)-steranes; (ii) the fragmentogram of shale Lo-14 does not match with those of either shale Yi-21 or Yi-18, since they differ in their carbon number distributions and therefore represent different depositional environments, although all three are of similar maturity; (iii) oil Cheng 15 and shale Lo-14 fail to correlate both on maturity and on carbon number distribution considerations.

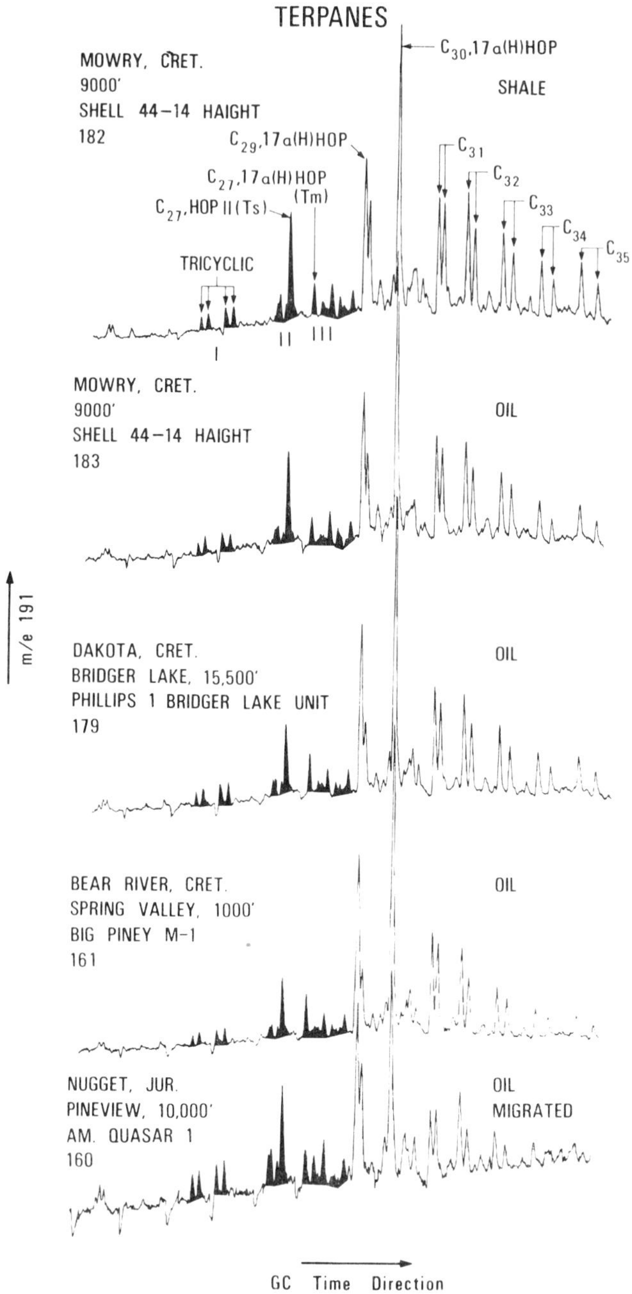

FIG. 33(*a*). Rocky Mountain Overthrust Belt: a good correlation of Mowry Cretaceous source shale with Type I oils. Taken from Seifert and Moldowan, 1981. *Reprinted with permission.*

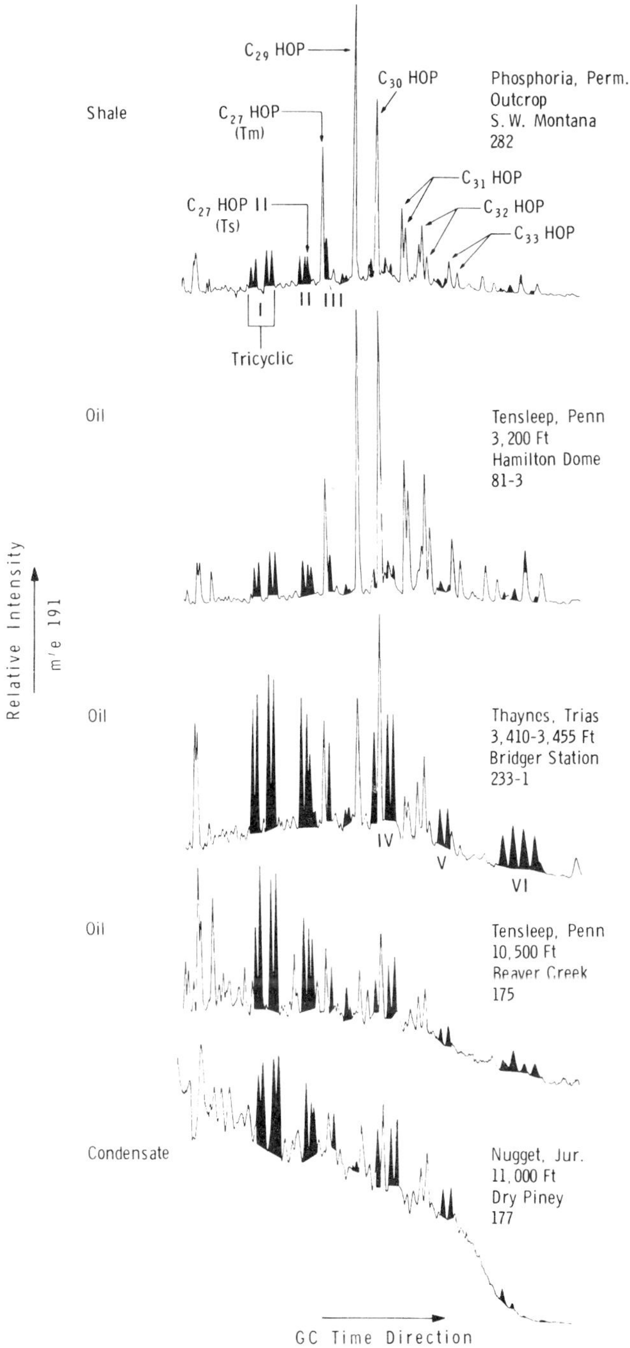

FIG. 33(*b*). Rocky Mountain Overthrust Belt: a good correlation of Type II oils with Phosphoria Permian source rock. Taken from Seifert and Moldowan, 1981. *Reprinted with permission.*

degradation relative to the other triterpanes present (hopanes) is not known. Grantham *et al.* (1980) have already shown in the North Sea, that triterpane distributions dominated by the modified hopanoids (I and II in Fig. 8) are altered to the typical hopane pattern by thermal maturation, because the modified hopanoids are preferentially degraded.

Despite these reservations, the *m/z* 191 fragmentograms can serve as excellent correlation tools, provided differences in maturity are not evident or can be allowed for and particularly if non-hopanoids are present in large relative amounts. This is because this fragmentogram can sample a relatively wide range of molecular types. Figure 33 shows one of the most successful published applications, where oils from the Rocky Mountain Overthrust Belt are divided into two groups: one group correlates with a Cretaceous source rock, the other with a Permian source rock (Seifert and Moldowan, 1981). The results of Fig. 33 are consistent with current geological thinking.

3. *Long-chain acyclic isoprenoid alkanes including "botryococcane-type" alkanes*

The use of these compounds for correlation purposes has recently been proposed (Albaiges, 1980a; Seifert and Moldowan, 1981) and applied to Spanish and Sumatran crude oils respectively. Different groups of oils can be distinguished by comparison of and ratios derived from the *m/z* 183 fragmentograms from the GC–MS analysis of branched and cyclic alkane fractions. Because of interference, *n*-alkanes must be removed first by molecular sieve, which is not normally necessary for the steranes and triterpanes.

4. *Alkyl porphyrins*

Despite the lengthy work-up procedure required to obtain the purified demetallated fractions, which are then analysed by high-performance liquid chromatography (HPLC), the distributions so generated (Eglinton *et al.*, 1980; Hajibrahim *et al.*, 1978, 1981) are often considerably more specific that those of the steroids and triterpanes. This means that the porphyrins can be used to further subdivide sets of oils and organic extracts grouped only on hydrocarbon data (e.g. Hajibrahim, 1978). It also means that differences in porphyrin distributions are often very difficult to rationalize with our current knowledge of oil-forming processes. One simplification is to analyse the nickel and vanadyl porphyrins separately, since it is known that the nickel porphyrins undergo the maturation effects (DPEP to aetio conversion, side chain cleavage) at a faster rate (Mackenzie *et al.*, 1980b) and that the ratio of nickel to vanadyl porphyrins in crude oils and sediments is highly variable (e.g. Baker and Palmer, 1978; Hajibrahim, 1978) for reasons that are not fully understood. There is every likelihood that as their structures and geological fate become better understood, and their analysis more routine, the alkyl porphyrins will make a major contribution to petroleum geochemistry.

5. *Concepts and limitations*

It is no problem for modern analytical chemistry to say whether two oils are similar, but petroleums can undergo many alteration processes in the reservoir (not just biodegradation) so that two samples from the same origin may have completely different bulk compositions. Biological markers are believed to be unaffected by all but the most extreme of such processes, and their distributions can be used to match subsurface oil accumulations with themselves and the organic extracts of sedimentary rocks. What, however, do these matches really mean? And to what extent can they be stressed in the face of other conflicting geochemical data?

One basic difficulty stems from using components, which often only comprise no more than one per cent of the complex sedimentary organic mixture of interest. This sobering observation is strikingly demonstrated in Fig. 34, taken from Tissot and Welte (1978). Immature rocks contain

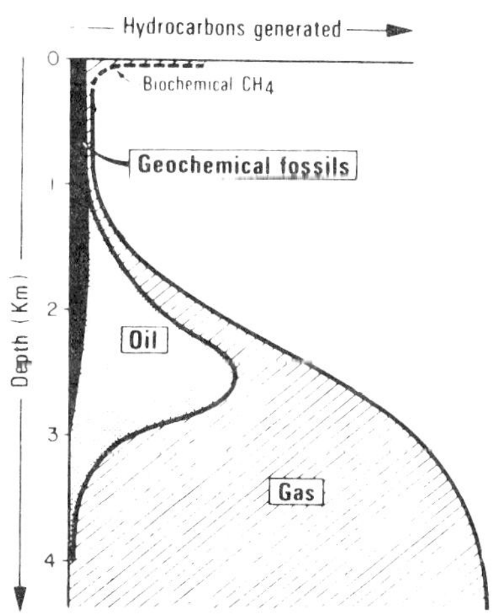

FIG. 34. General scheme of hydrocarbon formation as a function of burial, showing how the biological markers (called "geochemical fossils") are "diluted" with increasing maturity. Taken from Tissot and Welte, 1978. *Reprinted with permission.*

high relative concentrations of biological markers, yet the processes of maturation thermally degrade these components. By the stage of oil generation of interest for correlation studies they have also been diluted by the less specific species produced from the thermal breakdown of kerogen so that they form a small proportion of the total. In very mature oils and condensates, correlation with biological markers is occasionally not possible, because their concentrations are too low. Then correlation must revert either to more established methods (e.g. Tissot and Welte, 1978; Hunt, 1979) or to other recently developed techniques, such as the statistical treatment of the low-voltage GC–MS analysis of the aromatic hydrocarbon fraction (Rullkötter and Welte, 1980; Welte *et al.*, 1982).

An initial problem is that biological marker components present in such small quantities could be unrepresentative of the rest of the organic matter present in a sediment or a crude oil. In a recent pyrolysis study of the total organic matter of the Lower Toarcian shales of the Paris Basin (van Graas *et al.*, 1981), major differences in the chemical nature of the organic matter, which must have been caused by variations in the type of the original organic matter, were reported. These effects were wholly undetected by the biological markers (Mackenzie, 1980, and unpublished results).

A second complication is the inadequacy of our present understanding of the migration processes (see also part D below). It is often assumed that immediately after primary migration a continuous phase forms which will be divided and subtracted from, but never added to, as it proceeds towards the reservoir, perhaps as globules. Durand (1983) has warned that a biological-marker-poor oil could easily become significantly "impregnated" with biological markers from a quite different source. The picture that is emerging (Vandenbroucke *et al.*, 1983) is one of considerably more heterogeneous migration than previously imagined, where many components are behaving independently of each other, migrating depending on their mobility, and not always as a continuous phase. If this is at all the case, it would be foolish to base correlation entirely on biological markers, whose molecular weight range is normally untypical of the extract or the crude.

The tendency of the Chevron group to back their correlations with whole oil isotope measurements and bulk chemical measurements (e.g. Seifert and Moldowan, 1981) is to be applauded. The development of correlation with the carbon isotopic composition of individual *n*-alkanes (Vogler *et al.*, 1981) is welcomed. This gives an added specificity to components normally present in crude oils and mature extracts in much higher relative abundance than the biological markers.

The most acute correlation problem is deciding on representative samples of proposed source rocks. Extracting a powdered source rock with solvent bears little relationship to primary migration. This is not only worrying because of the fractionation effects which occur on "extraction" by primary migration (Brennemann and Smith, 1958) and which are avoided by solvent extraction, but also because of a recent survey by Sajgò *et al.*

(1983). Here, a number of samples were first extracted as chips (>1 cm), then powdered and re-extracted. This derives from the method of Beletskaya (1978). The first extract is called "coarse" and thought to represent "quasi-open" pores, and the second "fine" and represents "quasi-closed pores". The "coarse" extract is thought to be more amenable to migration and has quite different biological marker distributions from the "fine" and total extracts (Fig. 35). If such heterogeneity exists in a typical sample, then similar variation can be expected vertically and horizontally through a source sequence. Oils may represent an "average" of the migrated hydrocarbons of a source formation up to ca. 1000 m thick, so it would be foolish to emphasize an oil/source-rock correlation that was not based on a reasonable survey of the source beds in question and on more than just biological marker data.

Another problem relates to the heterogeneity of kerogen itself. Seifert (1978) and Seifert and Moldowan (1980) have discussed the value of pyrolysis of an extracted shale. One is to confirm that a bitumen (extract) is indigenous, the other is to test that a shale yields a pyrolysate as mature as the oil of which it is proposed to have been the source (on the molecular measurements). Both are valid approaches, but it does not follow that the extractable lipids which are generated by pyrolysis will reproduce the lipids that have already been generated by thermal breakdown in nature and migrated out. Huc (1978) and Mackenzie (1980) showed that pyrolysates yield much higher hopane-to-sterane ratios and sterane carbon number distributions more biased towards C_{27} than the corresponding extracts. In other words, as kerogens move through the oil-generation zone they may yield different biological marker distributions at different maturation stages. Thus the trend of enrichment of C_{27} in the triangular diagram of sterane carbon number abundance for the Chinese oils (Fig. 36) may merely reflect increasing maturity, as suggested by both the carbon and hydrogen isotopes (Schoell, 1982) and the molecular parameters (Shi Jiyang *et al.*, 1982), and not different sources for the different oils as previously proposed (Shi Jiyang *et al.*, 1982).

Frequently, the sampling during drilling does not allow the above complications to be adequately considered. Many geochemists will be aware of the problems associated with trying to judge a mature deep source area that lies off structure, with an immature shallow sample of the same age. Recent advances in the knowledge of the effects of maturation on biological marker distributions can help here. Also important is an ability to distinguish, for an oil, between maturity inherited from the source rock and that which derives from reservoir processes. Tentative evidence (Seifert and Moldowan, 1978; Price, 1980; Shi Jiyang *et al.*, 1982) suggests many maturational processes do not normally continue to any great extent, after

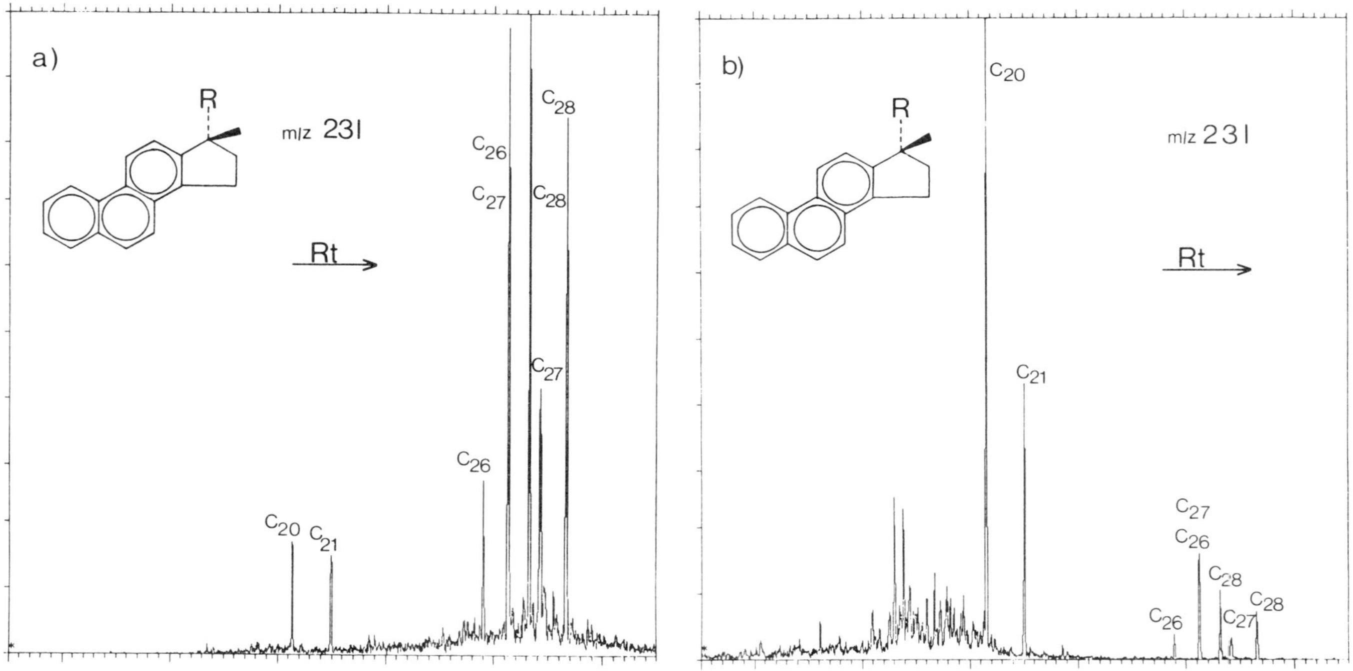

FIG. 35. The distributions of a series of triaromatic steroid hydrocarbons (a) in the more closed pores of a shale from the start of the zone of oil generation in the Pliocene of the Pannonian Basin (3480 m) and (b) in the more open pores of the same shale, as revealed by *m/z* 231 fragmentograms. Notice the relative enrichment of the C_{20} and C_{21} components ($R = C_2H_5$, $CH(CH_3)_2$) relative to the C_{26}–C_{28} species ($R = C_8H_{17}$, C_9H_{19} and $C_{10}H_{21}$) in (b) when compared to (a). The components in the closed pores will be less mobile, but the lower molecular weight components may migrate preferentially into the more open pores. A hexagon with an inscribed circle is an aromatic ring.

the finely dispersed oil has left the source rock. Such observations, if true, could assist petroleum geochemists.

In summary, the biological marker distributions represent the most specific "fingerprints" presently routinely available for the correlation of

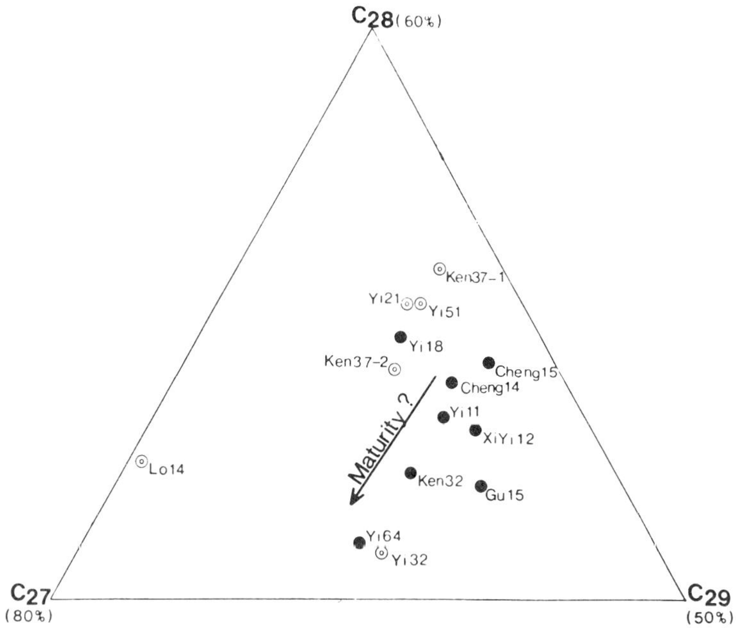

FIG. 36. Triangular diagram to show carbon number distributions of C_{27}–C_{29} steranes with a 5α(H), 14α(H), 17α(H), 20R stereochemistry for samples from the Zhanhua Depression (Shengli Oilfield), China. The solid dots are oils and the open dots with inscribed circles are shales. Although the position of a sample on this diagram was previously thought to be independent of maturity, circumstantial evidence suggests in certain cases that the steranes bound into kerogen are heterogeneous, and at later stages of kerogen breakdown (oil generation) more C_{27} species are produced. Oils derived primarily from later stages of generation may be displaced towards the C_{27} apex. Thus the positions of the oils, which increase in general maturity in the direction shown, could mean all are derived from a similar source as the immature Yi-18 oil, the Upper Shahejie (Ken-37, Yi-21, Yi-51) at higher maturity levels *or* are mixtures of Upper and Middle Shahejie (Yi-32).

oils and source rocks, and there are many published and unpublished examples of successful applications. Where possible, positive correlations should be confirmed with other techniques which are based on the chemistry of the whole oil or extract, such as stable isotopes. In many cases poor correlations can be explained by differences in maturity and depositional environments (undefined). Maturity assessment is best based on stereoisomer ratios, while depositional environment differences may be apparent from the carbon number distributions of compounds with the same stereochemistry or ratios of different types of carbon skeleton (hopanoid vs. steroid). There are many complications which could result in contradictory results. Biological markers represent a very small concentration (<1%) of a total oil or a mature extract. In theory they only speak for themselves and cannot purport to always represent the rest of the organics. It need not always follow that they have been in intimate association with each other and the rest of their present organic mixture since the primary migration of all the organic species of a given mixture.

6. *Future developments*

The most important measures required are maturity-independent ratios which reflect the original depositional environment. A small number have been suggested which attempt to measure the relative proportions of carbon skeletons synthesized in part by different organisms, for example: total hopanes/total steranes (Mackenzie *et al.*, 1982c), carbon number distribution of steranes of the same stereochemistry (Welte *et al.*, 1975; Leythaeuser *et al.*, 1977; Seifert and Moldowan, 1978; Shi Jiyang *et al.*, 1982), methylsteranes/steranes (Shi Jiyang *et al.*, 1982). In general, the actual palaeoenvironmental implications of the varition in such measures is not understood. Some trends are emerging, however; for example, terrestrial type (higher plant dominated) source rocks and crude oils often appear to have high amounts of C_{29} steranes relative to C_{26} and C_{28} steranes and high hopanoid/steroid ratios (>5). Even with these chosen ratios there are indications that small variations may result from maturity differences, because oils generated from a kerogen may contain different biological markers depending on the level of maturity (see above).

Despite the problems in using components present in low abundance relative to other components for correlation (see above), this line of research must continue, since it is only the structurally specific biological marker components which show promise of developing maturity-independent assessment of original organic input.

Pattern matching by eye can be very objective, and clearly computerized correlation with digitized descriptions of the fragmentograms may help here (Hohn *et al.*, 1981). Meanwhile, a number of diagnostic ratios should be incorporated into a computerized statistical correlation package.

A better appraisal of the problems should come if the absolute concentrations of biological markers were routinely measured for proposed source rocks and crudes, as suggested by Seifert and Moldowan (1979).

D. Migration

This section has been left until now because of the many uncertainties involved. For simplification, consideration of the effects of migration on the biological marker distributions was omitted from the section on correlation (part C above). The interested reader is encouraged to reread in the light of this discussion of migration.

The concept introduced for correlation still holds, i.e. that the biological markers, always present in unbiodegraded oils, in low relative abundance, may not always have experienced the same migration history as much of the rest of the crude oils. It is important, however, that the effects of migration on correlation and maturation parameters are considered—both to allow for them and possibly to use them to determine the migration history, at least of the biological markers, if not of the whole oil. Most of such documentation to date is based on a geological and geochemical consensus of which oils have migrated over longer distances. This exercise cannot be constrained with anything like the certainty possible, for example, when predicting maturity sequences for shales from geological information on burial history alone. Consequently, there remains much more doubt about the measures and effects of migration proposed. Furthermore, nearly all of the effects at present proposed to occur with increasing migration, are also observed with increasing maturity.

1. *Geochromatography*

Much of the interpretation depends on the model chosen for migration. All of the interpretations of Seifert and his co-workers (see section 2 below) depend on a picture of secondary migration where continuous oil phases move as globules through a sedimentary sequence and a large fraction remains in the rocks through which it passes, resulting in fractionation effects analogous to separation by, for example, liquid chromatography (cf. Nagy, 1960; see Tissot and Welte, 1978, for a discussion of the driving forces of secondary migration). If such a process exists, then the ability of both the inorganic and organic matter in the rocks through which it passes to absorb the components must vary with the nature of the component, and must also be sufficiently low, so that desorption can occur and so that the leading edge of the migrating plug saturates ("overloads") the active absorbing sites, thereby allowing some of those components with a high absorbing tendency to continue towards the reservoir. This process where the mobile phase, the oil, is less polar and the stationary phase, mineral matter (perhaps lined with structured water), is more polar corresponds to normal phase liquid chromatography and the absorbing tendency of the individual molecular components will increase with increasing polarity. If this highly simplified process were not so heterogeneous and the "overloading" and desorption effects described above did not occur relatively easily, then pure alkane oils could be produced! This process has been termed "geochromatography" by Seifert and Moldowan (1981). For the fractionation effects of secondary migration to be preserved in reservoired oils, as claimed by Seifert and co-workers (see below), implies either that large amounts of proportionally more polar material are permanently retained along the migration pathway and/or that in these cases much of the migrating oil has not yet reached the reservoir. If the mineralogy remains unchanged along the migration pathway, and no movement occurs along faults, then the extent of fractionation will depend on migration distance (Seifert and Moldowan, 1981) and the timing of migration and accumulation.

2. *Proposed effects of migration on the biological markers of oils*

(*a*) *Alkanes*

Seifert *et al.* (1980) believe that within the triterpanes the tricyclic components migrate more easily than the pentacyclic hopanes, and so are relatively rich in highly migrated oils. Thus, the bottom three "Permian derived" oils in Fig. 33(*b*) may be the result of relatively long-distance migration. The same authors have also proposed, partly from the results of liquid chromatography in the laboratory, that within the steranes 5α(H),14β(H),17β(H) components migrate faster than the 5α(H),14α(H),17α(H) species. All these effects can be seen in proposed highly migrated Prudhoe Bay oils, which show other effects observed with migration: increased aliphatic to aromatic hydrocarbon and *n*-alkane to cyclic alkane ratios (Vandenbroucke, 1972). Seifert and Moldowan (1978) had earlier proposed many other biological marker migration effects, but these have not been pursued in subsequent publications.

Unfortunately, all these effects could however also reflect increased maturity (see part B above for the biological markers and e.g. Tissot and Welte, 1978, for the bulk compositional changes). Seifert and Moldowan (1981) believe these can be allowed for in the steranes by plotting the isomerization at C-20 against that at C-14 and C-17 (Fig. 37). This, they believe, will follow the relatively narrow trend for "potential source rocks on the correct kinetic path towards petroleum generation" shown in Fig. 37. The meaning of this statement is not clear, but until the kinetics of the C-14 and C-17 isomerization are known and some of its problems (see above) better understood, the empirically established trend (the black line in Fig. 37) will remain unproven. The experience with plots of two different reaction extents against each other (cf. Figs. 29 and 31) suggests that not

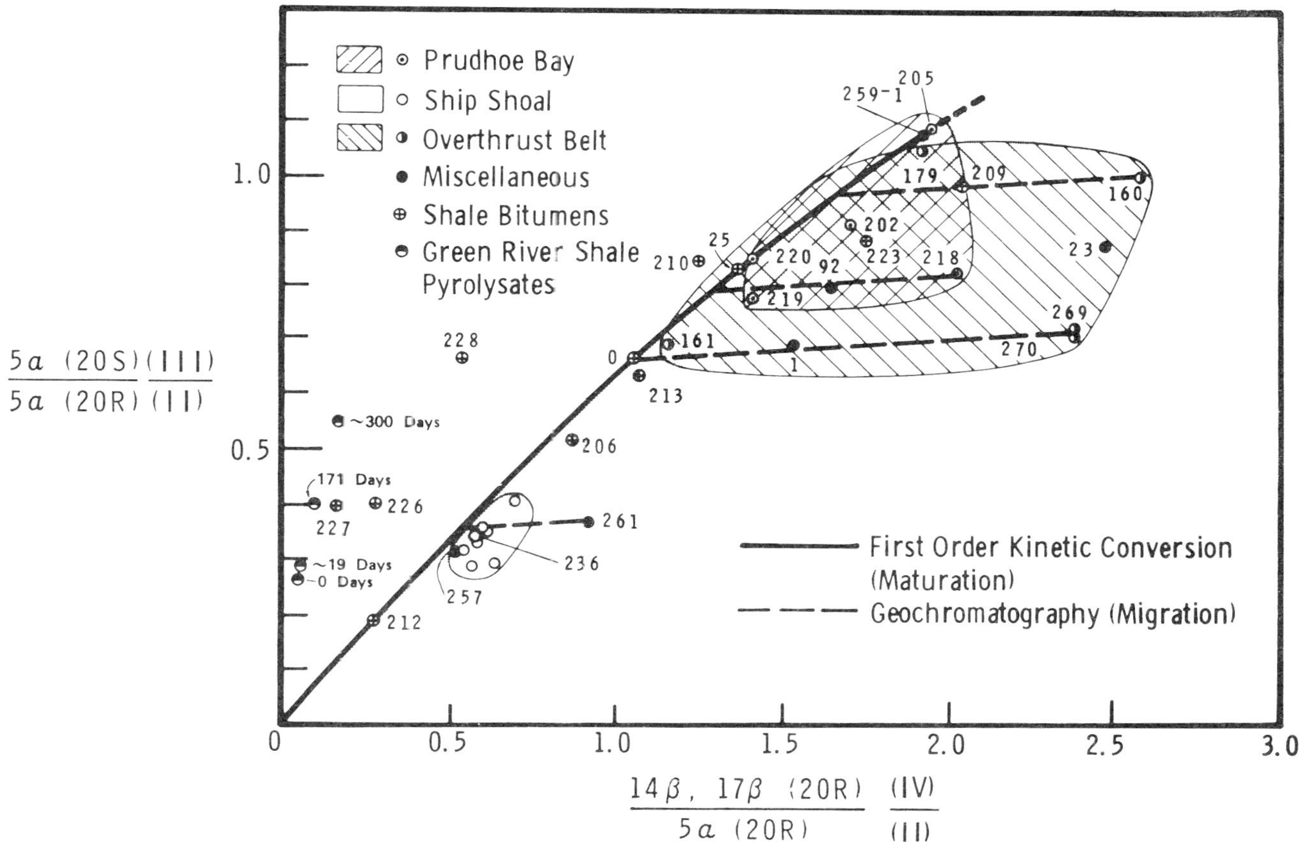

FIG. 37. The variations in the C-20 stereochemistry of steranes as a function of the variation in C-14 and C-17 stereochemistry. The solid line represents the position of extracts and oils unaffected by migration processes. The linear distance from the origin and along the curve represents maturation. Migration causes enrichment in 5α(H), 14β(H), 17β(H)-steranes, and can be assessed by the distance of displacement to the right of the curve along the dashed lines. The samples marked in days are the observations from laboratory heating experiments. Taken from Seifert and Moldowan, 1981. *Reprinted with permission.*

only catalytic effects (Seifert and Moldowan, 1981), but also variations in average sedimentary heating rates (witness the quite different trend described by laboratory heating in Fig. 37) could be responsible for major variations in the trend, so preventing universal application of the technique. Nonetheless, Seifert and Moldowan (1981) have shown that a number of crude oils lie to the right of the curve (Fig. 37), and have suggested that this displacement results from migration, which causes major enrichment of the 14β(H), 17β(H)- steranes, but only a very minor increase in the 20S- to 20R-ratio. Maturity is assessed by the distance from a point on the curve, or from its near horizontal projection onto the curve, to the origin, and the migration by the distance of its displacement to the right from the curve (e.g. the dashed lines in Fig. 37). The greater the distances the higher the maturity and the more extreme the migration fractionation respectively. Conditions for such fractionation and its interpretation in terms of migration distances are discussed below. Relative extents of source-rock maturation and subsequent migration proposed by Seifert and Moldowan (1981) for: Overthrust Belt; Prudhoe Bay, Shipshoal, Louisiana; McKitterick, California and other oils seem entirely consistent with current geological thinking.

(*b*) *Aromatic steroid hydrocarbons*

There is only one report for these compounds, and it is based on nine oils from the Handil field of the Mahakam Delta, Kalimantan, Indonesia. Here, Durand and Oudin (1980) proposed that all the oils from this field originated at depths greater than 3000 m, and that secondary migration was in a vertical direction. Therefore the reservoir depth can be related to migration distance: the shallower oils have migrated a greater distance. Durand and Oudin (1980) observed a number of fractionation effects which correlated with proposed migration distance. In general, the less polar components were relatively depleted in the source formations within the oil-generation zone and particularly enriched in the shallow, more migrated oils. Thus the shallower oils had higher relative abundances of alkanes and lower molecular weight hydrocarbons, resulting in lower densities and pour points (Durand and Oudin, 1980). In a survey of a series of nine oils, with reservoir depths ranging from about 400 to 2600 m, Hoffmann *et al.* (1984) did not see any of the changes in biological marker alkane distributions proposed to occur with migration by Seifert *et al.* (1980) and Seifert and Moldowan (1981). They did, however, see a general increase in the ratio of monoaromatic to triaromatic steroid hydrocarbons, previously suggested as a maturity measure (see above), with increasing reservoir depth (see Fig. 38). Such a decrease could mean that the monoaromatic components migrate more readily. The one anomalous oil (oil A in Fig. 38), which suggests a migration history more similar to that of the shallower oils, was also anomalous in the fractionation effects proposed by Durand and Oudin (1980). Either it has migrated over a longer distance than its present depth

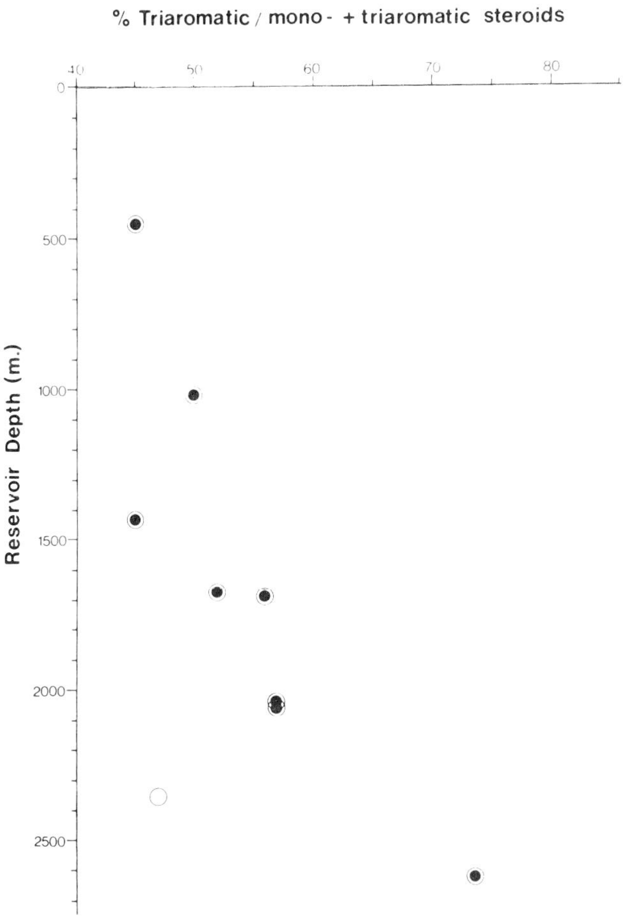

FIG. 38. Percentage of tri-/tri- + monoaromatic steroid hydrocarbons as a function of reservoir depth for oils from the Handil field of the Mahakam Delta, Kalimantan. The shallower, more migrated oils show relative enrichment in the less polar monoaromatics. Sample A is the unfilled circle.

suggests (more lateral migration than the other oils, perhaps) or the nature of its migration pathway has encouraged the absorption effects which result in fractionation to a greater extent than for the other oils.

Seifert and Moldowan (1981) suggest that the fractionation effects observed in the alkanes require a clay-rich migration pathway. Migration through clean sands and along faults will not occur with fractionation. The failure of Hoffmann *et al.* (1984) to detect the alkane fractionations in the well defined Mahakam sequence suggests, at least, that the changes in the aromatic steroid ratio and the enrichment of alkanes and low molecular weight components are more sensitive to migration effects. Indeed, the polarity difference between a mono- and triaromatic steroid hydrocarbon is considerably greater than, for example, that between 5α(H), 14β(H),17β(H)- and 5α(H),14α(H),17α(H)-steranes, and Durand and Oudin (1980) have proposed that migration in the Handil field is via faults. However, the enrichment in alkanes measured for the Mahakam Delta oils (Durand and Oudin, 1980) is more substantial than that seen in the Prudhoe Bay (Sag River) oil with the largest relative increase in 5α(H),14β(H),17β(H)-steranes proposed to derive from migration (Seifert *et al.*, 1980; Seifert and Moldowan, 1981). Some of these inconsistencies could be resolved if the aromatic steroid hydrocarbons ratios of the oils included in Fig. 37 were available.

3. *Possible effects of migration (redistribution) on the biological markers of sedimentary rock extracts*

The fractionation effects related above to secondary migration must equally apply to primary migration, and redistribution should cause the relative enrichment of the more mobile components in badly impregnated shales and possibly relative depletion in the corresponding "drained" shales. Such effects should be considered when using maturity ratios also thought to be sensitive to migration (e.g. the aromatization of steroid hydrocarbons).

Sajgò *et al.* (1983), in an attempt to understand primary migration, compared the organic material derived first by the solvent extraction of ca. 1 cm chips and secondly by the re-extraction of the same chips in a powdered form (ca. <100 μm). The first "coarse", extract is thought to contain the more mobile soluble organics in the more "open" pores, and the second, "fine", extract probably comprises the less mobile organic material of the more "closed" pores. The biological markers of the two extract types always showed consistent differences, despite originating from a single sample. This methodology was first recommended by Beletskaya (1978) and the implications of the biological marker results for correlation studies were discussed in section C.5 above. Some of the differences between the coarse and fine extracts could be explained by the fine extracts being more dispersed and more reactive (the molecular measurements of thermal maturation were advanced in one case when compared with the coarse extract) and by the coarse extracts containing more higher plant material (this concurs with the grain size distribution of the sedimentary organic matter of recent sediments (Thompson and Eglinton, 1978)). However, the biggest difference, always seen, was the much higher concentrations of lower molecular weight steranes (C_{21}, C_{22}) and triaromatic steroids (C_{20}, C_{21}) relative to their higher molecular weight homologues in the coarse extract (see Fig. 35). Similar effects were also observed for triterpanes and *n*-alkanes. There were no exceptions, among the eleven sediment samples analysed, to this observation. Fine-grained rocks showed the difference to a greater extent than coarser-grained rocks. Sajgò *et al.* (1983) concluded that the difference was caused by the lower molecular weight species having a greater ability to move from the more "closed" pores to the more "open" pores. Such a fractionation involving molecular shape and size differences to a greater extent than polarity differences is unlikely to arise from simple chromatography over very small distances, but could arise from molecular diffusion (cf. Leythaeuser *et al.*, 1983b). The authors suggest therefore that diffusion may be one important and initial mechanism for primary migration.

4. ***Summary***

This author is unwilling at present to accept the simplified models for migration, which involve the simple movement of a homogeneous oil from a source rock to a reservoir over a restricted time period. More realistic is that migration starts at early stages of oil generation and continues throughout generation. The contents of one reservoir may have derived from many source areas, all of which could correspond to the same formation. As such the migration history of the biological markers normally present in low relative abundance could be unrepresentative of the oil as a whole. Studies of the effect of migration on biological markers can improve their application to correlation problems, and the fractionation effects observed may help in the understanding of migration mechanisms.

V. Laboratory Simulation

Many of the geochemical processes proposed above can be further confirmed by their simulation in the laboratory.

Bailey *et al.* (1973b), Rubinstein *et al.* (1977) and Connan *et al.* (1980) were able to biodegrade many of the hydrocarbons of crude oils, which are proposed to be attacked by microbes in the reservoir, with laboratory cultures of bacteria. Only Goodwin *et al.* (1983) have succeeded in promoting the bacterial uptake of steranes in the laboratory. In general the stepwise uptake of different molecular types observed in natural systems (see above) was reproduced by the above authors in the laboratory.

There have been very few attempts to recreate the diagenetic and catagenetic fates of biological markers by laboratory heating. Sieskind *et al.* (1979) observed the conversion of 5α(H)-cholestanol to rearranged sterenes and steranes by heating the reactant on clays and other substrates at 160°C. The chemical treatment of these substrates prior to incubation suggested that Brønsted acid sites in the clays (silanol groups), of increased strength because of adjacent Lewis acid sites (trigonal and therefore not fully substituted aluminium ions at the edges of the clay structures), were required for the rearrangement of the steroid skeleton. Ensminger (1977), Seifert and Moldowan (1981) and Mackenzie *et al.* (1981c) heated immature shales at 200–300°C and reported many of the isomerization reactions observed in the biological marker alkanes of sedimentary sequences (see above). The biggest problem in the interpretation of the results was the generation of previously more protected, and therefore less isomerized, components by thermal breakdown of kerogen during the experiments (Mackenzie *et al.*, 1981c). Mackenzie *et al.* (1981c) also observed the aromatization of C-ring aromatic steroid hydrocarbons, to a far greater extent than the isomerization reactions. This was consistent with the greater temperature dependence of the aromatization reaction rate constant proposed in the discussion of kinetics in section IV.B.9. above.

The most worrying aspect of the high-temperature laboratory heating experiments is the extent to which the reactions proceed by the same mechanism as in sedimentary sequences. Mackenzie and McKenzie (1983) suggested this may be the case for the aromatization of steroids studied, but not for the alkane isomerization. Further evaluation of these phenomena awaits detailed analysis of sediments affected by medium-size igneous intrusions, which could form a link between sedimentary reaction temperatures (<130°C) and laboratory heating (>200°C), and more accurate laboratory maturation studies, perhaps involving the heating of standard compounds on clays. Didyk *et al.* (1975) heated a porphyrin-rich oil on an extracted shale at 210°C and observed the postulated conversion of DPEP to aetio porphyrins to an increasing extent with increasing time.

The one gap in laboratory simulation of the geological effects on biological marker distributions is published laboratory studies of possible migration mechanisms with respect to, for example, steranes and triterpanes. Clearly the technology of the reservoir engineer for flow tests could assist here, as could detailed investigations of biological marker distributions throughout the production of a well.

VI. Summary and Conclusions

The applications of biological marker organic geochemistry described here are only possible because such a large diversity of organic molecular structures which occur in the geosphere can be confidently assigned. This situation only exists because of years of painstaking research by groups prepared to use many of the skills and techniques developed for natural product chemistry. Mass spectrometry in itself is mainly only capable of providing approximate structures, and the determination of full structures including stereochemistry requires that sufficient amounts of the pure compound of interest can be obtained for analysis by, for example, nuclear magnetic resonance spectroscopy or X-ray crystallography. In many cases, structures can only be properly identified by the laboratory synthesis of reference compounds. Such an approach can often take years of research for a single compound type. There remain many more unidentified components, which in the future will provide the tools for a better determination of the geological fate of biogenic carbon, including the processes which lead to subsurface accumulations of petroleum.

Biological marker organic geochemistry involves the tracing of the chemical and physical pathways taken by organic compounds in the biosphere, subsequent to the death of the host organism. Of dominant importance are the controls on the distributions of compounds incorporated into bottom sediments at the time of deposition. Such compounds are only really of relevance to the petroleum geochemist, if their basic carbon skeletons remain intact, at least until a level of maturity corresponding to the early stages of petroleum generation by the thermal breakdown of kerogen. Many such compounds can now be recognized and their geochemical pathways followed with varying degrees of certainty. For example, triterpenoids, steroids and acyclic isoprenoids biosynthesized by organisms are steadily modified during early diagenesis by the chemical and microbial removal of oxygen-containing functional groups and carbon–carbon double bonds to yield saturated and aromatic hydrocarbons which retain the original carbon skeleton and which are observed in ancient sedimentary rocks and petroleums. Many of the large number of reactions and intermediates to these processes are now known. An analogous series of reactions converts the chlorophyll of plants and algae to the basic alkylated porphyrin macrocycle of ancient sediments and petroleums (Fig. 1).

Before these modified carbon skeletons are completely degraded by the thermal cleavage of the carbon–carbon covalent bonds which define the skeleton, a number of other temperature-dependent reactions occur. In general, this stage, which can include the earlier phase of oil generation, is less complicated with respect to biological markers than early diagenesis: the problems of many competing reactions are less severe and in favourable cases the reactants and products of certain single reactions can be recognized. This is possible because of the high structural specificity of the biological markers.

The reactions considered involve configurational isomerization (the alteration of the biologically inherited stereochemistry at certain chiral centres by the loss and regain of a hydrogen substituent) and aromatization. Ideally, the concentrations of the products of the reactions should be zero when the reaction starts. Only three of the reactions whose extents can routinely be calculated from GC–MS measurements of the concentrations of reactants and products fulfill this condition. They are the isomerization of hopanes at C-22 and steranes at C-20 and the aromatization of C-ring monoaromatic steroid hydrocarbons to triaromatic steroid hydrocarbons. These reactions should form the basis of maturity assessment of sedimentary rocks and immature petroleums, derived from the extent to which they have occurred. Many other similar reactions have been detected, but with one exception (the isomerization of pristane) the concentrations of the products can be significant when the reaction starts. They can, however, be used to supplement and extend maturity assessment with biological markers.

Recent progress has been made in determining the kinetic expressions for the derivation of the rate constants of the reactions for a given temperature. Aromatization of C-ring monaromatic steranes has a higher activation energy (i.e. shows a bigger increase in rate for a given temperature increase) than isomerization of steranes at C-20. Consequently, in sedimentary sequences which have not experienced uplift and where the temperature increase since deposition is rapid, the measured aromatization reaction extents appear promoted with respect to those for the isomerization of steranes at C-20, when compared to sequences where the temperature increase was more gradual. This increased understanding can be used to help determine past variations in heat flow and sedimentation rate, and perhaps provide comment on the extent to which a sequence has been uplifted.

The distributions of the biological marker compounds of sediments vary considerably with differences in the depositional environment and in the nature of their biological origin. They therefore respond in a sensitive way to changes in redox potential, salinity, water temperature and depth, nutrient supply and topography. The degree to which a diagenetic pathway is followed may strongly reflect the nature of the mineral matrix present. All these factors are superimposed on the maturation and diagenetic effects already discussed, and combine to impart a quite specific set of "fingerprints" to sedimentary rock extracts which are later inherited by crude oils when expulsion occurs. Thus, these "fingerprints" or biological marker distributions, revealed by GC–MS mass fragmentograms of diagnostic ions, are now routinely used for the classification of oils into groups and the correlation of these groups with potential source areas in a particular sedimentary basin. In an ideal situation the geochemist would have a number of biological marker ratios reflecting maturation, but entirely independent of variation in the nature of the original contributing organisms and the depositional environment, and others capable of assessing this variation, but entirely independent of maturity effects. As a very general rule for the hydrocarbons the ratios of different stereoisomers with the same carbon number can provide the necessary information on maturity, whilst the carbon number distributions over a restricted range of individual compound types with the same stereochemistry and ratios of different structural types (hopanes/steranes) may be relatively insensitive to changes in maturity, but rather reflect differences in the original depositional environment. With few exceptions, the palaeoenvironmental reasons for these differences cannot be defined. However, with the two types of measurements, poor correlations can often be rationalized as differences in maturity and/or in organic matter type (i.e. the nature of the precursor biomass).

Initial findings from the studies of crude oil sources and maturity suggest that many (but not all) of the maturity reactions which affect biological markers disseminated in a potential source rock may not continue to the same extent, when in a more accumulated state in a reservoir.

Only in rare, extreme cases of in-reservoir biodegradation are some of the biological markers of interest altered by the microbes. The sequence and results of the alteration are relatively well understood.

Despite the obvious advantages of the biological markers for correlation studies, much of their present application rests on a somewhat simplified model for the migration and accumulation of crude petroleum. This in no way invalidates many of the successful uses reported to date, but the extent of this simplification should always be borne in mind, particularly where the biological marker data contradict other geochemical information.

Recognizable biological markers are responsible for only a very small proportion of the total mass of a crude oil or the organic extract of sedimentary rocks. This and the fact that the reactions studied in detail bear little resemblance to those associated with oil generation is unimportant, because they can be regarded as trace indicators of thermal history—a very important consideration for all petroleum geochemists. Their low concentration is however more problematic when using them to correlate oils with each other and with the extracts of potential source rocks.

Most correlation studies assume a very static situation for most of the soluble organic matter in the geosphere. This is only disturbed by the movement of relatively homogeneous plugs of oil out of source rocks and into reservoirs over quite short time periods. Perhaps a more realistic model, supported by some observations, should involve a more dynamic picture, where a significant proportion of these soluble components moves and is redistributed continuously. This mobility will vary from one molecular type to another, so that, for example, light hydrocarbons ($\leq C_8$) will migrate much more readily than most biological markers (ca. C_{30}). If so-called "primary" migration out of relatively impervious sediments into more porous lithologies starts early and continues until the end of oil generation, then quite clearly situations could arise where the main origin of the biological markers differed from the main source of the rest of the oil. This danger exists partly because the biological marker compounds form such a small concentration of the total oil or mature extract, and means that these techniques should not be used in isolation from other less specific methods, which at least are based on the chemistry of the total mixture.

The other major problems relate to comparing the organic matter extracted from a powdered sedimentary rock with organic solvents to that which is removed by primary migration. Studies have shown that the extractable biological markers of the more open pores of a single sedimentary rock sample can differ considerably from those of the more closed pores. The material in the open pores will be selectively sampled by primary migration. Primary migration may sample quite large but heterogeneous (with respect to biological marker content) thicknesses of sediments. The resultant oil will represent a homogenized average and correlation studies should consider the implications of such a process. Finally, many of the biological markers now found in oils may be derived to a large extent by the thermal cracking of kerogen during oil generation, and not just from components that have remained "unbound" and extractable since early diagenesis. Evaluation of pyrolysis experiments suggests that those components could differ from the unbound material and vary in composition as generation progresses. These differences could produce small variations in the normal measurements of original organic matter for a single sediment as the zone of oil generation is crossed.

The assessment of migration with biological markers is still at an early stage. The extent to which the normal conditions of migration can alter biological marker distributions remains unclear. Enrichments of 14β(H),17β(H)-steranes relative to 14α(H),17α(H)-steranes and of monoaromatic steroid hydrocarbons relative to triaromatic steroid hydrocarbons have been proposed to occur under certain conditions with increased secondary migration distance.

The study of conclusively observed fractionations of biological markers by migration could provide useful clues to the mechanisms of migration.

Such is the complexity of geological systems, that in certain situations postulated processes can only be confirmed by their careful simulation in the laboratory. This approach has already had certain successes with maturation (laboratory heating) and biodegradation (laboratory cultures) studies. The evaluation of migration effects would clearly benefit from laboratory experiments, designed to recreate the movement of hydrocarbons in the subsurface.

Biological marker organic geochemistry can make a substantial contribution to the petroleum geochemist's understanding of the thermal history and evolution of sedimentary basins. Furthermore, it can greatly assist the correlation of oils with each other and the postulation of which area(s) of the basin they derived from. The present knowledge of the structures of biological marker hydrocarbons and the GC–MS technology currently available are sufficiently advanced, so that many of the important measurements can now be made routinely. As such, biological marker analyses can form an integral part of the geochemical analysis of oil and sediment samples from exploration wells.

Appendix

I and II below show the numbering systems for steranes and hopanes.

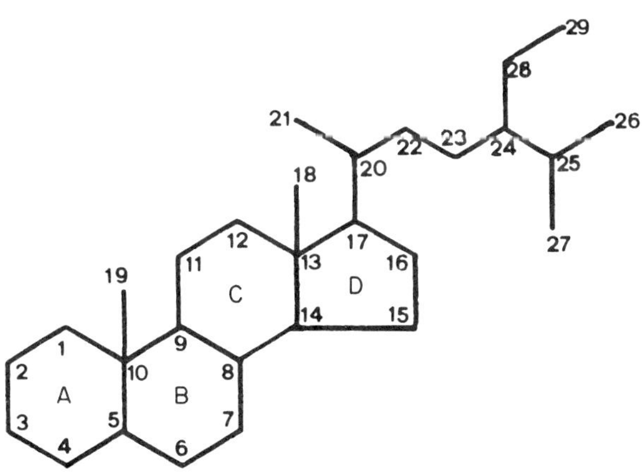

I

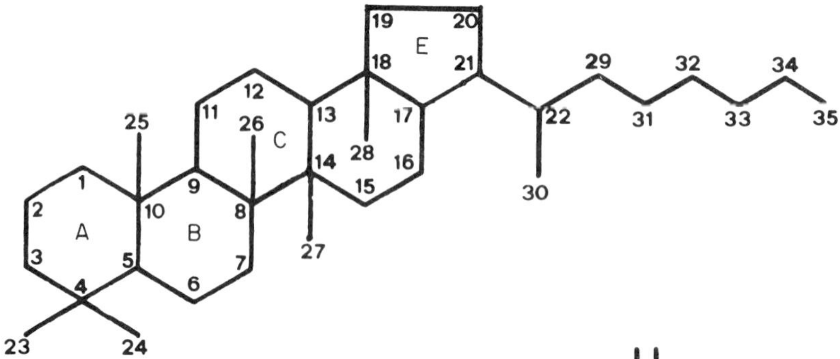

II

Acknowledgements

I thank Dr. J. R. Maxwell and Prof. G. Eglinton of Bristol University for continuous help and support during much of the research which laid the basis for this review. The support of Prof. D. H. Welte and Dr. D. Leythaeuser while writing this chapter is acknowledged. I thank the Alexander von Humboldt Foundation for a fellowship and Lydia Fuckardt for typing the manuscript with skill and patience.

References

Albaiges, J. (1980a). *In* "Advances in Organic Geochemistry 1979" (Douglas, A. G. and Maxwell, J. R., Eds.), 19–28. Pergamon Press, Oxford.

Albaiges, J. (1980b). "Analytical Techniques in Environmental Chemistry". Pergamon Press, Oxford.

Aldridge, A. K., Brooks, P. W., Eglinton, G. and Maxwell, J. R. (1977). *Proc. Inst. Petr. Microbiology Group Symp.* Oct. 1976, 4–21.

Alexander, R., Eglinton, G., Gill, J. P. and Volkman, J. K. (1980). *J. High Res. Chrom. and Chrom. Comm.* **3,** 521–522.

Alexander, R., Kagi, R. and Noble, R. (1983a). *J.C.S. Chem. Comm.*, 226–228.

Alexander, R., Kagi, A. V., Larcher, A. V. and Woodhouse, G. W. (1983b). *In* "Advances in Organic Geochemistry 1981" (Bjorøy, M. *et al.*, Eds.), 69–71. John Wiley, Chichester.

Allan, J., Bjorøy, M. and Douglas, A. G. (1977). *In* "Advances in Organic Geochemistry 1975" (Campos, R. and Gǫni, J., Eds.), 633–654. ENADIMSA, Madrid.

Anders, D. E. and Robinson, W. E. (1971). *Geochim. Cosmochim. Acta* **35,** 661–678.

Aquino Neto, F. R., Trendel, J. M., Restlé, A., Albrecht, P. and Connan, J. (1982). *Tetrahedron Lett.*, 2027–2030.

Bailey, N. J. L., Krouse, H. R., Evans, C. R. and Rogers, M. A. (1973a). *AAPG Bull.* **57**, 1276–1290.
Bailey, N. J. L., Jobson, A. M. and Rogers, M. A. (1973b). *Chem. Geol.* **11**, 203–221.
Baker, E. W. (1966). *J. Am. Chem. Soc.* **88**, 2311–2315.
Baker, E. W. (1969). *In* "Organic Geochemistry—Methods and Results" (Eglinton, G. and Murphy, M. T. J., Eds.), 464–497. Springer Verlag, Berlin.
Baker, E. W. and Louda, J. W. (1980). *In* "Initial Reports of the Deep Sea Drilling Project" (Lee, M. and Stout, L. N., Eds.), Vol. LVI, LVII, Pt. 2, 1397–1408. U.S. Government Printing Office, Washington.
Baker, E. W. and Louda, J. W. (1983). *In* "Advances in Organic Geochemistry 1981" (Bjorøy, M. *et al.*, Eds.), 401–421. John Wiley, Chichester.
Baker, E. W. and Palmer, S. E. (1978). *In* "The Porphyrins" (Dolphin, D., Ed.), Vol. I, 485–551. Academic Press, London, Orlando and New York.
Baker, E. W., Yen, T. F., Dickie, J. P., Rhodes, R. E. and Clarke, L. F. (1967). *J. Am. Chem. Soc.* **89**, 3631–3639.
Baker, E. W., Palmer, S. E. and Huang, W. Y. (1978). *In* "Initial Reports of the Deep Sea Drilling Project" (Lancelot, Y., Seibold, E. *et al.*, Eds.), Vol. XLI, 825–832. U.S. Government Printing Office, Washington.
Barwise, A. J. G. and Whitehead, E. V. (1980). *In* "Advances in Organic Geochemistry 1979" (Douglas, A. G. and Maxwell, J. R., Eds.), 181–192. Pergamon Press, Oxford.
Beletskaya, S. N. (1978). *In* "Fundamental Aspects of the Genesis and Accumulation of Oil and Gas" (Navlikin, V. D. and Aliev, M. M., Eds.), 52–70. Nauka, Moscow.
Bendoraitis, J. G. (1974). *In* "Advances in Organic Geochemistry 1973" (Tissot, B. and Bienner, F., Eds.), 209–224. Editions Technip, Paris.
Bjorøy, M. And Rullkötter, J. (1980). *Chem. Geol.* **30**, 27–34.
Bjorøy, M., Brooks, P. W. and Hall, K. (1983). *In* "Advances in Organic Geochemistry 1981" (Bjorøy, M. *et al.*, Eds.), 87–93. John Wiley, Chichester.
Bonnett, R. and Czechowski, F. (1980). *Nature* **283**, 465–467.
Bonnett, R., Brewer, P., Noro, K. and Noro, T. (1978). *Tetrahedron* **34**, 379–385.
Boon, J. J., de Leeuw, J. W. and Burlingame, A. L. (1978). *Geochim. Cosmochim. Acta* **42**, 631–644.
Boon, J. J., Rijpstra, W. I. C., de Lange, F., de Leeuw, J. W., Yoshioka, M. and Shimizu, Y. (1979). *Nature* **277**, 125–127.
Brassell, S. C. (1980). Ph.D. Thesis, Univ. of Bristol.
Brassell, S. C., Comet, P. A., Eglinton, G., Issacson, P. J., McEvoy, J., Maxwell, J. R., Thomson, I. D., Tibbetts, P. J. C. and Volkman, J. K. (1980). *In* "Initial Reports of the Deep Sea Drilling Project" (Lee, M. and Stout, L. N., Eds.), Vol. LVI, LVII, Pt. 2, 1367–1390. U.S. Government Printing Office, Washington.
Brassell, S. C., Wardroper, A. M. K., Thomson, I. D., Maxwell, J. R. and Eglinton, G. (1981). *Nature* **290**, 693–696.
Brennemann, M. C. and Smith, P. N. (1958). *In* "Habitat of Oil" (Weeks, L. G., Ed.), 818–849. AAPG, Tulsa.
Brooks, P. W., Cardoso, J. N., Didyk, B., Eglinton, G., Humberston, M. J. and Maxwell, J. R. (1977) *In* "Advances in Organic Geochemistry 1975" (Campos, R. and Gõni, J., Eds.), 433–453. ENADIMSA, Madrid.
Brunet, M. F. (1981). Thèse de Doctorat de troisième cycle, Univ. Pierre et Marie Curie.

Calvin, M. (1969). "Chemical Evolution". Oxford Univ. Press, Oxford.
Carman, R. M., Cowley, D. E. and Marthy, R. A. (1970) *Aust. J. Chem.* **23,** 1655–1665.
Caspi, E., Duax, W. L., Griffin, J. F., Moreau, J. P. and Wittstruck, T. A. (1975). *J. Org. Chem.* **40,** 2005–2006.
Chappe, B., Albrecht, P. and Michaelis, W. (1982). *Science* **217,** 65–66.
Claret, J., Tchikaya, J. B., Tissot, B., Deroo, G. and van Dorsselaer, A. (1977). *In* "Advances in Organic Geochemistry 1975" (Campos, R. and Gõni, J., Eds.), 509–522. ENADIMSA, Madrid.
Coleman, M. L., Curtis, C. D. and Irwin, H. (1979). *World Oil*, March 1979, 83–92.
Connan, J. (1974). *AAPG Bull.* **58,** 2516–2521.
Connan, J. (1981). *Bull. Centre Rech. Pau*, 151–171.
Connan, J., Restlé, A. and Albrecht, P. (1980). *In* "Advances in Organic Geochemistry 1979" (Douglas, A. G. and Maxwell, J. R., Eds.), 1–17. Pergamon Press, Oxford.
Corbet, B., Albrecht, P. and Ourisson, G. (1980). *J. Am. Chem. Soc.* **102,** 1171–1173.
Corwin, W. H. (1959). *Proc. Fifth World Petr. Cong.* **5,** 119–129.
Dastillung, M. and Albrecht, P. (1977). *Nature* **269,** 678–679.
Demaison, G. T. (1977). *AAPG Bull.* **61,** 1950–1961.
Deroo, G. (1967). *Internal Report 14427, Inst. Fran. Petr.*
de Rosa, M., de Rosa, S., Gambacorta, A., Bu'lock, J. D. (1977). *Phytochem.* **16,** 1961–1965.
Didyk, B. (1975). Ph.D. Thesis, Univ. of Bristol.
Didyk, B., Alturki, Y. I. A., Pillinger, C. T. and Eglinton, G. (1975). *Nature* **256,** 563–575.
Didyk, B., Simoneit, B. R. T., Brassell, S. C. and Eglinton, G. (1978). *Nature* **272,** 216–222.
Dow, W. G. (1977). *J. Geochem. Explor.* **7,** 79–99.
Durand, B. (1983). *In* "Advances in Organic Geochemistry 1981" (Bjorøy, M. *et al.*, Eds.), 117–128. John Wiley, Chichester.
Durand, B. and Oudin, J. L. (1980). *Proc. Tenth World Petr. Cong.* **2,** 3–11.
Eglinton, G. and Calvin, M. (1967). *Sci. Am.* **216,** 32–43.
Eglinton, G., Hajibrahim, S. K., Maxwell, J. R. and Quirke, J. M. E. (1980). *In* "Advances in Organic Geochemistry 1979" (Douglas, A. G. and Maxwell, J. R., Eds.), 193–203. Pergamon Press, Oxford.
Ekweozor, C. M. and Strausz, O. P. (1982). *Tetrahedron Lett.*, 2711–2714.
Ekweozor, C. M., Okogun, J. I., Ekong, D. E. U. and Maxwell, J. R. (1979a). *Chem. Geol.* **27,** 11–28.
Ekweozor, C. M., Okogun, J. I., Ekong, D. E. U. and Maxwell, J. R. (1979b). *Chem. Geol.* **27,** 29–37.
Ekweozor, C. M., Okogun, J. I., Ekong, D. E. U. and Maxwell, J. R. (1981). *J. Geochem. Explor.* **15,** 653–662.
Ensminger, A. (1977). Thèse de Doctorat ès-Sciences, Univ. Louis Pasteur.
Ensminger, A., Van Dorsselaer, A., Spyckerelle, C., Albrecht, P. and Ourisson, G. (1974). *In* "Advances in Organic Geochemistry 1973" (Tissot, B. and Bienner, F., Eds.), 245–260. Editions Technip, Paris.
Ensminger, A., Albrecht, P., Ourisson, G. and Tissot, B. (1977). *In* "Advances in Organic Geochemistry 1975" (Campos, R. and Gõni, J., Eds.) 45–52. ENADIMSA, Madrid.

Ensminger, A., Joly, G. and Albrecht, P. (1978). *Tetrahedron Lett.* **18,** 1575–1578.
Galbraith, M. N., Miller, C. J., Rawson, J. W. L., Ritchie, E., Shannon, J. S. and Taylor, W. C. (1965). *Aust. J. Chem.* **18,** 226–239.
Gallegos, E. J. (1971). *Anal. Chem.* **43,** 1151–1160.
Gallegos, E. J. (1975a). *Anal. Chem.* **47,** 1150–1154.
Gallegos, E. J. (1975b). *Anal. Chem.* **47,** 1524–1528.
Gallegos, E. J. (1976). *Anal. Chem.* **48,** 1348–1351.
Gaskell, S. J. and Eglinton, G. (1975). *Nature* **254,** 209–211.
Giger, W. and Schaffner, C. (1981). *Naturwissenschaften* **68,** 37–38.
Goodwin, N. S., Park, P. J. D. and Rawlinson, T. (1983) *In* "Advances in Organic Geochemistry 1981" (Bjorøy, M. *et al.*, Eds.), 117–128. John Wiley, Chichester.
Goy, G. (1979). Thèse de Doctorat ès-Sciences, Univ. Pierre et Marie Curie.
Grantham, P. and Douglas, A. G. (1980). *Geochim. Cosmochim. Acta* **44,** 1801–1810.
Grantham, P. J., Posthuma, J. and de Groot, K. (1980). *In* "Advances in Organic Geochemistry 1979" (Douglas, A. G. and Maxwell, J. R., Eds.), 29–38. Pergamon Press, Oxford.
Greiner, A. Ch., Spyckerelle, C. and Albrecht, P. (1976). *Tetrahedron* **32,** 257–260.
Gutsche, C. D. and Pasto, D. J. (1975). "Fundamentals of Organic Chemistry". Prentice Hall, New Jersey.
Hajibrahim, S. K. (1978). Ph.D. Thesis, Univ. of Bristol.
Hajibrahim, S. K., Tibbetts, P. J. C., Watts, C. D., Maxwell, J. R., Eglinton, G., Colin, H. and Guiochon, G. (1978). *Anal. Chem.* **50,** 549–553.
Hajibrahim, S. K., Quirke, J. M. E., Eglinton, G. (1981). *Chem. Geol.* **32,** 173–188.
Hills, I. R., Whitehead, E. V., Anders, D. E., Cummins, J. J. and Robinson, W. E. (1966). *J.C.S. Chem. Comm.* **20,** 752 754.
Hoffmann, C. F., Mackenzie, A. S., Lewis, C. A., Maxwell, J. R., Vandenbrouke, M., Durand, B. and Oudin, J. L. (1984). *Chem. Geol.* **42,** 1–23.
Hohn, M. E., Jones, N. W., Patience, R. L. (1981). *Geochim. Cosmochim. Acta* **45,** 1131–1140.
Huang, W. Y. and Meinschein, W. G. (1979). *Geochim. Cosmochim. Acta* **43,** 739–745.
Huc, A. Y. (1978). Thèse de Doctorate ès-Sciences, Univ. Louis Pasteur.
Hufnagel, H., Teschner, M. and Wehner, H. (1980). *In* "Advances in Organic Geochemistry 1979" (Douglas, A. G. and Maxwell, J. R., Eds.), 51–66. Pergamon Press, Oxford.
Hunt, J. M. (1979). "Petroleum Geochemistry and Geology". W. H. Freeman and Co., San Francisco.
Hussler, G., Chappe, B., Wehrung, P. and Albrecht, P. (1981). *Nature* **294,** 556–558.
Ikan, R., Baedecker, M. J. and Kaplan, I. R. (1975). *Geochim. Cosmochim. Acta* **39,** 187–194.
Jobson, A. M., Cook, E. D. and Westlake, D. W. S. (1979). *Chem. Geol.* **24,** 355–365.
Kates, M., Joo, C. N., Palameta, B. and Shier, T. (1967). *Biochem.* **6,** 3329–3338.
Laflamme, R. E. and Hites, R. A. (1978). *Geochim. Cosmochim. Acta* **42,** 289–303.
Leythaeuser, D., Hollerbach, A. and Hagemann, H. (1977). *In* "Advances in

Organic Geochemistry 1975" (Campos, R. and Gõni, J., Eds.), 3–20. ENA-DIMSA, Madrid.
Leythaeuser, D., Bjorøy, M., Mackenzie, A. S., Schaefer, R. G. and Altebäumer, F. J. (1983a). *In* "Advances in Organic Geochemistry 1981" (Bjorøy, M. *et al.*, Eds.), 136–146. John Wiley, Chichester.
Leythaeuser, D., Schaefer, R. G. and Yükler, A., (1983b). *AAPG Bull.* **67,** 889–895
Ludwig, B., Hussler, G., Wehrung, P. and Albrecht, P. (1981). *Tetrahedron Lett.*, 3313–3316.
Mackenzie, A. S. (1980). Ph.D. Thesis, Univ. of Bristol.
Mackenzie, A. S. and McKenzie, D. P. (1983). *Geol. Mag.* **120,** 417–470.
Mackenzie, A. S. and Maxwell, J. R. (1981). *In* "Organic Maturation Studies and Fossil Fuel Exploration" (Brooks, J., Ed.), 239–254. Academic Press, London, Orlando and New York.
Mackenzie, A. S., Patience, R. L., Maxwell, J. R., Vandenbroucke, M. and Durand, B. (1980a). *Geochim. Cosmochim. Acta* **44,** 1709–1721.
Mackenzie, A. S., Quirke, J. M. E., Maxwell, J. R. (1980b). *In* "Advances in Organic Geochemistry 1979" (Douglas, A. G. and Maxwell, J. R., Eds.), 239–248. Pergamon Press, Oxford.
Mackenzie, A. S., Hoffmann, C. F. and Maxwell, J. R. (1981a). *Geochim. Cosmochim. Acta* **45,** 1345–1355.
Mackenzie, A. S., Patience, R. L. and Maxwell, J. R. (1981b). *In* "The Origin and Chemistry of Petroleum" (Atkinson, G. and Zuckerman, J. L., Eds.), 1–32. Pergamon Press, Oxford.
Mackenzie, A. S., Lewis, C. A. and Maxwell, J. R. (1981c). *Geochim. Cosmochim. Acta* **45,** 2369–2376.
Mackenzie, A. S., Brassell, S. C., Eglinton, G. and Maxwell, J. R. (1982a). *Science* **217,** 491–504.
Mackenzie, A. S., Lamb, N. A. and Maxwell, J. R. (1982b). *Nature* **295,** 223–226.
Mackenzie, A. S., Patience, R. L., Yon, D. A., Maxwell, J. R. (1982c). *Geochim. Cosmochim. Acta*, 637–649.
Mackenzie, A. S., Li Renwei, Maxwell, J. R., Moldowan, J. M. and Seifert, W. K. (1983a). *In* "Advances in Organic Geochemistry 1981" (Bjorøv, M. *et al.* Eds.), 496–503. John Wiley, Chichester.
Mackenzie, A. S., Wolff, G. A., Maxwell, J. R. (1983b). *In* "Advances in Organic Geochemistry 1981" (Bjorøy, M. *et al.*, Eds.), 637–649. John Wiley, Chichester.
Mackenzie, A. S., Disko, U. and Rullkötter, J. (1983c). *Org. Geochem.* **5,** 55–63.
Mackenzie, A. S., Maxwell, J. R., Coleman, M. L. and Deegan, C. E. (1984). *Proc. Eleventh World Petrol. Cong.* **2,** in press.
McKenzie, D. P. (1978). *Earth Planet. Sci. Lett.* **40,** 25–32.
McKenzie, D. P. (1981). *Earth Planet. Sci. Lett.* **55,** 87–98.
McKirdy, D. M., Aldridge, A. K. and Ypma, P. J. M. (1983). *In* "Advances in Organic Geochemistry 1981" (Bjorøy, M. *et al.*, Eds.), 99–107. John Wiley, Chichester.
Mandava, N., Anderson, J. D., Dutky, S. R. and Thompson, M. J. (1974). *Steroids* **23,** 257–261.
Maxwell, J. R., Douglas, A. G., Eglinton, G. and McCormick, A. (1968). *Phytochemistry* **7,** 2157–2171.
Maxwell, J. R., Cox, R. E., Ackman, R. G. and Hooper, S. N. (1972). *In* "Advances in Organic Geochemistry 1971" (von Gaertner, H. R. and Wehner,

H., Eds.), 277–291. Pergamon Press, Oxford.
Maxwell, J. R., Mackenzie, A. S. and Volkman, J. K. (1980). *Nature* **286,** 694–697.
Michaelis, W. and Albrecht, P. (1979). *Naturwissenschaften* **66,** 420–422.
Moldowan, J. M. and Seifert, W. K. (1979). *Science* **204,** 169–171.
Moldowan, J. M. and Seifert, W. K. (1980). *J.C.S. Chem. Comm.*, 912–914.
Moldowan, J. M., Seifert, W. K. and Gallegos, E. J. (1983). *Geochim. Cosmochim. Acta* **47,** 1531–1534.
Mulheirn, L. J. and Ryback, G. (1975). *Nature* **256,** 301–302.
Mulheirn, L. J. and Ryback, G. (1977). *In* "Advances in Organic Geochemistry 1975" (Campos, R. and Gõni, J., Eds.), 173–192. ENADIMSA, Madrid.
Murphy, M. T. J., McCormick, A. and Eglinton, G. (1967). *Science* **157,** 1040–1042.
Nagy, B. (1960). *Geochim. Cosmochim. Acta* **19,** 289–296.
Nes, W. R. (1974). *Lipids* **9,** 596–612.
Nishimura, M. (1977). *Geochim. Cosmochim. Acta* **41,** 1817–1823.
Ourisson, G., Albrecht, P. and Rohmer, M. (1979). *Pure Appl. Chem.* **51,** 709–729.
Palmer, S. E. and Baker, E. W. (1978). *Science* **201,** 49–51.
Patience, R. L., Rowland, S. J. and Maxwell, J. R. (1978). *Geochim. Cosmochim. Acta* **42,** 1871–1875.
Patience, R. L., Yon, D. A., Ryback, G. and Maxwell, J. R. (1980). *In* "Advances in Organic Geochemistry 1979" (Douglas, A. G. and Maxwell, J. R., Eds.), 287–294. Pergamon Press, Oxford.
Patterson, G. W. (1971). *Lipids* **6,** 120–127.
Payzant, J. D., Rubinstein, I., Hogg, A. M. and Strausz, O. P. (1980). *Chem. Geol.* **29,** 73–88.
Petrov, A. A., Pustil'nikova, S. D., Arbiutina, N. N. and Kagramonova, G. R. (1976). *Neftekhimiga* **16,** 411–427.
Phillipi, G. T. (1977). *Geochim. Cosmochim. Acta* **41,** 33–52.
Philp, R. P., Gilbert, T. and Friedrich, J. (1981). *Geochim. Cosmochim. Acta* **45,** 1173–1180.
Price, L. C. (1980). *Chem. Geol.* **28,** 1–30.
Price, L. C., Clayton, J. L. and Rumen, L. L. (1981). *Org. Geochem.* **3,** 59–77.
Pym, J. G., Ray, J. E., Smith, G. W. and Whitehead, E. V. (1975). *Anal. Chem.* **47,** 1617–1622.
Quirk, M. M., Patience, R. L., Maxwell, J. R. and Wheatley, R. E. (1980). *In* "Analytical Techniques in Environmental Chemistry" (Albaiges, J., Ed.), 23–31. Pergamon Press, Oxford.
Quirke, J. M. E., Eglinton, G. and Maxwell, J. R. (1979). *J. Am. Chem. Soc.* **101,** 7693–7697.
Quirke, J. M. E., Shaw, G. J., Soper, P. D. and Maxwell, J. R. (1980). *Tetrahedron*, 3261–3267.
Radke, M., Sittardt, H. G. and Welte, D. H. (1978). *Anal. Chem.* **50,** 663–665.
Radke, M., Willsch, H. and Welte, D. H. (1980). *Anal. Chem.* **52,** 407–411.
Radke, M., Willsch, H., Welte, D. A. (1982). *Geochim. Cosmochim. Acta* **46,** 1–10.
Reed, W. E. (1977). *Geochim. Cosmochim. Acta* **41,** 237–247.
Rhead, M. M., Eglinton, G. and Draffan, G. H. (1971). *Chem. Geol.* **8,** 277–297.
Roberts, W. H. and Cordell, R. J. (1980). *In* "Problems of Petroleum Migration"

(Roberts, W. H. and Cordell, R. J., Eds.), AAPG Studies in Geology No. 10. AAPG, Tulsa.
Rubinstein, E. and Albrecht, P. (1975). *J.C.S. Chem. Comm.*, 957–958.
Rubinstein, I., Sieskind, O. and Albrecht, P. (1975). *J. Chem. Soc. Perkin 1*, 1833–1835.
Rubinstein, I., Strausz, O. P., Spyckerelle, C., Crawford, R. S. and Westlake, D. W. S. (1977). *Geochim. Cosmochim. Acta* **41**, 1341–1353.
Rubinstein, I., Spyckerelle, C. and Strausz, O. P. (1979). *Geochim. Cosmochim. Acta* **43**, 1–6.
Rullkötter, J. and Philp, P. (1981). *Nature* **292**, 616–618.
Rullkötter, J. and Welte, D. H. (1980). *In* "Advances in Organic Geochemistry 1979" (Douglas, A. G. and Maxwell, J. R., Eds.), 93–102. Pergamon Press, Oxford.
Rullkötter, J. and Welte, D. J. (1983). *In* "Advances in Organic Geochemistry 1981" (Bjorøy, M. *et al.*, Eds.), 438–448. John Wiley, Chichester.
Rullkötter, J. and Wendisch, D. (1982). *Geochim. Cosmochim. Acta* **46**, 1545–1553.
Rullkötter, J., Cornford, C., Flekken, P. and Welte, D. H. (1980). *In* "Initial Reports of the Deep Sea Drilling Project" (Lee, M. and Stout, L. N., Eds.), Vol. LVI, LVII, Pt. 2, 1291–1304. U.S. Government Printing Office, Washington D.C.
Rullkötter, J., Leythaeuser, D. and Wendisch, D. (1982). *Geochim. Cosmochim. Acta* **46**, 2501–2509.
Sajgò, Cs., Maxwell, J. R. and Mackenzie, A. S. (1983). *Org. Geochem.* **5**, 65–73.
Sajgò, Cs., Mackenzie, A. S. and Maxwell, J. R. (1984). Unpublished results.
Schaeflé, J. (1979). Thèse de Doctorat ès-Sciences, Univ. Louis Pasteur.
Schaeflé, J., Ludwig, B., Albrecht, P. and Ourisson, G. (1977). *Tetrahedron Lett.*, 3673–3676.
Schmitter, J. M., Arpino, P. and Guiochon, G. (1978). *J. Chromatog.* **167**, 149–158.
Schmitter, J. M., Vajta, Z. and Arpino, P. (1980). *In* "Advances in Organic Geochemistry 1979" (Douglas, A. G. and Maxwell, J. R., Eds.), 67–76. Pergamon Press, Oxford.
Schmitter, J. M., Arpino, P. and Guiochon, G. (1981). *Geochim. Cosmochim. Acta* **45**, 1951–1955.
Schmitter, J. M., Sucrow, W. and Arpino, P. J. (1982). *Geochim. Cosmochim. Acta* **46**, 2345–2350.
Schoell, M. (1982). Unpublished results.
Schoell, M., Teschner, M., Wehner, H., Durand, B., Oudin, J. L. (1983). *In* "Advances in Organic Geochemistry 1981" (Bjorøy, M. *et al.*, Eds.), 156–163. John Wiley, Chichester.
Seifert, W. K. (1977). *In* "Advances in Organic Geochemistry 1975" (Campos, R. and Gõni, J., Eds.), 21–44. ENADIMSA, Madrid.
Seifert, W. K. (1978). *Geochim. Cosmochim. Acta* **42**, 473–484.
Seifert, W. K. (1981). *In* "The Impact of the Treibs' Porphyrin Concept on the Modern Organic Geochemistry" (Prashnowsky, A. A., Ed.), 1–23. Springer Verlag, Berlin.
Seifert, W. K. and Moldowan, J. M. (1978). *Geochim. Cosmochim. Acta* **42**, 77–92.
Seifert, W. K. and Moldowan, J. M. (1979). *Geochim. Cosmochim. Acta* **43**, 111–126.

Seifert, W. K. and Moldowan, J. M. (1980). *In* "Advances in Organic Geochemistry 1979" (Douglas, A. G. and Maxwell, J. R., Eds.), 229–237. Pergamon Press, Oxford.
Seifert, W. K. and Moldowan, J. M. (1981). *Geochim. Cosmochim. Acta* **45,** 783–794.
Seifert, W. K., Gallegos, E. J., Teeter, R. M. (1972). *J. Am. Chem. Soc.* **94,** 5880–5887.
Seifert, W. K., Moldowan, J. M., Smith, G. W. and Whitehead, E. V. (1978). *Nature* **271,** 436–437.
Seifert, W. K., Moldowan, J. M. and Jones, R. W. (1980). *Proc. Tenth World Petr. Cong.* **2,** 425–438.
Seifert, W. K., Carlson, R. M. and Moldowan, J. M. (1983). *In* "Advances in Organic Geochemistry 1981" (Bjorøy, M. *et al.*, Eds.), 710–724. John Wiley, Chichester.
Shaw, G. J., Quirke, J. M. E. and Eglinton, G. (1978). *J. Chem. Soc. Perkin 1,* 1655–1659.
Shaw, G. J., Franich, R. A., Eglinton, G., Allan, J. and Douglas, A. G. (1980). *In* "Advances in Organic Geochemistry 1979" (Douglas, A. G. and Maxwell, J. R., Eds.), 281–286. Pergamon Press, Oxford.
Shi Jiyang, Mackenzie, A. S., Alexander, R., Eglinton, G., Wolff, G. A. and Maxwell, J. R. (1982). *Chem. Geol.* **35,** 1–31.
Sieskind, O., Joly, G. and Albrecht, P. (1979). *Geochim. Cosmochim. Acta* **43,** 1675–1680.
Simoneit, B. R. T. (1977). *Geochim. Cosmochim. Acta* **41,** 463–476.
Smith, G. W. (1970). *Acta Cryst.* **B26,** 1746–1755.
Smith, G. W., Fowell, D. T. and Melsom, B. G. (1970). *Nature* **228,** 355–356.
Snowdon, L. R. (1980). *Can. Soc. Petr. Geol. Mem.* **6,** 421–446.
Spyckerelle, C. (1975) Thèse de Doctorat ès-Sciences, Univ. Louis Pasteur.
Spyckerelle, C., Greiner, A. C., Albrecht, P. and Ourisson, G. (1977). *J. Chem. Res.* (M) 3746–3777, (S) 330–331.
Thompson, S. and Eglinton, G. (1978). *Geochim. Cosmochim. Acta* **42,** 199–207.
Tibbetts, P. J. C. (1980). Ph.D. Thesis, Univ. of Bristol.
Tissot, B. P. and Welte, D. H. (1978). "Petroleum Formation and Occurrence". Springer Verlag, Berlin.
Tökes, L. and Amos, B. A. (1972). *J. Org. Chem.* **37,** 4421–4429.
Treibs, A. (1934). *Ann. Chem.* **510,** 42–62.
Treibs, A. (1936). *Angew. Chem.* **49,** 682–686.
Trendel, J. M., Restlé, A., Connan, J. and Albrecht, P. (1982). *J.C.S. Chem. Comm.*, 304–306.
Tsuda, K., Hayatsu, R., Kishida, Y. and Akagi, S. (1958). *J. Am. Chem. Soc.* **80,** 921–925.
Vandenbroucke, M. (1972). *In* "Advances in Organic Geochemistry 1971" (von Gaertner, H. R. and Wehner, H., Eds.), 547–565. Pergamon Press, Oxford.
Vandenbroucke, M., Durand, B. and Oudin, J. L. (1983). *In* "Advances in Organic Geochemistry 1981" (Bjorøy, M. *et al.*, Eds.), 147–155. John Wiley, Chichester.
van Dorsselaer, A., Ensminger, A., Spyckerelle, C., Dastillung, M., Sieskind, O., Arpino, P. Albrecht, P., Ourisson, G., Brooks, P. W., Gaskell, S. J., Kimble, B. J., Philp, R. P., Maxwell, J. R. and Eglinton, G. (1974). *Tetrahedron Lett.* **14,** 1349–1352.

van Dorsselaer, A., Albrecht, P. and Ourisson, G. (1977). *Bull. Soc. Chim. France*, 165–176.
van Graas, G., de Leeuw, J. W., Schenck, P. A., Haverkamp, I. (1981). *Geochim. Cosmochim. Acta* **45**, 2465–2474.
van Graas, G., de Lange, F., de Leeuw, J. W. and Schenk, P. A. (1982a). *Nature* **296**, 59–61.
van Graas, G., Baas, J. M. A., van de Graaf, B. and de Leeuw, J. W. (1982b) *Geochim. Cosmochim. Acta* **46**, 2399–2402.
Vogler, E. A., Meyers, P. A. and Moore, W. A. (1981). *Geochim. Cosmochim. Acta* **45**, 2287–2293.
Volkman, J. K. and Maxwell, J. R. (1984). *In* "Biological Markers—A Monograph" (Johns, R. B., Ed.). Elsevier, Amsterdam, in press.
Volkman, J. K., Alexander, R., Kagi, R. I. and Rullkötter, J. (1983). *Geochim. Cosmochim. Acta* **47**, 1033–1040.
Wade, S. W., Mackenzie, A. S. and Maxwell, J. R. (1982). Unpublished results.
Wakeham, S. G., Farrington, J. W., Gagosian, R. B., Lee, C., de Baar, H., Nigrelli, G. E., Tripp, B. W., Smith, S. O. and Frew, N. M. (1980). *Nature* **286**, 798–800.
Walker, J. D., Colwell, R. R. and Petrakis, L. (1975). *Can. J. Microbiol* **21**, 1760–1667.
Wardroper, A. M. K. (1979). Ph.D. Thesis, Univ. of Bristol.
Watts, C., Maxwell, J. R. and Kjosen, H. (1977). *In* "Advances in Organic Geochemistry 1975" (Campos, R. and Goni, J., Eds.), 391–413. ENADIMSA, Madrid.
Welte, D. H. and Yükler, M. A. (1981). *AAPG Bull.* **65**, 1387–1396.
Welte, D. H., Hagemann, H. W., Hollerbach, D., Leythaeuser, D. and Stahl, W. (1975). *Proc. Ninth World Petr. Cong.* **2**, 179–191.
Welte, D. H., Kratochvil, H., Rullkötter, J., Ladwein, H. and Schaefer, R. G. (1982). *Chem. Geol.* **35**, 33–68.
Whitehead, E. V. (1973). *In* "Proc. Symp. Hydrogeochemistry and Biochemistry" (Ingerson, E., Ed.), Vol. II, 158–211. The Clarke Company, Washington, D.C.
Winters, J. C. and Williams, J. A. (1969). *Am. Chem. Soc. Div. Petr. Chem., New York*, 7–12 September, Preprints **4** (14), E22–31.
Yen, T. F. and Silverman, S. R. (1969). *Am. Chem. Soc. Div. Petr. Chem., New York*, 7–12 September, Preprints **4** (14), E32.
Yon, D. A. Maxwell, J. R. and Ryback, G. (1982). *Tetrahedron Lett.*, 2143–2146.

Stable Isotopes in Petroleum Research

Martin Schoell*
Bundesanstalt für Geowissenschaften und Rohstoffe, Hannover, West Germany

I. Introduction

Stable isotope investigations applied to petroleum research have increased considerably since the early investigations of Silverman and Epstein (1958) and Eckelmann *et al.* (1962). With regard to questions of the origin of petroleum, stable isotopes have been used as conservative tracers that are not appreciably altered during the thermal evolution of hydrocarbons from sedimentary organic matter or kerogen which eventually leads to oil formation. Stable isotope investigations in petroleums and kerogens have mainly been applied to questions of oil/oil and oil/source-rock correlation.

Stable isotope investigations have also proved to be important in natural gas research. Here stable isotope concentrations in gaseous compounds are

* Currently with Chevron Oil Field Research Co., Le Habra, California, U.S.A.

ADVANCES IN PETROLEUM GEOCHEMISTRY Vol. 1
ISBN 0-12-0320[illegible]0

properties which vary and are indicative of the type of gas and its origin. Again, work was started twenty years ago, and stable isotopes have since developed as an important tool in natural gas geochemistry.

This paper summarizes developments since 1977 and is restricted to carbon and hydrogen isotopes. It does not claim to cover all published material in this field; rather it reviews various topics which the author believes are important to the field. Emphasis is given to new developments and questions arising with these developments. The status of the work before 1977 has been reviewed in papers by Fuex (1977) and Stahl (1977). Other useful reviews of the isotope geochemistry of organic carbon have been compiled by Deines (1980) and Galimov (1980). For basic data and principles, the reader is referred to these papers.

II. Reference Materials

Stable isotope ratios of $^{13}C/^{12}C$ as well as D/H are directly proportional to the concentration of the heavy, rare isotope when reported in the usual δ-notation as per mil deviation from a standard:

$$\delta = \frac{R_{\text{sample}} - R_{\text{standard}}}{R_{\text{standard}}} \times 1000 \text{ (p.p.t.)}$$

where R is the isotope ratio of the respective element. One of the prerequisites for accurate measurements is to have reliable isotope reference materials. These are well established for carbonates, but not for organic materials. The adopted reference scale is the PDB scale, with the only calibrated sample on the scale being NBS 21 graphite, which has been determined to −28.79 per mil versus PDB by Craig (1957). The isotopic reference material most commonly used in stable isotope laboratories working with organic materials is, however, NBS 22 lubricating oil, which has been established as a standard by Silverman (1964) with a value of −29.4 per mil on the PDB scale. In a later calibration, Mook (1968) suggested that NBS 21 should be changed to a $\delta^{13}C$ value of −29.19 per mil on the PDB scale. Comparison of isotope values based on the NBS 21 calibration on the one hand and the NBS 22 calibration on the other revealed discrepancies of the order of 0.4 per mil. A new inter-laboratory comparison of NBS 22 consequently led to a new value for NBS 22 lubricating oil of −29.8 per mil versus PDB (Schoell *et al.*, 1983a). It is essential when comparing literature data to take into consideration the value of the reference material which has been used. NBS 22 is also a suitable reference material for hydrogen isotopes. It revealed a δD value of −119 per mil versus SMOW.

III. Coals and Kerogens

There is surprisingly little systematic data available on isotope variations in coals and kerogens. They are initially (i.e. in their immature stage) heterogeneous materials, and the bulk isotopic composition of a coal or a kerogen is the result of its various constituents. Rigby *et al.* (1981) found considerable variations, for example, in different macrolithotypes and brown coals (see Fig. 1), which were not found in coals in higher stages of coalification. This indicates that an isotopic homogenization occurs during coalification due to the rearrangement of organic matter. Experimental evidence as well as natural examples support the early finding of

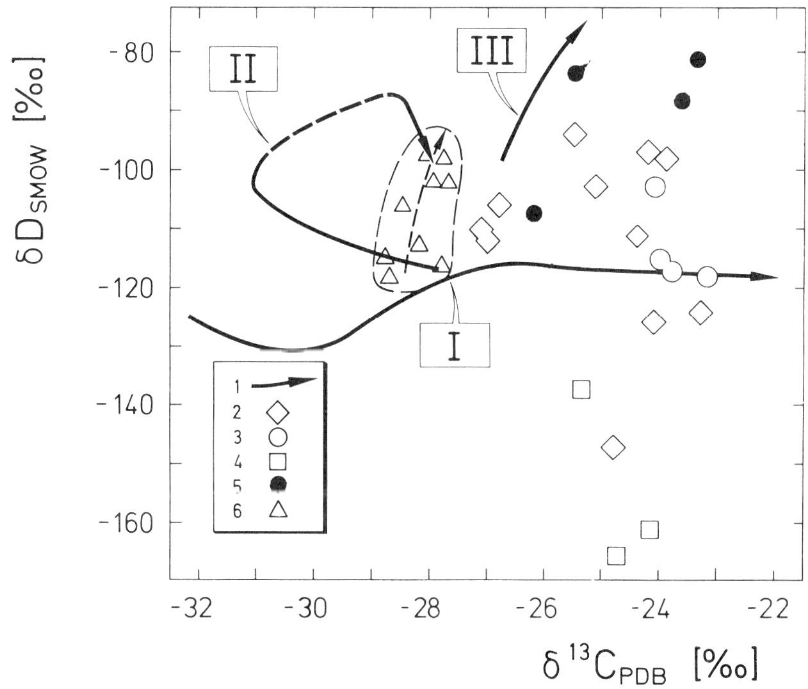

FIG. 1. C and H isotope analyses on coals and kerogens.

1. Isotopic variation in Type kerogens (I Uinta Basin, II Paris Basin, III Douala Basin; the arrows point to increasing maturity of the samples).
2. Coals from Europe and North America.
3. Coals, Ruhr area, Germany.
4. Permian coals, Australia.
5. Various macrolithotypes in Australian coals.
6. Type III kerogens, Mahakam Delta.

Data from Redding *et al.* (1980), Rigby and Smith (1981) and Schoell *et al.* (1983b).

Wickmann (1953) that kerogens and coals do not change their carbon isotope composition during carbonization within the catagenesis stage (Redding *et al.*, 1980; Lewan, 1982).

The isotopic composition of kerogens varies without obvious regularity (Fig. 1). Type kerogens have been analysed by Redding *et al.* (1980), and showed different isotope patterns from basin to basin. For example, kerogens from the Uinta basin, which are all of Type I, exhibited large variations in carbon isotopes but little variability in hydrogen isotopes. Type III kerogens from the Donala Basin and the Mahakam Delta were isotopically uniform for carbon but variable in hydrogen isotopes. There is no current concept to explain the isotopic variability in kerogens. The most simple assumption is that kerogens are isotopically heterogeneous mixtures of the various components of which they are comprised. Variations in any of these components would lead to isotope changes in the kerogen. It is not, however, possible to relate C-isotope values of ca. −20 to −23 per mil to a marine and values of −27 to −30 per mil to a terrestrial origin respectively, as it is for living organisms (Galimov, 1980). Kerogens, however, which are composed of uniform algal material, are isotopically enriched in ^{13}C. Hollerbach *et al.* (1977), for example, have reported $\delta^{13}C$ values around −18% for algal kerogens from Miocene lacustrine algal kerogens in the Ries area.

Coals are also very variable in their isotopic composition. Tertiary coals from the Mahakam Delta are relatively low in ^{13}C and high in D, whereas Permian coals from Australia are high in ^{13}C and low in D. The deuterium content in coals is possibly climatically controlled (Smith, personal communication).

More information is required on the causes of isotope variations in kerogens and coals. In particular, petrographic characterization of kerogens combined with isotope studies is lacking in the literature. Such studies would allow a deeper insight into the interdependence of isotopic and kerogen composition, which would in turn allow a more accurate interpretation of isotope data.

A. Isotope variations in source rocks

The above considerations suggest that isotope variations in coals and kerogens primarily reflect the various inputs of organic material. Therefore, isotope concentrations in kerogens may vary in a sedimentary basin or in an organic-rich unit with changing organic facies. The variations in the kerogens from the Green River Basin (Type I, Fig. 1) reflect primary facies changes within the basin. Liassic kerogens from Luxembourg are another example (Fig. 2). Here the isotopic composition is similar to that of coals from the Ruhr area. Detrital coal particles are found in the Liassic sediments near the old coastline and contribute to the organic matter in the Lias in Luxembourg.

Completely different processes have been found to cause isotope variations in kerogens from a Liassic oil shale in south-west Germany (Küspert, 1982). In a vertical distance of a few metres the ^{13}C concentration in both the carbonate fraction and the organic carbon species changed notably (Fig. 3). A clear correlation was found between organic richness of the

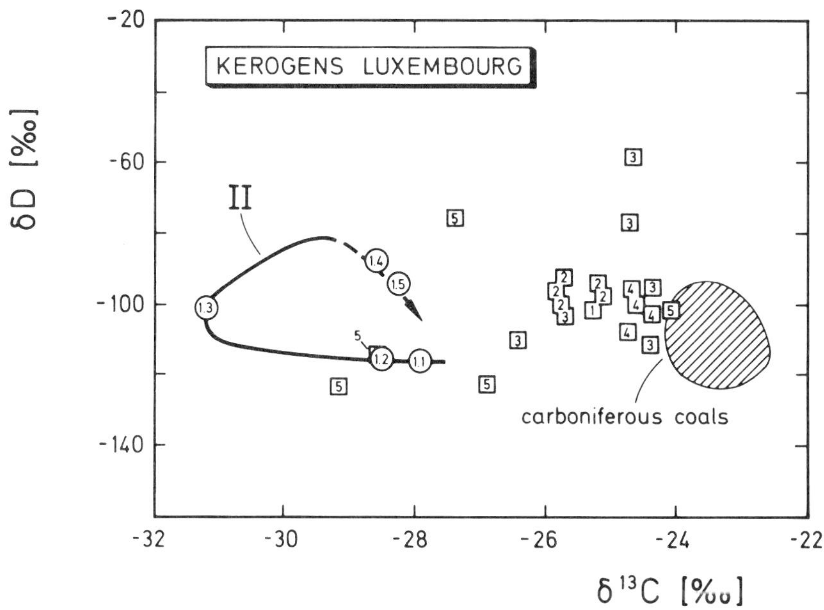

FIG. 2. Isotope variations in kerogens from Toarcian shales in Luxembourg. Note that the Luxembourg shales are close to coals from the Ruhr area. This is attributed to the input of coaly material from near-shore areas during shale deposition. Arrow indicates kerogens from the Paris Basin. (Schoell, 1982.)

shale and all isotopic species. With increasing organic carbon content the kerogen and extracts became depleted in ^{13}C. Within the zone of highest organic carbon content and highest extract content the ^{13}C-concentration was almost 4 per mil lower than in the units below and above. The isotopic variations in the kerogens have been interpreted as the result of different stages of the euxinic system in which the organic matter was deposited. Decreasing ^{13}C concentrations both in inorganic and organic carbon indicate an increasing incorporation of reworked organically derived carbon by biota.

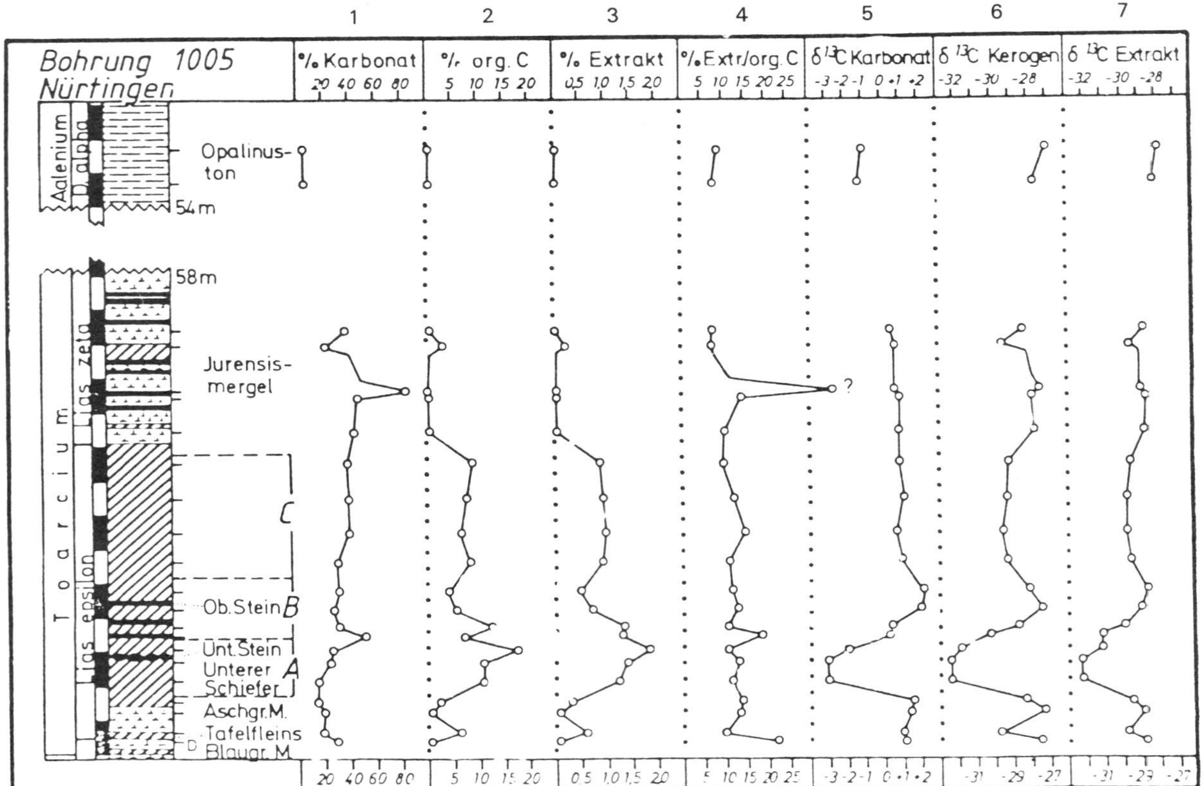

FIG. 3. Chemical and isotopic variations in a lower Jurassic shale section in S. Germany (Küspert, 1982). On the left-hand side is the stratigraphy with local names. Data from left to right: (1) carbonate content; (2) organic carbon content; (3) extractables; (4) extract/TOC ratio; (5) $\delta^{13}C$ of carbonate fraction; (6) $\delta^{13}C$ of organic carbon; (7) $\delta^{13}C$ of extract.

Whatever the ultimate reason for the changes, the fact that there are possibly geochemically heterogeneous source rocks is most important for petroleum geochemistry and requires further study. Often single samples are picked and are inferred to be representative of the whole unit. In the case of the Lias this technique would lead to erroneous results.

IV. Extracts and Oils

Based on findings of Galimov (1973), the isotopic composition of compound classes in petroleums is routinely used for oil correlation purposes, either as "type curves" (Stahl, 1977) or as cross-plots of two compound classes (Fuex, 1977). Isotope concentrations are applied in a pattern recognition mode to define genetically related oils. The prerequisite for such applications is that the properties are primarily source-controlled.

Recent studies both in natural systems and in the laboratory indicate that isotope patterns in compound classes in extracts and petroleums are also dependent on maturation levels.

Figure 4 shows results from the Mahakam Delta, where isotopically relatively uniform kerogens have been investigated at different stages of carbonization. The isotopic differences between kerogens and extracts decrease with increasing maturity. In the immature section, extracts tend to be isotopically depleted in ^{13}C compared with the kerogens, whereas in the mature section extracts become isotopically more similar to the kerogens. This trend has been interpreted to be the result of formation of increasing amounts of indigeneous extracts during rearrangement of kerogen (Schoell *et al.* 1983b).

Results from hydrous pyrolysis experiments allow a detailed insight into the maturation-dependent isotope variations in extractable organic matter (Lewan, 1983). Figure 5 is an example in which the isotopic composition of all compound classes change throughout the course of maturation. The saturates become increasingly enriched in ^{13}C, the aromatics tend to remain relatively unchanged, whereas the polar compounds show a wide spread in ^{13}C concentrations. These variations probably indicate transformations of molecules with internal isotopic heterogeneities. Similar effects have been found in petroleums, where saturated hydrocarbons have been observed to be more susceptible to secondary processes compared with aromatics. Application to oil correlation work is discussed below.

Chung *et al.* (1981) found considerable isotope variations in thermally altered oils from the Big Horn Basin. The ^{13}C concentration in *all* compound classes increased with increasing maturity. Also, an increase in the amounts of saturates and a decrease in aromatic and polar compounds was observed with increasing maturity (Fig. 6). These results clearly demonstrate the

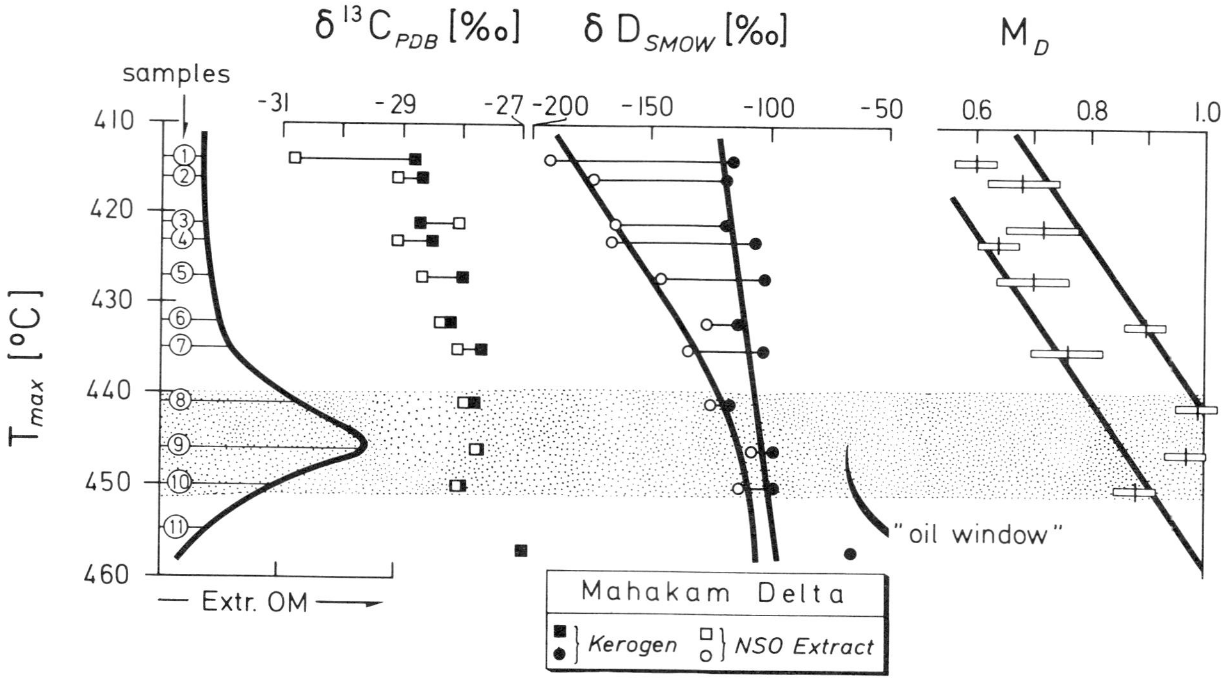

FIG. 4. Change of isotopic composition of extracts and kerogens in a section of increasing carbonization of the organic matter (410 subbituminous, 460 high volatile bituminous). M_D is a maturity index and is the ratio of δD (Kerogen)/δD (NSO extract). (Schoell *et al.*, 1983.)

non-conservative character of ^{13}C concentrations in petroleums with respect to maturation. The data shown in Fig. 6 would normally be interpreted as being indicative of isotopically different sources. The author's unpublished

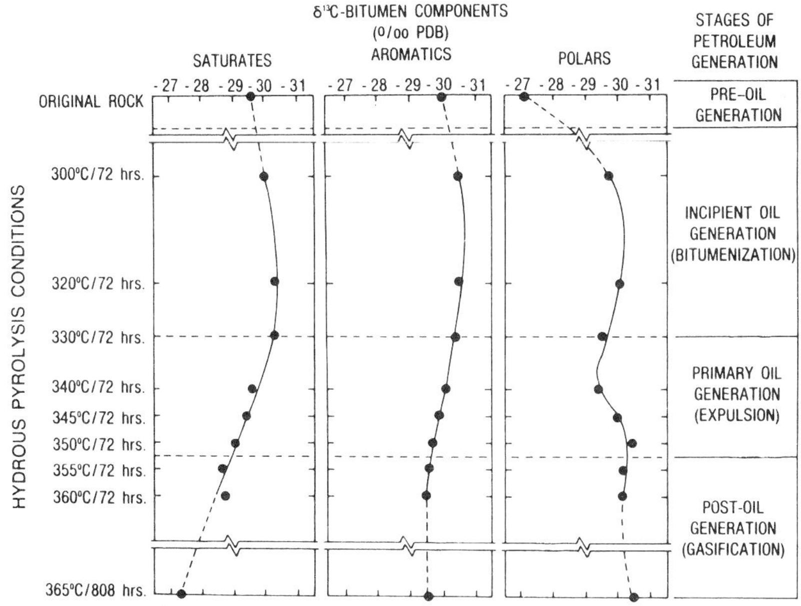

FIG. 5. Isotope variations in extracts during experimentally promoted maturation of a source rock by means of hydrous pyrolysis. The pyrolysis time increases from top to bottom. Note that the saturates change with maturation, whereas the aromatics remain constant. Data from Lewan (1983).

data show that this is not a unique case. Oils from the Arabian Penninsula, for example, revealed an inverse relationship between the content of saturates and aromatics, paralleled by an increase in ^{13}C in all compound classes.

A. Oil–oil correlation

Consequences of the above findings may have some future application in oil correlation work. Internal isotope patterns in extracts and probably in oils may in some areas be used as maturation indices. Schoell *et al.* (1983b) used the ratio of the δD values of kerogen and polar compounds, which approaches unity with increasing maturity. Lewan (1982) proposed the isotopic difference of polar and saturate compounds as a maturation parameter—the Bitumen Isotope Index (BII):

$$BII = \delta^{13}C_{saturates} - \delta^{13}C_{polars}.$$

Although this has not yet been shown to be generally applicable, future research is certainly justified to fully evaluate the information which can be derived from internal isotope patterns of compound classes.

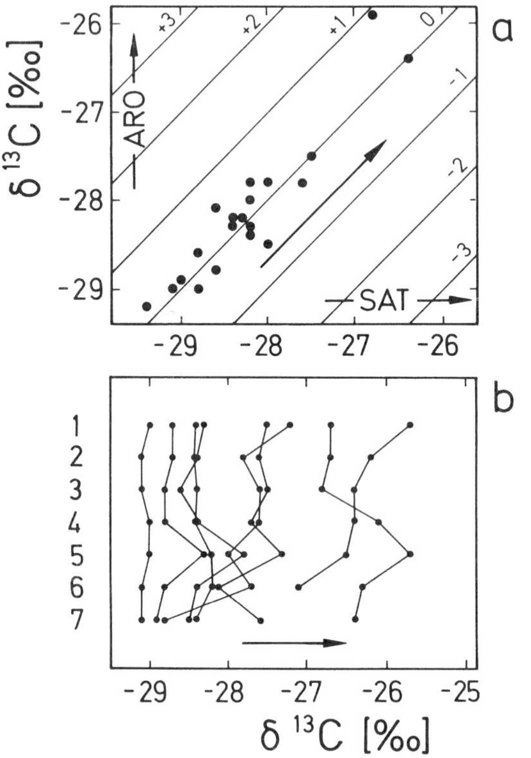

FIG. 6. Isotope patterns of compound classes in petroleums from Big Horn Basin (Chung *et al.*, 1981).

(a) Cross-plot of $\delta^{13}C$ values of saturates and aromatics.

(b) Isotope variations in various compound classes: (1) whole crude; (2) topped crude; (3) saturates; (4) monoaromatics; (5) diaramatics; (6) polyaromatics and polar compounds; (7) asphaltenes. The arrow gives the direction of increasing API gravity (i.e. increasing thermal alteration). Note that these oils are supposedly from one source unit and that the isotope variations are believed to be due to thermal alteration.

Combined carbon and hydrogen isotope analyses have proved useful for oil correlation (Schoell and Redding, 1978; Yeh and Epstein, 1981). Because petroleums can be isotopically changed by thermal processes, it is preferable to use those compound classes which are least affected by such processes. As discussed above, experimental evidence (Lewan, 1982) and natural examples (Schoell, 1982) indicate that the aromatic fraction of oils is the most suitable for oil correlation purposes. A $\delta^{13}C$ versus δD

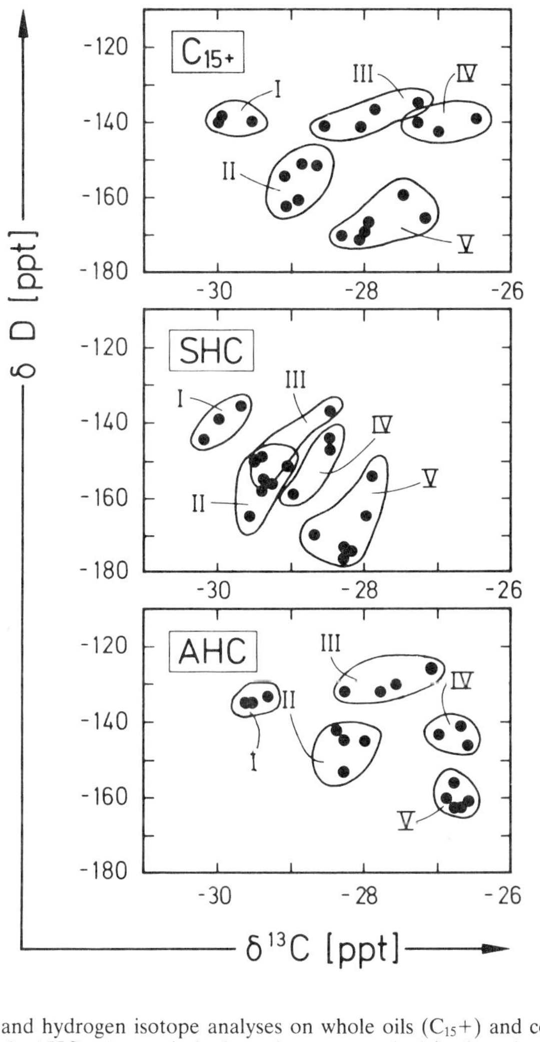

FIG. 7. Carbon and hydrogen isotope analyses on whole oils (C_{15}+) and compound classes (SHC—Saturated, AHC—Aromatic hydrocarbons respectively) of crude oil families from various sedimentary basins. I and II respectively: Triassic and Tertiary oils without secondary alteration, S. German W. Molasse Basin. III: Tertiary oils, S. German Tertiary Basin (E. Molasse) partly biodegraded. IV: Biodegraded oils Beaufort Basin. V: Thermally altered oils Viking formation (W. Canada Basin).

The saturated hydrocarbons of the various groups of oils exhibit a large scatter within the groups so that the groups cannot be clearly differentiated. The aromatics revealed less scatter and a better discrimination. This is attributed mainly to the fact that aromatics are not that much affected by secondary processes like thermal alteration and biodegradation. Note the differences between aromatic and saturates for the thermally altered Viking oils (V).

crossplot from aromatic compound classes is often the most appropriate to identify genetic relationships of oils. Figure 7 gives an example of oils from various oil-producing basins (Schoell, 1982). Increasing depth of burial in one oil family (group V in Fig. 7), for example, results in a higher degree of thermal alteration, specifically in the deepest sample. The saturate fraction seems to reflect these differences by slight enrichment in ^{13}C and deuterium, whereas the aromatic fraction remains unchanged, indicating a very uniform source.

For many oils, secondary processes may often be ruled out from geological and compositional information. Genetic correlation of oils and oil families can then be achieved by carbon and hydrogen analyses on whole oils. Figure 8 is an example using data from Yeh and Epstein (1980). Comparison of the groups plotted in Fig. 8 reveals that, while some are

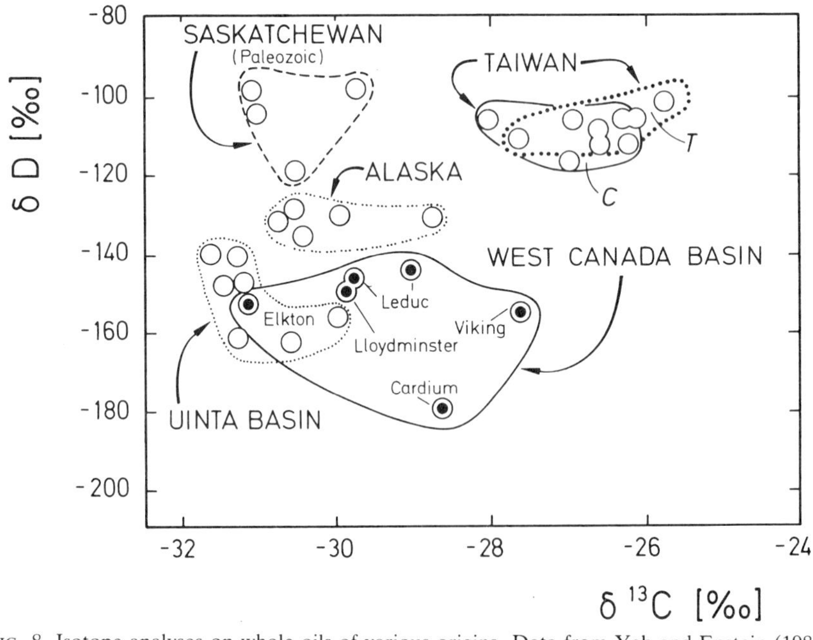

FIG. 8. Isotope analyses on whole oils of various origins. Data from Yeh and Epstein (1981).
C = Changchihkeng
T = Talu

quite uniform in D-concentration but variable in carbon-13 (Alaska, Taiwan), others vary both in carbon-13 and deuterium. The few analyses from the Western Canadian basin provide a good demonstration of the type of information that can be obtained from H and C isotopes in oils when considered together: the Devonian Leduc oils are reservoired in a deep play in the basin and are, in some areas, suspected to have migrated into younger reservoirs, such as the Cretaceous (Deroo *et al.*, 1977). Two samples of Cretaceous oils are shown in Fig. 8, from the Upper Cretaceous Cardium Sandstone and from the Lower Cretaceous Lloydminster. Both Cardium and Lloydminister oils must be suspected to be Leduc-type oils because both are similar to the Leduc oils in C-isotope values. The information provided by the δD analyses results in a clear distinction between Leduc and Cardium oils, but still allows Lloydminster to be grouped with the Leduc oils. Although perhaps these observations do not solve an actual problem in this area, they demonstrate the type of information which can be drawn from these analyses. Another example in Fig. 8 is the distinction between a Palaeozoic oil in the more southern part of the Western Canadian basin (Elkton) and a Saskatchewan Palaeozoic oil. Again, the two oils are not differentiated by their $\delta^{13}C$ values but by their deuterium concentra-

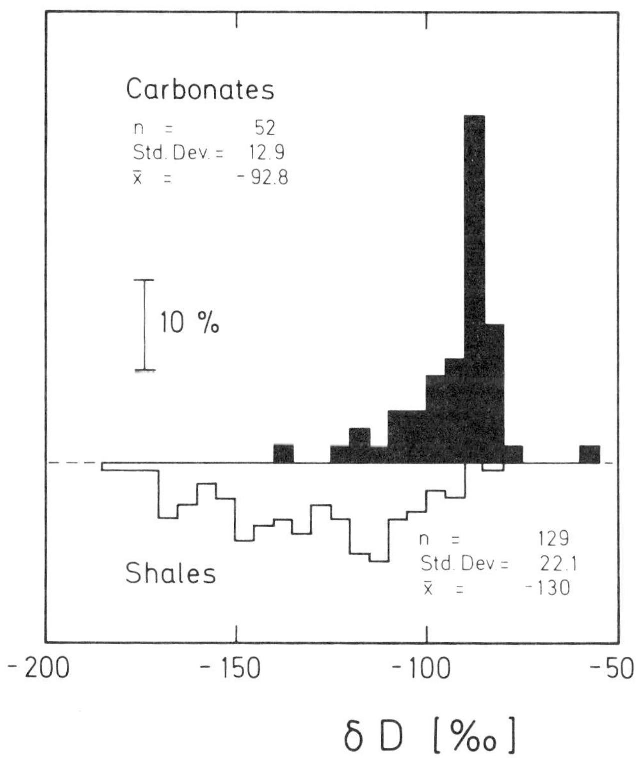

FIG. 9. Hydrogen isotope variations in petroleums from carbonate and shale source rocks (Berner, 1982; Schoell, 1983c).

tions. A last example is provided by two petroleum groups from Taiwan (Changchihkeng and Talu), showing nearly identical scatter in both C and H isotopes, suggesting either a close relationship between the sources of even identical sources.

In using C and H isotopes for correlation purposes, it should be noted that the causes for variation in D- and ^{13}C-isotope concentrations are not yet fully understood. For H-isotope variations in petroleums, Yeh and Epstein (1980) have discussed a latitude effect, which they assumed to be caused through the incorporation of cellulose in the respective source rocks. There are many arguments against such a latitudinal variation of H isotopes in oils. The type of source rock, for example, has been found to be an important consideration. Analyses on approximately 200 oils are displayed in Fig. 9 and indicate that petroleums from carbonate source rocks are rich in deuterium compared with those from shale source rocks. (Schoell and Berner, unpublished results). It should be noted that most of the oils in Fig. 9 are derived from the Arabian peninsula; however, analyses on carbonate sourced oils from the U.S. indicate this to be a general pattern. Certainly more work is needed to further the understanding of causes for isotope variations in oils.

V. Natural Gases

The predominant product of the maturation of organic matter beyond the oil-generation stage is natural gas. The cracking of petroleum and/or the breakdown of kerogen structures are processes in gas formation. Many factors, including the processes and complex interactions of gases during migration, account for the ultimate composition of a gas (e.g. the type of source, maturation, mixing during migration or in the reservoir, trap timing etc.).

A. Genetic characterization of natural gases

The molecular composition of natural gases, as well as the isotope ratios in the hydrocarbons, are controlled by processes during the formation of the gases (Galimov, 1968, 1973; Colombo *et al.*, 1969; Stahl, 1974b). Bernard (1978) presented a useful diagram in which the molecular composition is correlated with the C-isotope concentration of the methane (Fig. 10). This diagram permits genetic characterization of gases of unknown origin. Seep gases from the Gulf of Mexico (see Fig. 19) could be recognized as mixtures of biogenic and thermogenic gases (Brooks *et al.*, 1974). Schoell (1980, 1981) used a genetic diagram in which C and H isotopes of methanes are related to each other. This concept has been extended by a crossplot similar to that of Stahl (1974b), adding C and H isotopes of the methane in a combination of diagrams as shown in Fig. 11 (Schoell, 1983a). These diagrams have been designed as a guide to interpretation, not only of biogenic and thermogenic gases, but also of associated and mixed gases and of even more complex origins.

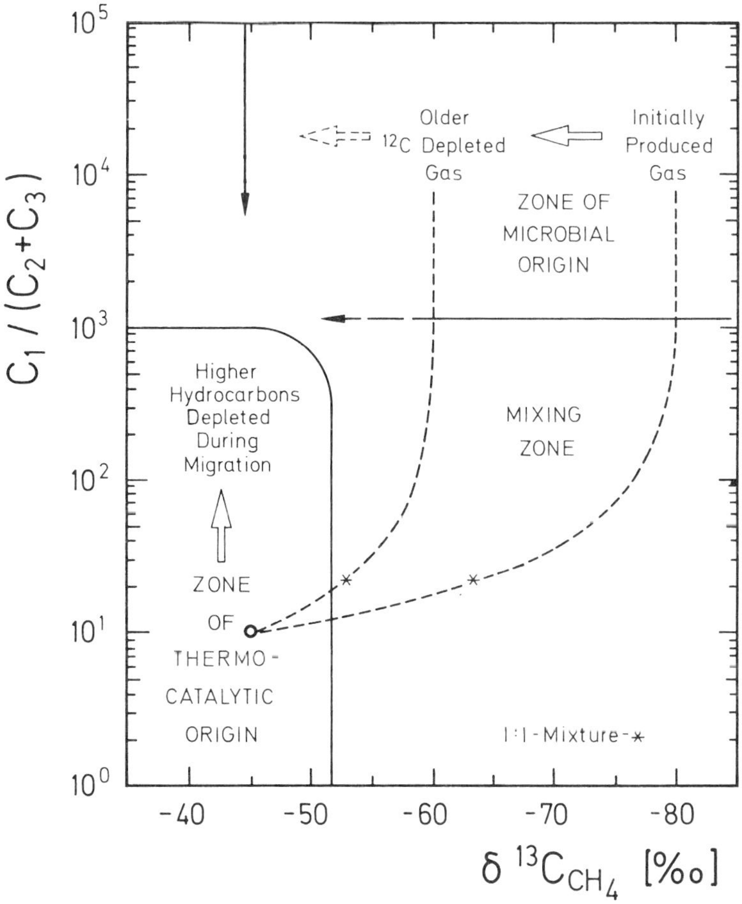

FIG. 10. Relationship of molecular versus carbon isotopic composition of methane in natural gases (Bernard, 1978).

An example of the use of genetic interpretive diagrams is given in Figs. 12 and 13, where data of gases from the North Italian Po Basin have been

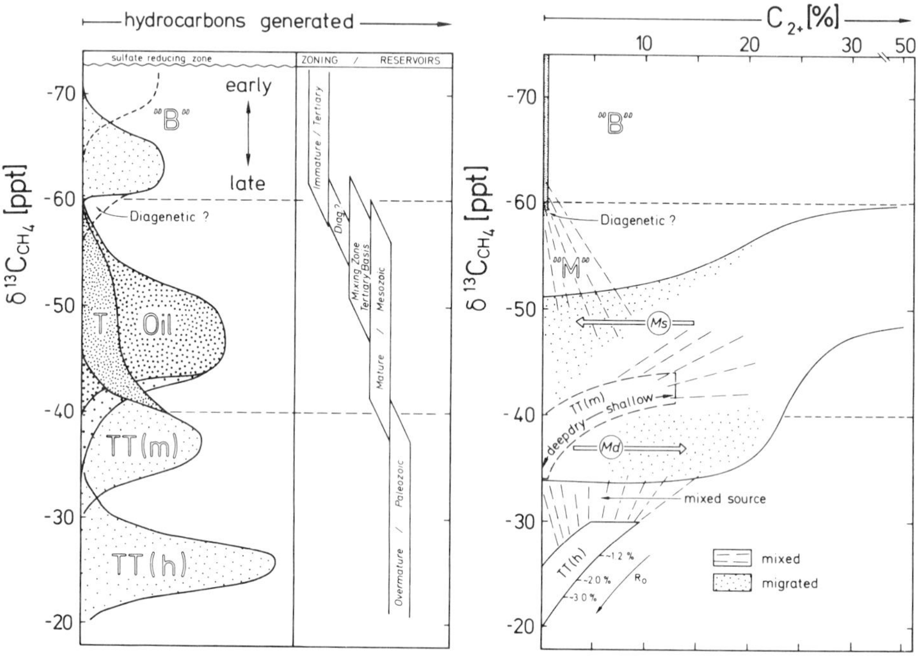

FIG. 11. Combined diagrams for the genetic characterization of natural gases (Schoell, 1983a).

Left: Schematic display of zoning of natural gas and petroleum formation.
Middle: Variations of molecular composition in natural gases related to the isotope variations in methane.
Right: Carbon and hydrogen isotope variations in methanes.

The principle for the genetic characterization of natural gases is that the primary gases (B—biogenic gas, T—associated gas, TT—non-associated gas) are defined by fields of compositional variations. These primary gases may become mixed and form various mixtures "M" of intermediate composition. "TT(m)" and "TT(h)" are non-associated gases from marine source rocks and coal gases from N.W. Germany respectively, compositional shifts due to migration are indicated by arrows *Md* (deep migration) and *Ms* (shallow migration) respectively. "T_O" are gases associated with petroleum in an initial phase of formation. "T_C" are gases associated with condensates.

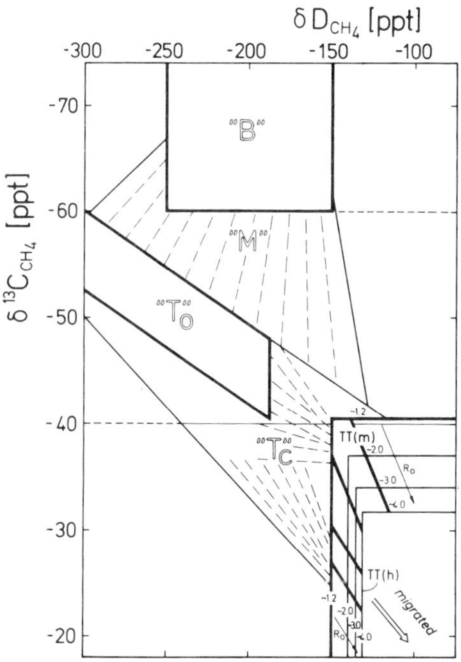

δ D$_{CH_4}$ [ppt]
-300
-250
-200
-150
-100
δ ^{13}C$_{CH_4}$ [ppt]
-70
-60
-50
-40
-30
-20
"B"
"M"
"T$_O$"
"T$_C$"
TT(m)
TT(h)
R$_o$
-1.2
-2.0
-3.0
-4.0
migrated

selected (Mattavelli *et al.*, 1983). In the Po Basin biogenic as well as thermogenic associated gas and mixtures of both have been found. Migrated gases (Medesano) and deep gases associated with condensates (Malossa) indicate a basin of complex gas origins. Comparison of the same data in the Bernard and Schoell diagrams in Figs. 12 and 13 respectively shows that a cross-plot of molecular versus carbon isotopic composition allows

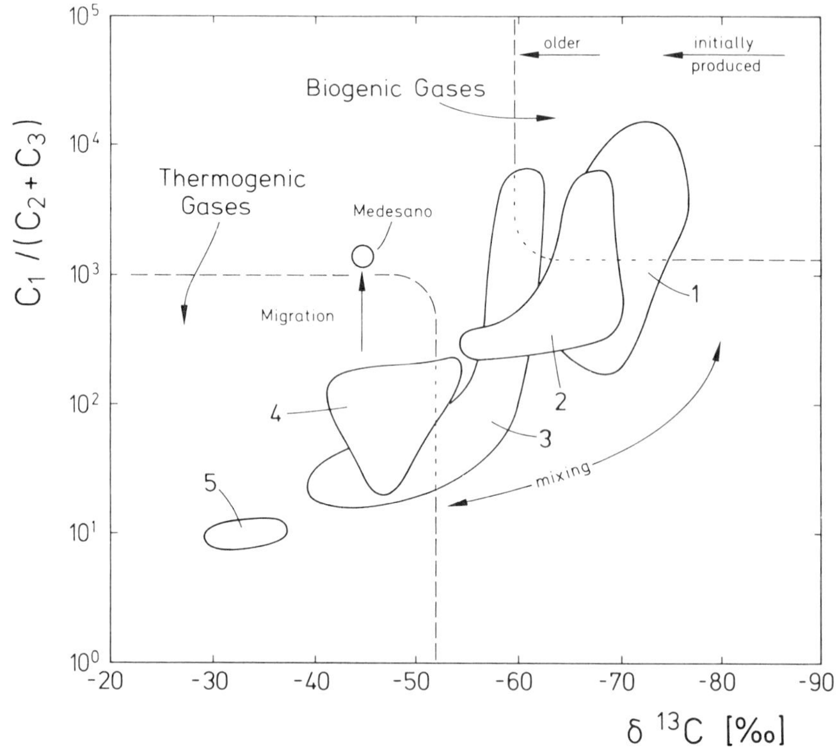

FIG. 12. Molecular and isotope variation of natural gases from the Po Basin displayed in a diagram after Bernard.
Reservoirs: (1) Pleistocene and Upper/Middle Pliocene, (2) Lower Pliocene, (3) Messinian, (4) Middle Miocene, (5) Pretertiary.

differentiation of gases of various origins. Specific problems of mixing and migration are more readily solved using the $\delta^{13}C/\delta D$ diagram. For example, the gas of Medesano (Figs. 12 and 13) can be recognized as a migrated associated gas (Fig. 13, right-hand diagram) whereas the compositional change in the molecular composition of a gas during migration (see arrow labelled "Migration" in Fig. 12) does not indicate a precise genetic source of such a migrated gas.

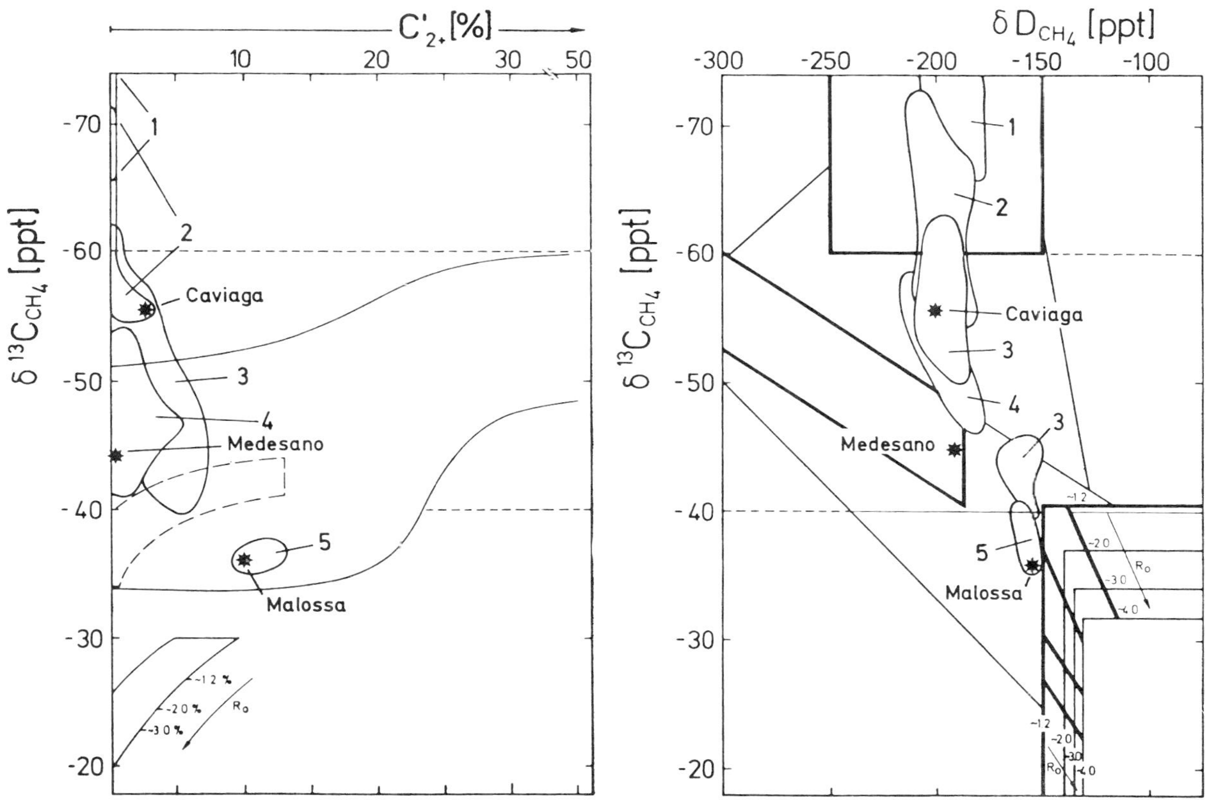

FIG. 13. Genetic characterization of natural gases from the Po Basin using diagrams after Schoell. Reservoirs—see Fig. 12.

B. Maturity-controlled properties in natural gases

Since the early investigations of Galimov (1968), it has been known that the carbon-isotope composition of the methane changes with the stage of carbonization (maturation) of organic matter.

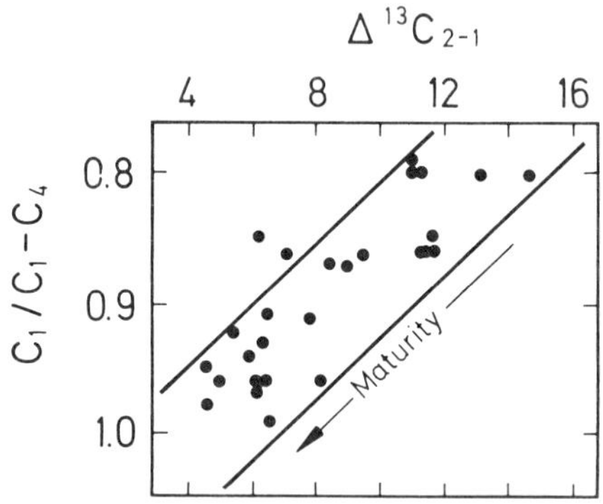

FIG. 14. Difference in isotopic composition of methane and ethane ($\Delta^{13}C$) in relation to wetness for gases from the Cooper Basin, Australia (Rigby and Smith, 1981).

In addition, pyrolysis experiments indicate that the isotopic fractionation between methane and precursor carbon is dependent on maturity (Chung and Sackett, 1979).

A general relationship between maturity and isotopic composition of gaseous constituents, if universally applicable, would be extremely useful in hydrocarbon exploration, as the isotopic composition of a gas could then be used to estimate the maturity of its source.

Stahl (1977) has published a quantitative relationship between $\delta^{13}C$ of the methane and the maturity for gases from north-west Germany. This relationship is different for (i) gases from marine sources, as compared with (ii) gases from coal beds:

$$\text{(i) } \delta^{13}C\ 17 \log R_o - 42 \text{ for } -30 \leqslant \delta^{13}C \leqslant -50$$

$$\text{(ii) } \delta^{13}C\ 14 \log R_o - 28 \text{ for } -20 \leqslant \delta^{13}C \leqslant -30$$

Despite the usefulness of such a quantitative relationship, few case histories are available in the published literature. Parameters other than maturity-controlled isotope effects probably play a role, e.g. in associated gases methane may be as low in ^{13}C as -60 per mil (Fuex, personal communication). The relationship found for German coal gases is presumably not representative of Type III source rocks in general. Gases from the Cooper Basin in Australia have methanes with $\delta^{13}C$ values between −27 and −42 per mil, a range which far exceeds that of the German coal gases (Rigby and Smith, 1981). Gases from the Mahakam delta, where only Type III kerogens have been found, have methane $\delta^{13}C$ values around −45 per mil (Durand and Oudin, 1979). Gases from coal beds in the San Juan Basin have $\delta^{13}C$ values around −41 per mil (Rice, 1983). Moreover, a quantitative relationship can be obscured if different precursors act as source materials, or if gases from different sources become mixed. Despite these limitations the $\delta^{13}C/R_o$ (vitrinite reflectance) relationship has been found to be useful in specific geological situations (Stahl and Carey, 1975), but more case histories are necessary in order to better establish the constraints of the concept, specifically with regard to the influence of the type of organic matter.

Recent developments suggest that an improvement of maturity-dependent isotope parameters may be better achieved by using isotopic differences between molecular compounds rather than $\delta^{13}C$ values of single constituents (Rigby and Smith, 1981; Sundberg and Bennet, 1983; James, 1983). Rigby and Smith (1981) showed for gases from the Cooper Basin, Australia, that there is a clear relationship between $\delta^{13}C$ (i.e. the C-isotopic difference of methane and ethane) and wetness, which suggests that the isotope concentrations in molecular species other than methane are also controlled by maturity.

James (1983) and Sundberg and Bennet (1983) developed this concept systematically for low molecular compounds in natural gases up to pentanes. Here the isotopic differences among C_1, C_2, C_3, C_4 and C_5 are supposed to be independent of source. The procedure depends on thermodynamic calculations of quasi-equilibrium fractionations between methane, ethane, propane and butane which are temperature dependent. James (1983) related these fractionations to a LOM scale using the time–temperature scale of Hood *et al.* (1975). Figure 15 is the result of these calculations, giving the isotopic differences of C_1–C_5 alkanes against the maturity scale. James (1983) presented examples which demonstrate the applicability of this entirely theoretical diagram.

Alternative diagrams can be developed from Fig. 15 by correlating the isotopic differences of the various alkanes. Figure 16 is an example for the isotopic differences of C_1, C_2 and C_3. In this diagram a linear relationship is shown for those gases which exhibit $\delta^{13}C$ isotope differences in their gaseous constituents which are controlled only by the maturity of the source. Some published data have been added to this diagram and show considerable scatter: the three groups of gases shown in Fig. 16 occupy areas of different maturity but do not fall along the equilibrium line. In particular, the data of coal gases are scattered, supporting the finding of James (1983) that this concept does not apply to coal gases. The scattered data as such are not evidence against the applicability of this diagram, but the geochemical significance of scattered data as compared with quasi-equilibrium data has to be investigated. James has pointed out that many

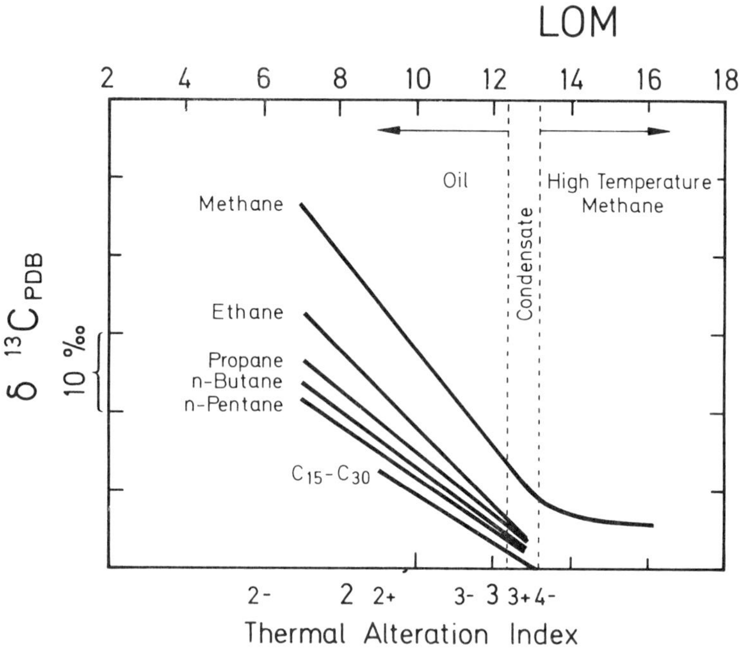

FIG. 15. C-isotopic composition of molecular species in gases related to maturity of its source (James, 1983).

processes may obscure a quasi-equilibrium composition of a gas, e.g. mixing with biogenic gas or formation of gases in coals or from liquid hydrocarbons and cracking of higher hydrocarbons in a gas. The examples quoted by James indicate, however, that in specific areas gases approach equilibrium composition, which allows estimation of source maturity.

C. Effect of migration on isotopic composition

The use of isotope values as maturity parameters implies that the isotopic composition of a gas is not considerably changed during migration and accumulation of gases. Theoretical and experimental results as well as

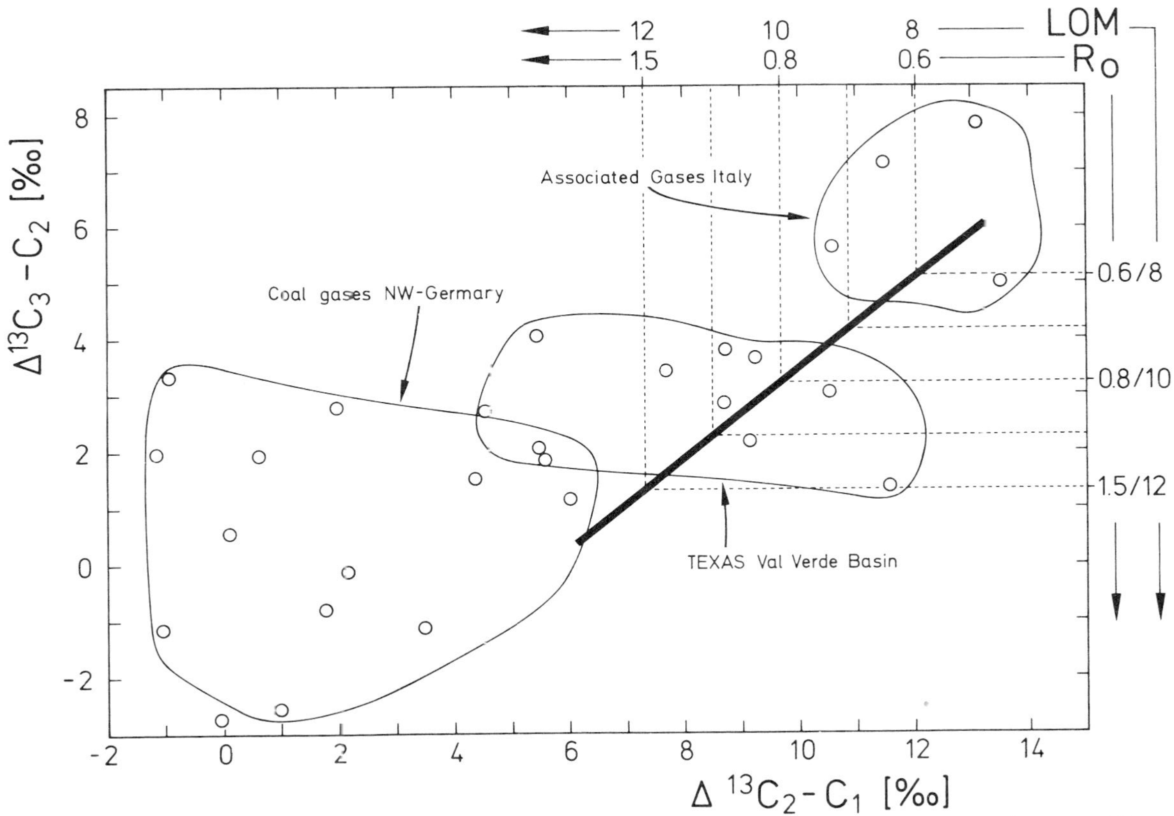

FIG. 16. Cross-plot of differences in isotopic composition of molecular species, as derived from Fig. 15. $\Delta^{13}C_2–C_1 = \delta^{13}C_{ethane} - \delta^{13}C_{methane}$; $\Delta^{13}C_3–C_2 = \delta^{13}C_{propane} - \delta^{13}C_{ethane}$. Data for Texas from Stahl and Carey (1975) and Schoell (1983c).

observation in natural systems provide evidence against a significant change of the isotopic composition of hydrocarbons during migration. The main argument is that in natural systems migration is a quasi-steady-state process in which isotope fractionation does not occur. Further discussions of this topic may be found in Galimov (1973), Fuex (1980), James (1983) and Schoell (1983b).

VI. Isotopes in Exploration

The genetic significance of the isotope variations in gases facilitate their application in exploration projects. The two main applications are isotope analyses on headspace gases from wells and gases in surface exploration.

A. Headspace gas analyses

Traditionally, headspace gas analyses are applied to investigate the compositional changes in gases with increasing depth in a well, which often enables identification of immature and mature sections in a well (Evans *et al.*, 1971).

The analysis of the isotopic composition of methane in headspace gases provides some additional information: the isotopic composition of the gases encountered can be related to the maturity of the sections drilled, and thereby it can be ascertained whether the gas is indigenous or migrated.

The principle of this concept is given in Fig. 17, in which it is assumed that the maturity in a well increases regularly. The isotopic composition of the indigenous methane which is expected in the well can be estimated. If gas migrates from deeper, more mature zones, the isotopic composition should be more positive than the expected value.

An application of this has been presented by Reitsema *et al.* (1981). Figure 18 displays the results: $\delta^{13}C$ values of methane from headspace gas samples became continuously isotopically heavier with increasing depth. This linear trend, which is in accordance with the increasing maturity, and is assumed to be the trend of the indigenous gas, was extrapolated to greater depth. Isotope values in deeper parts of the well were significantly more enriched in ^{13}C, as would be expected from the maturity trend of the sediments encountered; therefore a deeper, more mature source was assumed. Other informative examples have been given by Faber *et al.* (1977) and Reitsema *et al.* (1981), and for further information the reader is referred to these publications.

B. Gas analyses in surface exploration

The application of isotope analyses for surface geochemical exploration has been the subject of many studies in recent years. Brooks *et al.* (1974) and Bernard (1978) used isotope analyses in seepage gases to differentiate their biogenic versus thermogenic origin. Similar isotope analyses were applied by Kvenvolden *et al.* (1980, 1981a, b). In these cases gases were seeping in relatively large quantities and the interpretation of the data is the same as for data from natural gases.

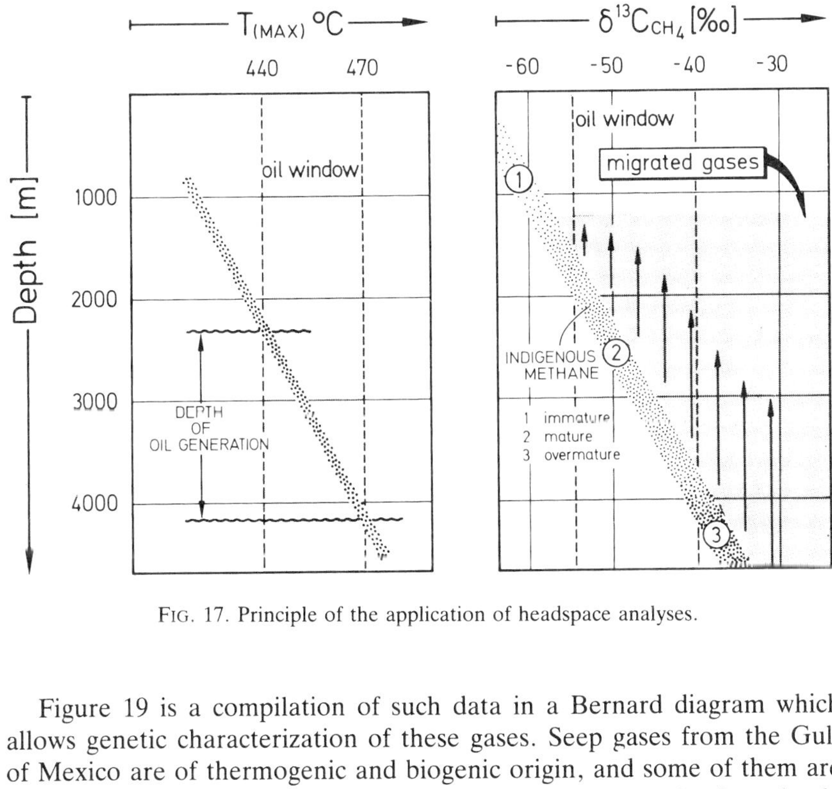

FIG. 17. Principle of the application of headspace analyses.

Figure 19 is a compilation of such data in a Bernard diagram which allows genetic characterization of these gases. Seep gases from the Gulf of Mexico are of thermogenic and biogenic origin, and some of them are mixtures from both these sources. Vents in the Gulf of Mexico funnel only thermogenic gases to the surface. In the Norton Sound and Western Gulf of Alaska most of the seep gases are of biogenic origin; only one has been found to have a thermogenic signature.

The difficulties of surface exploration arise in drawing conclusions from such results. Thermogenic gas seeps almost certainly indicate processes of thermal gas formation, or may even result from a leaking reservoir. The finding of bacterial gases, on the other hand, does not allow conclusions about the absence of thermogenic gas in the subsurface. Claypool and Kvenvolden (1983) have compiled isotope data of interstitial gases from 23 DSDP wells from all major shelf areas of the world, showing, without exception, bacterial gases to be present at shallow depths. Overwhelming evidence has accumulated that the top 500 m of shelf sediments in the

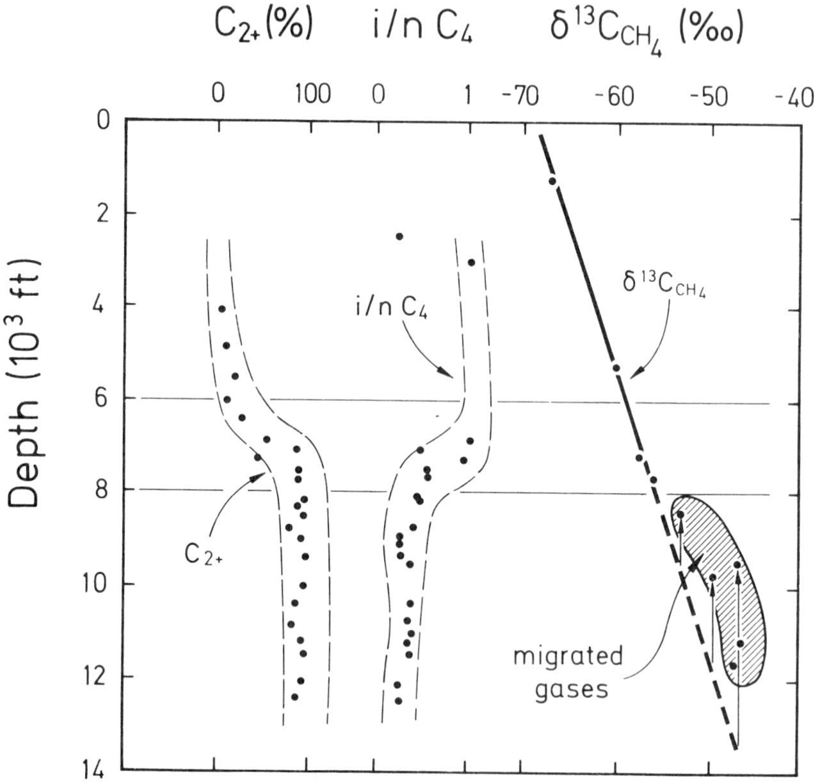

FIG. 18. Isotope analyses on headspace gases, Antelope Creek well. After Reitsema *et al.* (1981).

oceans almost invariably contain bacterial gases, in which methane is the main constituent and C_2+ hydrocarbons are present in traces with C_2/C_1 ratios of 10^{-5} to 10^{-2}. Thermogenic gases, if they are to appear at the surface, have to pass through these bacterial gases in shallow sediments. This is conceivable if channels and conduits are assumed to exist, which is probably the case for vent and seep gases.

A further recent development in surface geochemical prospecting has been the application of isotope analyses on traces of gases adsorbed in sediments (Stahl *et al.*, 1981). Adsorbed gases (normally between 20 and 500 p.p.b. by weight, i.e. between 4 and 140 μl CH_4 from 200 g of sediment!) are released from the sediments by an acid treatment and analysed for their isotopic composition (Faber and Stahl, in preparation). Stahl *et al.* (1981) have shown examples where similar isotope values have been found at the surface as in subsurface gas accumulations, which was inferred to be evidence for migration of gas from the subsurface.

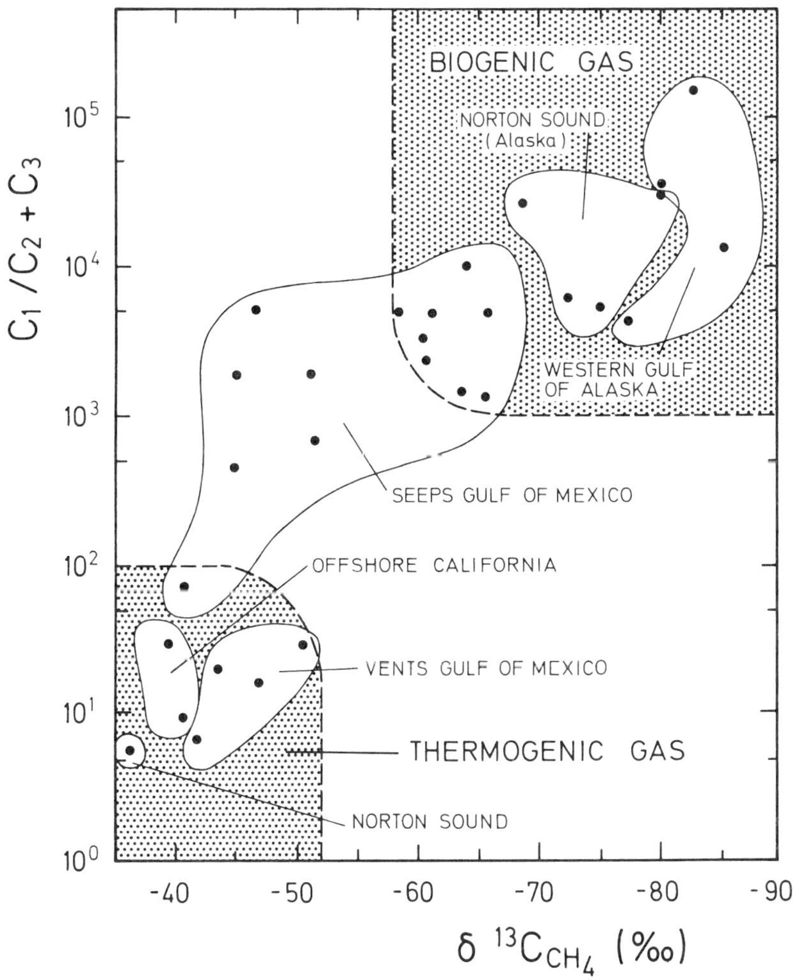

FIG. 19. Genetic characterization of gases from seeps and vents in various offshore shelf areas. After Claypool and Kvenvolden (1983).

With regard to the application of this method in marine sediments, there are at present controversial discussions as to what extent traces of hydrocarbons are derived from the deep subsurface, or whether they are produced more or less *in situ* (Hunt, 1981; Fuex, 1981; Stahl and Faber, 1981; Bernard 1981a). Bernard (1981b) and Claypool and Kvenvolden (1983) have summarized observations, which call for great caution in interpretation:

(i) Microbial processes, especially methane oxidation, can change the isotopic composition. Oxidation of bacterial gas may result in isotope values which are typical for thermogenic gases (Coleman *et al.*, 1981).

(ii) There is direct evidence for formation of C_2+hydrocarbons in microbial processes (Hunt, 1974; Oremland, 1981).

(iii) Diffusion of hydrocarbons from large accumulations through unconsolidated sediments would not be detectable in near surface sediments (Bernard, 1979).

(iv) Concentration profiles of C_2 to C_4 hydrocarbons with depth in near-surface sediments do not support a derivation of interstitial gases from great depth (Bernard, 1978; Kvenvolden and Redden, 1980).

It is, however, beyond the scope of this review to go into the details of these observations. In this field of study further research is definitely required to investigate, among other questions, the origin of hydrocarbons in interstitial gases and in adsorbed gases in near-surface sediments. In the author's opinion it is at present premature to apply this method to exploration for oil and gas.

VII. Summary

This review has summarized some recent important developments in stable carbon and hydrogen isotope studies, as applied to hydrocarbon and fossil fuel research.

Stable isotope concentrations in kerogens and coals are independent of the stage of their thermal evolution and are hence conservative properties which remain unchanged during oil formation. Primary isotope variations in kerogens and coals are independent of the type of kerogen or coal, and little is known on the causes of variations. Occasionally source rock units have been found to be isotopically heterogeneous due to changing depositional environments, a fact which is very important in correlation work.

Oils are related to their source by isotopic similarity. Investigation of thermally altered oils, however, has shown that the isotopic composition of oils may vary by more than 3 per mil due to thermal alteration. Compound classes also change isotopically with thermal alteration.

For secondarily unchanged petroleums, combined carbon and hydrogen isotope analyses allow an easy genetic correlation of oils.

In natural gases, stable isotope concentrations in methane are controlled by the process of gas formation. Using molecular and isotopic compositions allows genetic characterization of natural gases. Application of combined carbon and hydrogen isotope analyses in methane facilitates elucidation of any complex origins of gases. Recent research aims to use isotopic variations in gases as source-independent maturity parameters. The isotopic differences among C_2, C_3 and C_4 hydrocarbons decrease with increasing maturity of organic matter from which the gas is formed.

Isotope analyses on gases from headspace methane in exploration wells allow determination of whether gases have formed *in situ* or migrated from deeper strata. Latest developments make use of stable isotopes in gas traces from sediments for surface exploration, but must be viewed with great caution as an exploration method.

References

Bernard, B. B. (1978). Ph.D. Dissertation, Texas A and M Univ.

Bernard, B. B. (1979). *Deep Sea Res.* **26** A, 49–43.

Bernard, B. B. (1981a). *AAPG Bull.* **65,** 900.

Bernard, B. B. (1981b). "Basic Research in Organic Geochemistry Applied to National Energy Needs", University of South Florida Workshop Report OPI/6-81, B.7–B.33.

Berner, U. (1982). Unpublished thesis, Univ. Clausthal.

Brooks, J. M., Gormly, J. R. and Sackett, W. M. (1974). *Geophys. Res. Lett.* **1** (5), 312–316.

Chung, H. M. and Sackett, W. M. (1979). *Geochim. Cosmochim. Acta* **43,** 1979–1988.

Chung, H. M., Brand, S. W. and Grizzle, P. L. (1981). *Geochim. Cosmochim. Acta* **45,** 1803–1815.

Claypool, A. E. and Kvenvolden, K. A. (1983). *In* "Methane and Other Hydrocarbon Gases in Marine Sediments", Annual Review of Earth and Planetary Science Vol. 11, in press.

Coleman, D. D., Risatti, J. B. and Schoell, M. (1981). *Geochim. Cosmochim. Acta* **45,** 1033–1037.

Colombo, U., Gazzarrini, F., Gonfiantini, R., Tongiorgi, E. and Caflish (1969). *In* "Advances in Organic Geochemistry 1968" (Schenck, P. A. and Havenaar, I., Eds.), 499–516. Pergamon Press, Oxford.

Craig, H. (1957). *Geochim. Cosmochim. Acta* **12,** 133–149.

Deines, P. (1980). *In* "Handbook of Environmental Isotope Geochemistry" (Fritz, P. and Fantes, J. Ch., Eds.), Vol. 1, 329–406. Elsevier, Amsterdam.

Deroo, G., Powell, T. G., Tissot, B., McCrossan, J. G. and Hacquebard, P. A. (1977). *Alberta. Geol. Surv. of Can. Bull.* **262.**

Durand, B. and J. L. Oudin (1979). *Tenth World Petr. Cong., Panel Discussion PD 11–9.* Heyden and Son, London.

Eckelmann, W. R., Broecker, W. S., Whitlock, D. W., Allsupp, J. R. (1962). *AAPG Bull.* **46,** 699–704.

Evans, C. R., Rogers, M. A., Bailey, N. J. L. (1971). *Chem. Geol.* **8,** 147–170.

Faber, E., Stahl, W. and Carey, B, D. (1977). *Erdöl und Kohle–Erdgas–Petrochemie, Compendium 77/78,* 380–391.

Fuex, A. N. (1977). *J. Geochem. Expl.* **7,** 155–188.

Fuex, A. N. (1980). *In* "Advances in Organic Geochemistry 1979" (Douglas, A. G. and Maxwell, J. R., Eds.), 725–732. Pergamon Press, Oxford.

Fuex, A. N. (1981). *AAPG Bull.* **65,** 928.

Galimov, E. (1968). *Izv. Akad. Nauk SSSR, Ser. Geol.* (5).

Galimov, E. (1973). "Carbon Isotopes in Oil and Gas Geology. National Aeronautics and Space Administration, Washington D.C. [Translation (1974) from Russian.]

Galimov, E. (1980). *In* "Kerogen, Insoluble Organic Matter from Sedimentary Rocks" (Durand, B., Ed.), 271–299. Editions Technip, Paris.

Hollerbach, A., Hufnagel, H. and Wehner, H. (1977). *Geol. Bav.* **75,** 139–153.

Hood, A., Gutjahr, C. C. M. and Heacock, R. L. (1975). *AAPG Bull.* **59,** 986–996.

Hunt, J. M. (1974). *In* "The Black Sea—Geology, Chemistry, and Biology" (Degens, E. T. and Ross, D. A., Eds.), Mem. 20, 499–504. AAPG, Tulsa.

Hunt, J. M. (1981). *AAPG Bull.* **65,** 939.

James, A. T. (1983). *AAPG Bull.* **67,** 1176–1191.

Küspert, W. (1982). *In* "Cyclic and Event Stratification" (Einsele, A. and Seilacher, A., Eds.), 482–501. Springer Verlag, Berlin.

Kvenvolden, K. A. and Redden, G. D. (1980). *Geochim. Cosmochim. Acta* **44,** 1145–1150.

Kvenvolden, K. A., Redden, G. D., Thor, D. R., Nelson, C. H. (1981a). *In* "The Eastern Bering Sea Shelf: Oceanography and Resources" (Hood, D. W. and Calder, J. A., Eds.), Vol. 1, 411–24. Univ. of Washington Press, Seattle.

Kvenvolden, K. A., Vogel, T. M., Gardner, J. V. (1981b). *J. Geochem. Expl.* **14,** 209–219.

Lewan, M. D. (1983). *Geochim. Cosmochim. Acta* **47,** 1471–1480.

Mattavelli, L., Ricchiuto, T. Grignani, D. and Schoell, M. (1983). *AAPG Bull.* **67**(12), in press.

Mook, W. G. (1968). Proefschrift, Univ. of Groningen.

Oremland, R. S. (1981). *Applied and Envir. Microbiol.* **42,** 122–129.

Redding, C. E., Schoell, M., Monin, J. C. and Durand, B. (1980). *In* "Advances in Organic Geochemistry 1979" (Douglas, A. G. and Maxwell, J. R., Eds.), 711–723. Pergamon Press, Oxford.

Reitsema, R. H., Kaltenback and Lindberg, F. A. (1981). *AAPG Bull.* **65,** 1536–1542.

Rice, D. D. (1983). *AAPG Bull.* **67,** 1199–1218.

Rigby, D. and Smith, J. W. (1981). *APEA J.* **21,** 222–229.

Rigby, D., Batts, B. D. and Smith, J. W. (1981). *Org. Geochem.* **3,** 29–36.

Schoell, M. (1980). *Geochim. Cosmochim. Acta* **44,** 649–661.

Schoell, M. (1981). *Erdöl und Kohle* **34,** 537–544.

Schoell, M. (1983a). *AAPG Bull.* **67**(12), in press.
Schoell, M. (1983b). *J. Geol. Soc. Lond.* **140,** 415–422.
Schoell, M. (1983c). *Geol. Jahrbuch* **D67,** in press.
Schoell, M. and Redding, C. E. (1978). *Short Papers of the Fourth International Conference, Geocheomology, Cosmochromology, Isotope Geology 1978, Geological Open File Report* 78–101, 384–385.
Schoell, M., Faber, E. and Coleman, M. L. (1983a). *Org. Geochem.* **5,** 3–6.
Schoell, M., Teschner, M., Wehner, H., Durand, B. and Oudin, J. L. (1983b). *In* "Advances in Organic Geochemistry 1981" (Bjorøy, M., *et al.*, Eds.), 156–163. John Wiley, Chichester.
Silverman, S. R. (1964). *In* "Isotopic and Cosmic Chemistry", 92–102. North Holland Publishing Co., Amsterdam.
Silverman, S. R. and Epstein, S. (1958). *AAPG Bull.* **42,** 998–1012.
Stahl, W. J. (1968). Dissertation, Technische Hochschule Clausthal.
Stahl, W. J. (1974a). *In* "Advances in Organic Geochemistry 1973" (Tissot, B. and Bienner, F., Eds.), 453–462. Editions Technip, Paris.
Stahl, W. J. (1974b) *Nature* **251,** 134–135.
Stahl, W. J. (1977). *Chem. Geol.* **20,** 121–149.
Stahl, W. J. and Carey, B. D. (1975). *Chem. Geol.* **16,** 257–267.
Stahl, W. J. and Faber, E. (1981). *AAPG Bull.* **65,** 997.
Stahl, W. J., Faber, E., Carey, B. D. and Kirksey, D. L. (1981). *AAPG Bull.* **65,** 1543–1550.
Sundberg, K. R. and Bennet, C. R. (1983). *In* "Advances in Organic Geochemistry 1981" (Bjorøy, M., *et al.*, Eds.), 769–774. John Wiley, Chichester.
Wickmann, F. E. (1953). *Geochim. Cosmochim. Acta* **3,** 244–252.
Yeh, S. W. and Epstein, S. (1981). *Geochim. Cosmochim. Acta* **45,** 753–762.

Pyrolysis Studies and Petroleum Exploration

Brian Horsfield*

Conoco Inc., Ponca City, Oklahoma, U.S.A.

I. Introduction

Pyrolysis is the process whereby solid, liquid, and gaseous materials are thermally degraded, in the absence of oxygen, into smaller molecular fragments. This technique has been widely used by earth scientists, primarily in the study of complex, naturally-occurring geopolymeric materials by breaking them down into more readily analysable, volatile fragments. Thus, for example, humic acids and kerogen isolated from soils and coals (e.g. Girling, 1963; Kimber and Searle, 1970a, b), and from sediments of

* Currently with ARCO Oil and Gas Company, Dallas, Texas, U.S.A.

ADVANCES IN PETROLEUM GEOCHEMISTRY Vol. 1.
ISBN 0-12-032001-0

Pre-Cambrian through to Recent age (e.g. Scott *et al.*, 1970; Dungworth and Schwartz, 1972; Leventhal *et al.*, 1975; Douglas *et al.*, 1970, 1977) have been analysed by a variety of pyrolysis techniques, as have native bitumens, asphalts and petroleum fractions (e.g. Connan, 1972; Connan and van der Weide, 1974; Le Plat, 1967; Poxon and Wright, 1971; George *et al.*, 1977; Moschopedis *et al.*, 1978).

Results from these and other pyrolysis experiments have been applied to specific problems in petroleum exploration (see Fig. 1). Thus, long-term heating experiments have been used to simulate the processes by which petroleum is generated in source rocks. As part of this work, the roles played by various precursors, reaction mechanisms and mineral catalysts have been examined in detail. Simulation experiments continue to be important in petroleum geochemistry, particularly when used to artificially mature sediments for oil-to-source correlations, but in recent years the primary role of pyrolysis has shifted from simulation to evaluation. High-temperature, evaluative pyrolysis systems have been used extensively to provide quantitative source-rock data rapidly, and continue to play an important role in both the screening of rock samples and in their subsequent, more detailed analysis.

This review illustrates how pyrolysis has been used in petroleum geochemistry, and how it might be used in the future, by reference to specific problem areas such as maturation, migration and source-rock evaluation. The reader is referred to Samer (1972) and Barker (1978) for more comprehensive listings of publications relating to pyrolysis.

II. Techniques

Many different pyrolysis systems have been developed to examine a diverse array of problems in petroleum geochemistry (see Table I and Fig. 2). The essential feature of all pyrolysis systems is a pyrolyser, which is used to heat samples under controlled conditions. While many different types of pyrolyser can be used (Walker, 1972), the emphasis here is placed on the types of separation and/or detection devices that may be attached to the pyrolyser, as these attachments govern how and where the pyrolysis system can be used. Each pyrolysis system given in Table I is briefly described below.

A. Static pyrolysis

Static pyrolysis is intended to simulate progressive burial and maturation. The reactants are heated in a closed container at relatively low temperatures

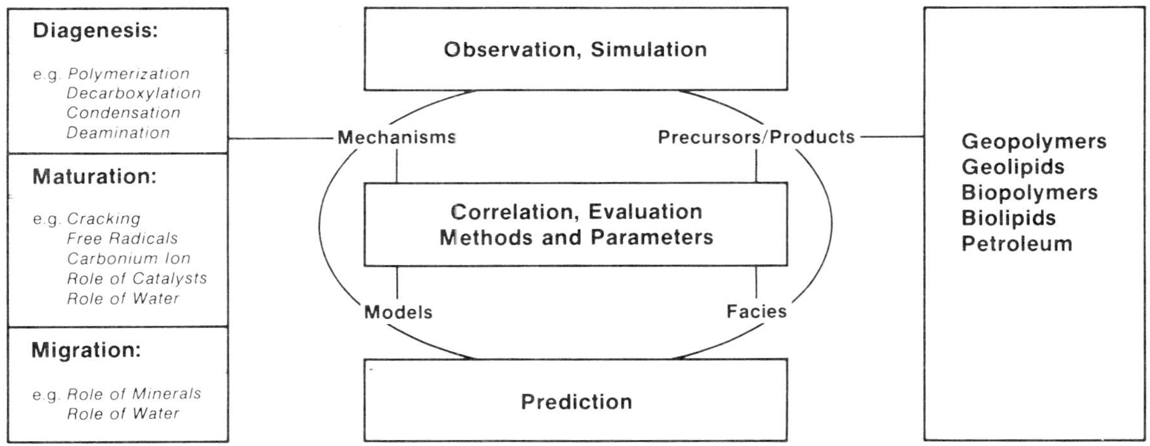

FIG. 1. The scope of pyrolysis in petroleum geochemistry.

TABLE I. *Potential applications of pyrolysis to various aspects of petroleum geochemistry.*

System	Laboratory	Well-Site	Evaluation	Simulation	Analytical	Screening
Autoclave	√			√		
Bulk Flow						
P-FID	√	√	√			√
P-FID/-TCD	√	√	√			√
P-MS	√		√		√	
C_R/C_T	√		√			√
P-FL		√	√			√
Chromatographic						
P-GC	√		√		√	
P-BF/-GC	√	√	√		√	√
P-GC-MS	√		√		√	
P-TLC	√		√		√	

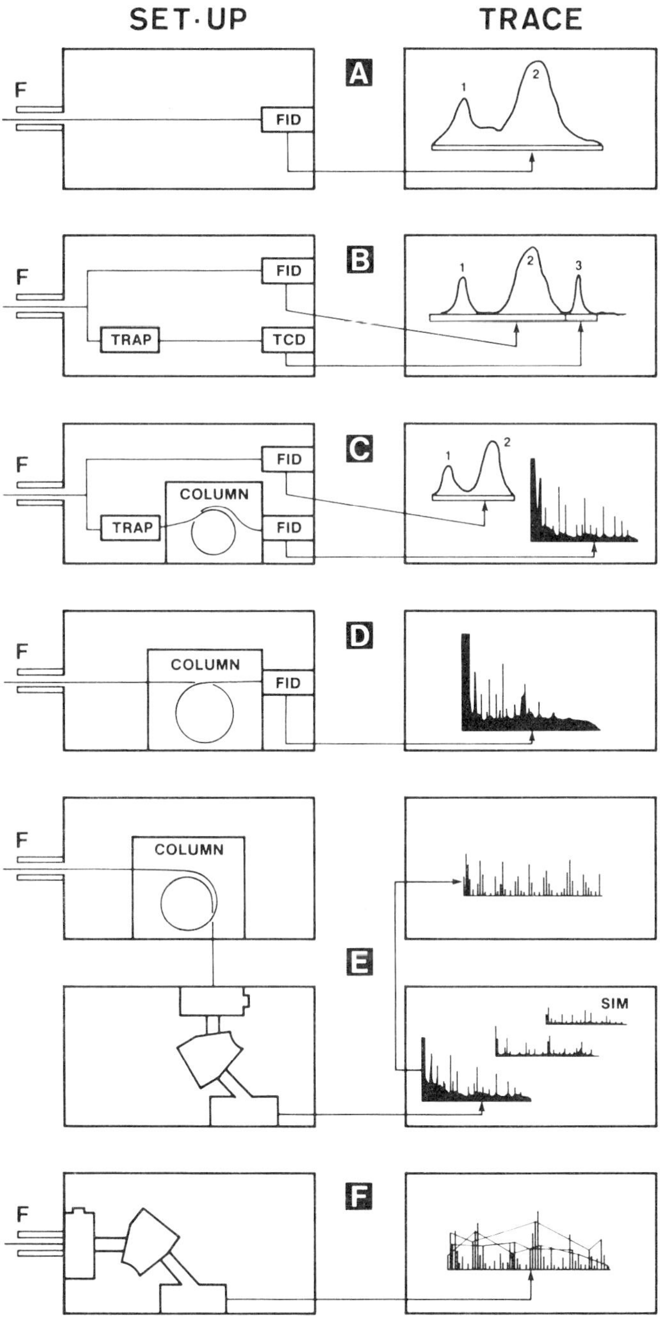

FIG. 2. Schematic analytical configuration and data trace of (A) pyrolysis–FID, (B) pyrolysis–FID/–TCD, (C) pyrolysis–FID/–GC, (D) pyrolysis–GC, (E) pyrolysis–GC–MS, (F) pyrolysis–MS.

for long periods of time, after which the pyrolysis products are extracted, fractionated and analysed (e.g. Murphy, 1969). For example, Recent sediment samples have been heated at temperatures ranging between 35°C and 550°C for periods of 1 to 15,000 hours, and the progressive changes in sedimentary organic matter composition studied by a variety of analytical techniques (Rohrback and Kaplan, 1978; Peters *et al.*, 1981).

A major question concerning the use of static pyrolysis is whether or not it realistically simulates geological processes (Snowdon, 1979). The similarity in carbon-isotopic composition of C_1–C_8 hydrocarbons in pyrolysates and petroleum (Hoering and Abelson, 1963a), the reduced yield of unsaturated compounds at lower pyrolysis temperatures (Giraud, 1970) and the related observation that alkenes are absent in most crude oils (Hunt, 1979a, p. 38) all support the contention that such experiments closely approximate to geological systems. Nevertheless, significant quantities of alkenes have been generated even at relatively low temperatures (ca. 100°C) from sedimentary organic matter (Rohrback and Kaplan, 1978). Moreover, the maturation tracks on van Krevelen diagrams (cf. Dormans *et al.*, 1957) of artificially matured Recent and older kerogens can be significantly different from those of naturally matured samples (Ishiwatari *et al.*, 1977; Peters *et al.*, 1980; Monin *et al.*, 1980). The discrepancy probably exists, in the case of Recent sediments, because it is impossible to simulate early diagenesis, including microbial history, by heat under sterile conditions. In the case of older sediments, both good and poor agreements between the compositions of naturally and artificially matured kerogens have been documented (e.g. Durand *et al.*, 1976; Monin *et al.*, 1980). Where anhydrous pyrolysis conditions are used, humic kerogens appear to show a greater discrepancy than lipid-rich kerogens. This is because the elimination of, in particular, oxygen from organic matter in nature and in the laboratory takes place by different reaction mechanisms (Monin *et al.*, 1980; cf. Yellow, 1965); the higher temperatures used in the laboratory appear to favour loss of oxygen as water instead of carbon dioxide. Excess water in the pyrolysis chamber appears to result in a close compositional similarity between naturally and artificially matured humic kerogens (cf. Lewan *et al.*, 1979; Winters *et al.*, 1983) but this remains to be fully tested.

B. Bulk-flow pyrolysis techniques

In bulk-flow pyrolysis (Larter *et al.*, 1977) samples are heated rapidly in a flow of carrier gas and the yield and composition of the pyrolysate, or, less commonly, the composition of the residue, are monitored using detector responses, mass-spectral data or combustion techniques. No chromatographic separation of the pyrolysate is attempted.

A bulk-flow pyrolysis technique that has been used in petroleum geochemistry for several years involves heating a sample at 900°C for 1.5 h in a stream of nitrogen carrier gas. The ratio of the residual organic carbon (C_R) to the original total organic carbon (C_T), as determined by combustion (LECO), can be used to evaluate source-rock potential (Gransch and Eisma, 1970; Le Tran and van der Weide, 1969). A bulk-flow method that has potential well-site applications is pyrolysis–fluorescence (Heacock and Hood, 1970). This inexpensive method determines "live carbon" or source richness (see Hunt, 1979a, p. 463) by measuring the fluorescence of pyrolysate driven from cuttings when they are heated in a test tube.

Of the bulk-flow techniques commonly used today, pyrolysis–FID is the simplest and least expensive way to evaluate, in a screening mode, source-rock potential (e.g. Barker, 1974a; Claypool and Reed, 1976). Samples are heated from ambient to high temperature (e.g. 700°C) at a pre-set rate in a flow of oxygen-free carrier gas. The products evolved during programmed heating are swept to a flame ionization detector (FID) which responds only to readily combustible products, such as hydrocarbons. The recorded output from heating a whole-rock sample consists of two peaks (Levy *et al.*, 1970). The first peak, comprising products evolved at temperatures up to about 300°C, has an area which corresponds to the yield of free organic matter (bitumen) distilled from the rock. The area of the second, higher-temperature peak corresponds to the yield of hydrocarbon-like products from the thermal decomposition of kerogen.

Rock Eval is a pyrolysis–FID/–TCD system, in which the flow of products from the pyrolyser is split and directed on one line to an FID, and on the other line to a thermal conductivity detector (TCD) via a series of traps (Espitalié *et al.*, 1977a, b). The TCD can detect both combustible and non-combustible materials. The recorded output from a whole-rock sample consists of three peaks. The first two peaks (from the FID) are equivalent to the output from the pyrolysis–FID system described above. The area of the third peak corresponds to the yield of carbon dioxide (as measured by the response of the TCD) that is generated during pyrolysis from oxygen-containing functional groups in the kerogen. Rock Eval is now widely used both in the laboratory and at the well-site (Clementz *et al.*, 1979a, b) to evaluate source-rock potential.

Pyrolysis–mass spectrometry has been used as an analytical tool in many disciplines (Jones and Cramers, 1977) but only in recent years has it been employed to study sedimentary organic matter. In pyrolysis–mass spectrometry, all volatile pyrolysis products are fed from the pyrolyser directly to a mass spectrometer without prior separation by gas chromatography (e.g.

Hummel, 1977). Low voltages (ca. 15 eV) are often used to ionize the products so that maximum structural information is obtained and secondary fragmentations are minimized. The spectra may be statistically processed using a computer (van Graas *et al.*, 1980a, b, 1981) and data interpreted in conjunction with that from other pyrolysis techniques (Maters *et al.*, 1977; Larter *et al.*, 1983). Alternatively, selective ion monitoring of given masses at higher voltages (e.g. 70 eV) allows the evolution of specific compound types from kerogen to be monitored during programmed heating (Gallegos, 1975).

C. Pyrolysis–chromatographic techniques

Pyrolysis–gas chromatography (P–GC), pyrolysis–gas chromatography–mass spectrometry (P–GC–MS) and pyrolysis–bulk flow/–gas chromatography (P–BF/–GC) are the most commonly applied chromatographic pyrolysis techniques, although a pyrolysis–thin-layer chromatography technique also has been reported (Stahl, 1977).

In pyrolysis–GC systems, products from an isothermal, high-temperature (ca. 600°C) pyrolysis are swept directly onto the front of a gas-chromatographic column. In the case of whole-rock samples, this high-temperature pyrolysis is usually preceded by thermal (ca. 300°C) or solvent extraction. The composition of the thermal extract or kerogen pyrolysate is determined from the gas-chromatographic trace.

Pyrolysis–BF/–GC systems (e.g. Scrima *et al.*, 1974; Hunt, 1979b) employ a stream splitter, directing products to a flame ionization detector or thermal conductivity detector on one line, and to a trap (e.g. Dembicki and Woods, 1982) plus gas chromatograph on the other. The yield and composition of thermal distillate and pyrolysate are thus obtained. Pyrolysis–BF/–GC systems are very versatile and can be used in a screening or analytical mode in the laboratory or at the well-site (Horsfield *et al.*, 1983).

Mass spectrometry has been used to characterize compounds that elute from the pyrolysis–gas chromatograph. Thus hydrocarbon and NSO-containing compounds have been identified in pyrolysates from coals and kerogens (Maters *et al.*, 1977; Larter, 1978; van de Meent *et al.*, 1980b; van Graas *et al.*, 1980b). A recent advance has been the use of the mass spectrometer as a specific detector whereby only ions of selected mass, corresponding to known compound classes, are scanned. In this way, single or multiple ion chromatograms are produced, and used to fingerprint kerogen pyrolysates (Gallegos, 1975; Larter *et al.*, 1978; Philp and Saxby, 1980; Philp *et al.*, 1982; Solli *et al.*, 1980a).

D. Reproducibility and comparability

Poor reproducibility and comparability of data have been identified as two major potential problems in short-term/high-temperature analytical pyrolysis (see Jones and Cramers, 1977). Nevertheless, when rock samples are analysed the reproducibility and comparability of data obtained from furnace, Curie point and filament pyrolysers (see Walker, 1972, for review) are good (Horsfield and Larter, 1981). Indeed, Horsfield and Larter found that major differences in analytical data from standard samples were associated more with the efficiency and resolution of hardware used than with the means by which pyrolysis was carried out. Thus, comparisons made in the present review are considered valid when analytical pyrolysis data refers to ancient, lithified sediments, and where gas-chromatographic resolution, for example, is good.

III. Formation of Petroleum Hydrocarbons

The formation of petroleum hydrocarbons has been a major research topic in petroleum geochemistry for many years. Static pyrolysis experiments have contributed to this work by providing important information concerning the conversion of biological organic matter to geological organic matter and the roles played by various hydrocarbon precursors. Thus, heating experiments undertaken by Treibs (1934) and later workers (e.g. Eglinton, 1972; van Dorssalaer *et al.*, 1977; Seifert and Moldowan, 1980; Mackenzie *et al.*, 1981) have shown that chemically and stereochemically specific carbon compounds found in crude oils and rock extracts can be related to biological precursors, demonstrating conclusively the organic origin of petroleum.

Moreover, the conversion of biological to geological organic matter, including petroleum, is gradual, as evidenced by normal alkane distributions in plants, sediments and petroleum (Eglinton and Hamilton, 1963; Bray and Evans, 1961). The most substantial changes in *n*-alkane distribution actually precede intense oil generation (Brooks and Smith, 1967; Durand *et al.*, 1977), and pyrolysis data has shown that these changes are readily explicable by decarboxylation of fatty acids and dehydrogenation and reduction of alcohols (Mitterer and Hoering, 1968; Henderson *et al.*, 1968; Brooks and Smith, 1969; Douglas *et al.*, 1970, 1977; Johns and Shimoyama, 1972; Connan, 1974).

On a quantitative basis, however, the conversion of lipids, such as free fatty acids, to hydrocarbons cannot account for known petroleum reserves. Rather, the sheer mass of kerogen dispersed in fine-grained sedimentary rocks, both in absolute terms (1.3×10^{18} tons, compared with 1.5×10^{15} tons of coal (Hunt, 1972)) and relative to other organic constituents of sedimentary rocks (Tissot and Welte, 1978), logically makes it the major potential petroleum precursor (Hoering and Abelson, 1963a; Abelson, 1964). There is also compelling geochemical evidence indicating that most petroleum is generated from kerogen as it matures during progressive burial under the cumulative effects of time and temperature (e.g. Philippi, 1965; McIver, 1967; Vassoyevich *et al.*, 1969).

A. Maturation of kerogen

Simulated maturation experiments (Louis and Tissot, 1967; Bajor *et al.*, 1969; Connan and van der Weide, 1974; Allan and Douglas, 1974; Rohrback, 1978; Ishiwatari *et al.*, 1976, 1977; Allan, 1975) have played an important role in demonstrating that the generation of hydrocarbons and other compounds from kerogen is a kinetic process, and that only over an optimum temperature–time interval are hydrocarbons generated and preserved in significant quantities. Insufficient heating gives insignificant conversion of kerogen, whereas overheating results in the breakdown of primary products (liquid and solid hydrocarbons) to lighter liquids and gases. Such experimental findings concur with geochemical data from many of the world's sedimentary basins, where with increasing depth of burial, corresponding to increasing time and temperature, immature, mature and over-mature liquid hydrocarbon generation zones have been delineated in the subsurface (e.g. Philippi, 1965; Albrecht and Ourisson, 1969; Le Tran *et al.*, 1974; Tissot *et al.*, 1978). Thus the originally empirical concept of the "oil floor" (Landes, 1967) has been verified using simulation experiments, and is now used to define economic basement in many areas of the world.

At maturation levels that precede the oil phase-out or oil floor, the simulation of geological thermal processes by pyrolysis, in conjunction with field data, has provided a fruitful basis for the development of both predictive (e.g. Tissot, 1969; Tissot and Welte, 1978) and evaluative geochemistry (e.g. Hunt, 1979a). The conversion of kerogen to petroleum-like compounds during maturation results in a disproportionation of hydrogen, as shown schematically in Fig. 3. The kerogen becomes progressively more hydrogen-deficient and aromatic (Dormans *et al.*, 1957; McIver, 1967; Tissot *et al.*, 1974), whereas the petroleum products represent hydrogen-rich moieties. This increase in kerogen aromaticity with increasing maturity, and the associated decrease in the concentration of thermally labile substituents have been investigated by a number of pyrolysis techniques. Gransch and Eisma (1970) and Le Tran and van der Weide (1969) showed that the residual or "dead" carbon (C_R) in kerogens increases relative to total carbon (C_T) as thermally labile, volatilizable carbon species (C_V) are lost as petroleum-like compounds, This increase in fixed carbon during maturation had actually been observed in coals and related to petroleum occurrence many years earlier (White 1915).

Changes in kerogen composition during natural and artificial maturation

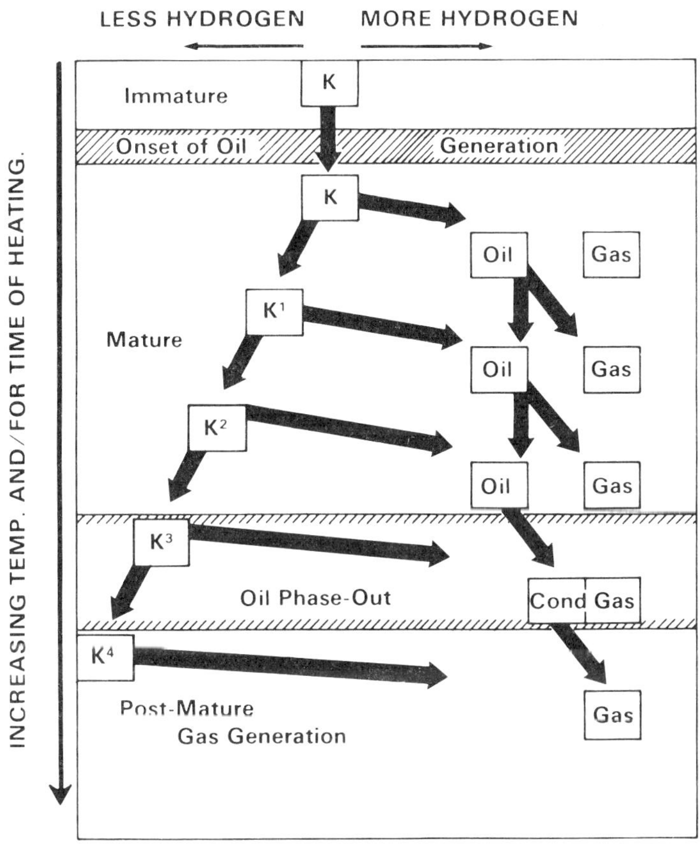

FIG. 3. Disproportionation of hydrogen during maturation.

have been studied by high-temperature pyrolysis. Aromatic compounds generated from kerogens during these experiments are mainly low molecular weight hydrocarbons such as benzene, toluene, xylenes, ethyl benzene and styrene (Holden and Robb, 1958; Girling, 1963). However, oxygenated species (e.g. phenol, cresols, xylenols) may also be present, especially in immature, humic samples (van de Meent *et al.*, 1980b). Romovacek and Kubat (1968) showed that coals of increasing rank yielded pyrolysates that were progressively enriched in total and low molecular weight aromatic compounds. This feature, which relates to the increasing aromaticity of the kerogen, has also been documented by other workers using pyrolysis–gas chromatography and pyrolysis–hydrogenation–gas chromatography (Larter and Douglas, 1980b; McHugh *et al.*, 1976, 1978).

Progressive decreases in the ratios of aromatic or branched compounds versus straight-chain hydrocarbons in pyrolysates may occur during early stages of maturation. For example, a relative increase in *n*-alkenes and *n*-alkanes was observed by van Graas *et al.* (1981) in the $C_{10}+$ pyrolysate of immature through to early mature Toarcian shale kerogen. This is readily explained by the release from the kerogen of thermally labile moieties, such as certain isoprenoid and phenolic compounds, during early maturation. The decrease in relative concentration of phenolic compounds from coal pyrolysates with increasing rank has also been documented by pyrolysis–MS (van Graas *et al.*, 1980b).

The cracking reactions during both simulated and natural maturation result in the generation of two free electrons (or radicals), one on the parent kerogen and one on the volatile product, for each covalent bond that is ruptured. The vast majority of free radicals are generally short-lived, reactive species that quickly alter to more stable forms. Thus the volatile product may abstract a hydrogen radical from kerogen or water, implied or demonstrated in pyrolysis experiments carried out by Rhead *et al.* (1971) and Hoering and Abelson (1963b). The free radical on the kerogen nucleus is stable because it is delocalized on a complex, polyaromatic ring system which acts as an electron sink.

As maturation proceeds, the number of free radicals increases (Ho, 1978; Pusey, 1973). ESR data from the laboratory heating experiments of Ishiwatari *et al.* (1976, 1977), Durand *et al.* (1977) and Peters *et al.* (1981), are entirely consistent with these field observations and, in addition, demonstrate that a reversal in the spin density trend occurs in over-mature sediments. These and other data indicate that structural rearrangements in a maturing kerogen involve the progressive elimination of thermally labile moieties. After these have been lost, pairing of electrons to form carbon–carbon bonds takes place. The reversal in spin density, also observed in coals (Retcofsky *et al.*, 1968), sediments affected by contact metamorphism (Peters *et al.*, 1978) and sedimentary rocks from deep gas wells (Fig. 4), corresponds to the decrease in aromatic C–H vibrations noted by Rouxhet and Robin (1978) in the infra-red spectra of naturally and artificially matured kerogens. This structural rearrangement produces a rapid shift in kerogen T_{max}, as determined by Rock Eval pyrolysis (Peters *et al.*, 1980). Austen *et al.* (1958) described this rearrangement in terms of aromatic clusters that join together, saturating broken edge bonds. The aggregation, rather than growth, of individual aromatic stacks or bricks has actually been observed in naturally and artificially metamorphosed kerogens using electron microscopy (Oberlin *et al.*, 1974a, b; Boulmier *et al.*, 1977). Larter and Douglas (1978) found that the amount of aromatic

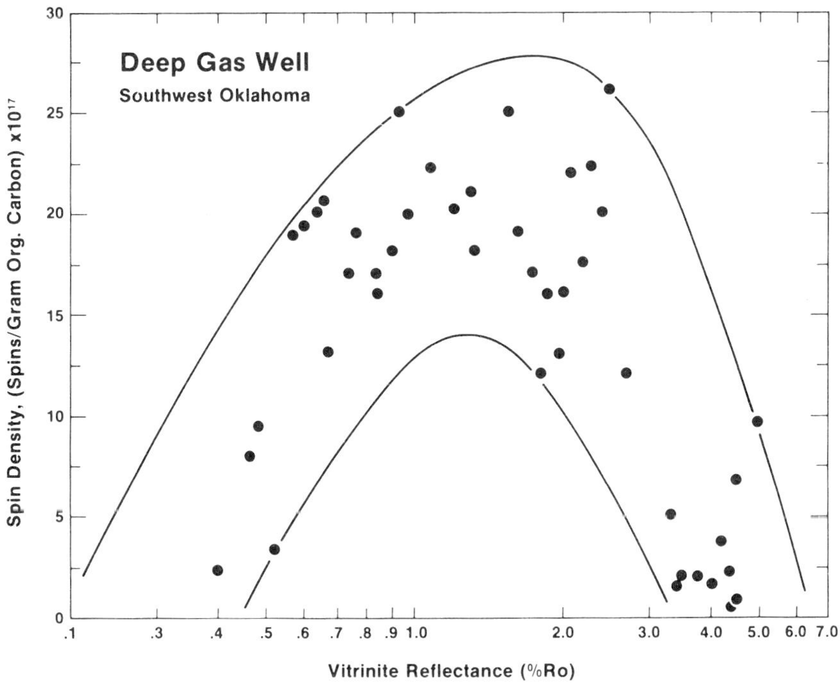

FIG. 4. Relationship between spin density and maturity in a deep gas well from Oklahoma.

hydrocarbons in the low molecular weight fraction of flash (500°C) pyrolysates was anomalously high for vitrinites of medium–low volatile bituminous rank. This was attributed to a structural hiatus (cf. Hirsch, 1954) whereby the development of a mobile, lamellar structure gave rise to thermally labile aromatic moieties.

B. Intermediate compounds

Pyrolysis experiments, particularly those designed to simulate maturation, have provided important information not only on gross transformations of kerogen structure but also on the way that the first-formed products may be converted into petroleum hydrocarbons. For instance, Louis and Tissot (1967) found that the amount of very heavy polar compounds, extracted by a methanol–acetone–benzene (MAB) mixture from naturally and artificially metamorphosed sedimentary organic matter, decreases as the amount of chloroform-solubles, including hydrocarbons, increases. Based on these observations, compounds in the MAB extract have been proposed as intermediates between kerogen and hydrocarbons (Tissot, 1969; Tissot *et al.*, 1971). Pyrolysis experiments on coal macerals, recent sediments and various sedimentary organic fractions, conducted by Allan *et al.* (1980), Peters *et al.* (1981) and Ishiwatari *et al.* (1976, 1977), tend to substantiate this hypothesis regarding hydrocarbon generation. For instance, in the heating of Tanner Bay sediment (150 to 410°C, 5 to 120 h) it was found that optimum yields of soluble organic matter were obtained at kerogen H/C atomic ratios in the range 0.80 to 0.95, whereas maximum yields of *n*-alkanes occurred when the H/C ratio was close to 0.8. NSO and asphaltene fractions of oils and rock extracts can produce hydrocarbons when heated in the laboratory. Douglas *et al.* (1977) found that polar organic eluates of Recent sediment extracts yielded hydrocarbons under varying autoclave conditions. Rubenstein *et al.* (1979) demonstrated that both aliphatic and aromatic hydrocarbons are produced when asphaltenes are heated. Similarly, certain asphaltic oils can be upgraded, by heating, to more paraffinic and naphthenic products (Connan, 1972; Connan *et al.*, 1975). The increase in the C_R/C_T ratio noted for the asphaltenes in this work further supports the contention that "MAB" extractable compounds may be intermediates in the natural formation of petroleum hydrocarbons.

IV. Evaluation of Kerogen Type and Source-Rock Potential

In recent years pyrolysis has emerged as the major means for rapidly acquiring information on kerogen type and other key aspects of source-rock geochemistry, namely source richness and maturity. An advantage of using pyrolysis to evaluate potential petroleum source rocks is that meaningful results can be obtained from wells drilled with oil-based mud systems. Other methods for evaluating source-rock potential (e.g. $C_{15}+$ hydrocarbon yield and composition) may be rendered useless under such circumstances. Moreover, pyrolysis allows the source quality of immature samples to be determined, unlike parameters based on light hydrocarbon and rock extract concentration data, which can be ambiguous. Thus, in regional studies, for example, immature, mature and over-mature control samples can be related using the same frame of reference. Also, only small samples (ca. 1–100 mg) are required, and total analysis time is rapid (less than about 45 min), making pyrolysis a viable analytical technique at the well-site (Clementz *et al.*, 1979a, b).

In the following sections, two working procedures, namely sample screening and detailed analysis, are identified, and the role played by pyrolysis in both these procedures is discussed.

A. Screening analysis

Any screening tool should handle many analyses in a day, and generate reliable data from whole-rock samples. Bulk-flow pyrolysis systems such as pyrolysis–FID, pyrolysis–FID/–TCD and pyrolysis–BF/–GC are mainly used for this purpose. Using pyrolysis–FID, *present source richness* is obtained from the concentration of thermally distilled hydrocarbon-like compounds, S_1, that evolve as Peak 1 (Barker, 1974a; Claypool and Reed, 1976; see Fig. 2). *Genetic richness* (Espitalié *et al.*, 1977a, b), or ultimate generating potential, corresponds to the sum of the concentration of free, hydrocarbon-like materials present in the rock (S_1) plus the concentration of thermally labile organic matter (S_2) which is still part of the kerogen, but which would, during any subsequent thermal maturation, be released as petroleum (Barker, 1974a). The area S_2 of Peak 2 (Fig. 2) is related to organic carbon concentration (Claypool and Reed, 1976; Anders *et al.*, 1978). The presence of oil or gas shows and seeps is detected by high S_1 and/or $S_1/(S_1 + S_2)$ values, as migrated hydrocarbons are detected, primarily, in the lower-temperature peak. The pyrolysis temperature at which the second peak reaches maximum detector response (T_{max}) has been shown (van Krevelen *et al.*, 1951; Barker, 1974b; Espitalié *et al.*, 1977a) to be related to maturity for any given heating rate (Juntgen and van Heek, 1968). Less mature samples tend to have lower T_{max} values than more mature samples.

In contrast with the basic pyrolysis–FID system, the dual-detector configuration of the pyrolysis–FID/–TCD system is designed to obtain information on source quality using whole-rock samples. The concentration of the hydrocarbon-like materials and CO_2 released during kerogen pyrolysis as S_2 and S_3 respectively, when related to total organic carbon concentrations, are termed the "hydrogen index" and "oxygen index" (Espitalié *et al.*, 1977a). These indices are analogous to the atomic H/C and O/C ratios of kerogens, and therefore allow kerogen types and source quality to be assessed. In general, oil-prone kerogens have high hydrogen indices and low oxygen indices, whereas gas-prone kerogens have lower hydrogen indices and higher oxygen indices.

In the case of pyrolysis–BF/–GC, gas-chromatographic data such as the relative concentrations of long-chain versus short-chain hydrocarbons and aliphatic versus aromatic hydrocarbon components may be used to help define kerogen composition. Similarly, the composition of thermally extracted indigenous or migrated bituminous organic matter can be obtained chromatographically (Jonathan *et al.*, 1975; Martin, 1977; Saint-Paul *et al.*, 1980). The pyrolysis–BF/–GC instrument can, in addition, determine S_1, S_2 and T_{max}.

1. *Applications and limitations*

The screening of samples is usually made according to source richness, and can conceivably be carried out either at the well-site or in the laboratory. However, for well-site applications, the detection and characterization of migrated hydrocarbons might be of primary concern (Horsfield *et al.*, 1983). Non-indigenous hydrocarbons are readily recognized in bulk-flow pyrolysis as an unusually large concentration of thermal distillate relative to the concentration of compounds in the kerogen pyrolysate (see Huc and Hunt, 1980). Pyrolysis–bulk-flow data alone cannot, however, unequivocally differentiate refined hydrocarbons (e.g. drilling mud additives) from naturally migrated species (oil shows/seeps). Pyrolysis–BF/–GC systems allow heavy, low-API-gravity crude oils to be readily distinguished from lighter oils (see Fig. 5), and distillation cuts of refined petroleum products to be identified as such by their narrow boiling range (Horsfield *et al.*, 1983; Dembicki *et al.*, 1983). Being able to distinguish bona fide migrated oil from refined products allows oil shows and seeps to be collected for later use in detailed oil correlation studies.

When evaluating source-rock potential, Peaks 1 and 2 are good measures of richness (Claypool and Reed, 1976). However, when samples are stained by oil or contaminated by petroleum distillates (e.g. diesel), not only can thermal distillate data not be used to determine present richness, but pyrolysate (Peak 2) yield may also be suspect, especially where the indigenous organic carbon concentration is low. This is because volatile compounds (including hydrocarbons) can be held over from Peak 1 into Peak 2 because of their interaction with the mineral matrix of the rock (Horsfield *et al.*, 1983; Tarafa *et al.*, 1983). If Peak 1 is very large compared with Peak 2, as in the case of contamination, the hold-over effect can be significant. In the case of oil-stained samples, pyrolysate yield (S_2) is not a reliable measure of genetic richness. This is because petroleum asphaltenes are not volatilized during thermal distillation, and instead reside in the heated zone until their pyrolytic breakdown occurs in the same temperature range over which kerogen degradation takes place (Clementz, 1978).

Preliminary assessments of maturity can be made using T_{max} if samples are uncontaminated (cf. Clementz, 1978), and especially if the data are used to define maturation *trends*. However, substantial variation in T_{max} is common at any given level of maturity (e.g. Claypool *et al.*, 1977) and a single T_{max} value is not definitive of maturity unless kerogen composition is known (cf. Peters *et al.*, 1980; Horsfield *et al.*, 1983; Price *et al.*, 1981).

The determination of kerogen type by pyrolysis screening techniques

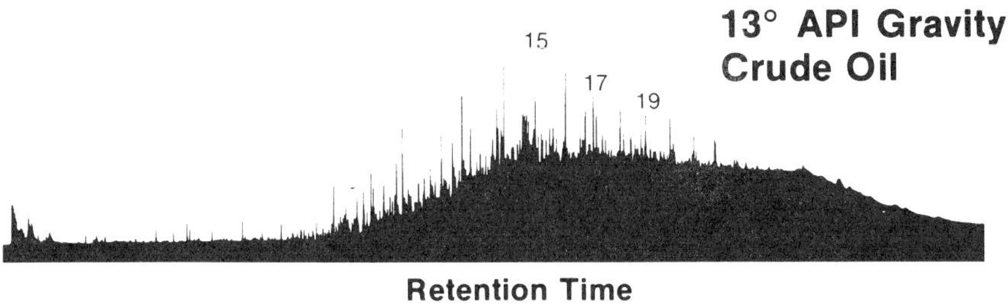

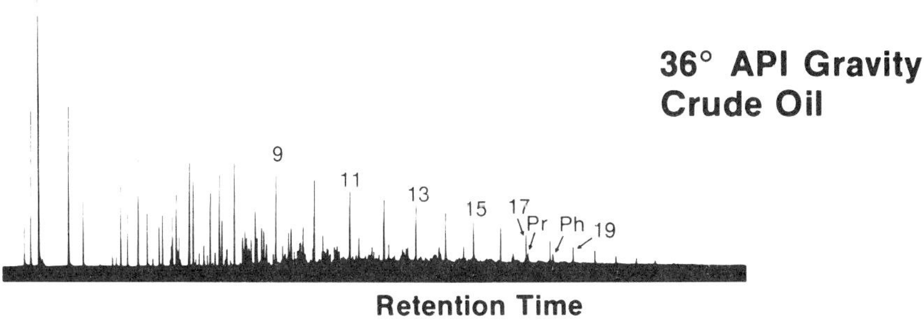

FIG. 5. Gas chromatograms of two crude oils, thermally distilled (300°C) from whole-rock samples.

is not entirely reliable because of mineral matrix effects. In this regard, oxygen index data obtained from Rock Eval are very suspect (Katz, 1981), especially for rocks from carbonate provinces (cf. Palmer and Zumberge, 1981). The rationale of the Rock Eval technique assumes that CO_2 evolved during pyrolysis is derived solely from the kerogen in the rock, because the heating range over which CO_2 is collected lies below the temperature (400°C) at which carbonate minerals begin to breakdown (Espitalié *et al.*, 1977a). It has since been shown (Larter, 1981; Horsfield *et al.*, 1983) that CO_2 is generated from mineral phases at much lower temperatures. A linear relationship between CO_2 yield and organic carbon content was obtained, indicating that CO_2 generation was controlled by the availability of organic carbon. The high CO_2 yields obtained suggest that organic carbon catalyses the thermal decomposition of CO_2 precursors in carbonates and clays.

Mineral matrix effects such as retention and cracking of primary pyrolysate into secondary pyrolysate and a condensed, residual char (Espitalié *et al.*, 1980; van de Meent *et al.*, 1980b; Orr, 1983; cf. Thomas 1970) also affect pyrolysis–BF–GC data. In the case of a carbonate matrix the composition of the resultant pyrolysate may be similar to that from the isolated kerogen, whereas for montmorillonitic sediments it is usually enriched in light (C_1–C_6) and aromatic hydrocarbons (Horsfield and Douglas, 1980; Espitalié *et al.*, 1980; Horsfield *et al.*, 1983, Davis and Stanley, 1982).

Schemes for source quality assessment based on the relative abundance of light versus heavy (Larter *et al.*, 1977; Bailey 1981) or aromatic versus aliphatic (Romovacek and Kubat, 1968) compounds must be thus interpreted in light of possible mineral matrix effects. The latter tend to result in source quality assessments that are more gas-prone (enhanced low molecular weight and aromatic pyrolysate) than those based on isolated kerogen pyrolysates.

2. *Examples*

In the light of the foregoing discussion, it is not surprising that rapid screening has been based mainly on richness, and sometimes maturity, criteria. Bulk-flow pyrolysis has been used extensively in the Deep Sea Drilling Project (Baker and Louda, 1980). Anders *et al.* (1978) correlated S_2 to organic carbon concentration in Black Sea sediments. Claypool and Baysinger (1978) used T_{max} to show that certain sediments in the Blake-Bahama Basin are immature. Similar work in the Moroccan Basin (Claypool and Baysinger, 1980) demonstrated that the kerogen pyrolysate (Peak 2) of an immature chalk sample was bimodal. Peters *et al.* (1980) reported similar results for an immature algal kerogen, and suggested that the first half of the peak might represent thermally labile substituents. Tissot *et al.* (1980) and Summerhayes (1981) summarized Rock Eval data for Cretaceous sediments in the North Atlantic, and showed that wide variations in kerogen type exist in the study area.

Source richness, source quality and maturity of sediments in Continental Offshore Stratigraphic Test (C.O.S.T.) Wells have been determined using pyrolysis data (Claypool *et al.*, 1976, 1977; Huc and Hunt, 1980; Whelan *et al.*, 1980; Huc *et al.*, 1981). Fisher (1980, 1982) reported low hydrogen indices and high T_{max} values for sediments on the Kodiak Shelf and in the Norton Basin, Alaska. These data, which indicated that petroleum generation had already taken place, were used to resolve conflicting vitrinite reflectance and spore colouration data. Magoon and Claypool (1981) presented S_2 and T_{max} data for four wells in Cook Inlet Basin, and used a value of 30% "live" carbon (S_2/organic carbon as percent) to distinguish poor from good potential petroleum source rocks. Bulk-flow pyrolysis data has also been published on sediments from basins in Europe (Palmer and Zumberge, 1981) and the Mid-Continental United States (Cardwell, 1977; Tissot *et al.*, 1978; Merewether and Claypool, 1980; Price *et al.*, 1981).

B. Detailed analysis

Pyrolysis and other degradative techniques, such as oxidation and saponification, have played an important role in the detailed characterization of various kerogens (e.g. Burlingame *et al.*, 1969; Allan *et al.*, 1980) but this information is not routinely applied in petroleum exploration.

Classification schemes currently used for kerogen suffer from being either overly simplistic or too complex for ready use in defining source quality. The frames of reference by which kerogens are described, and to which quite detailed analytical data commonly are related, are based on very simple chemical or microscopic nomenclature (Tissot *et al.*, 1974; Staplin, 1969), despite the known physico-chemical complexity of kerogen (Durand, 1980). An alternative classification scheme based on published pyrolysis and other data is presented below.

1. *Kerogen classification by elemental and microscopy data*

The chemical composition of kerogen is most commonly defined in terms of its elemental H/C and O/C composition (Tissot *et al.*, 1974; cf. Dormans *et al.*, 1957). Thus, kerogen Types I, II and III at early stages of maturation describe organic matter that is progressively more depleted in hydrogen and enriched in oxygen. Types I and II are oil-prone, whereas Type III is gas-prone. Harwood (1977) defined an additional kerogen type (IV) that is even more depleted of hydrogen and richer in carbon than the others. Using simulated maturation, he showed that its most likely maturation product is dry gas. These four kerogen types can be readily distinguished by pyrolysis techniques at maturation levels up to about 1.2% vitrinite reflectivity (Horsfield *et al.*, 1983; Larter and Douglas, 1980). At higher rank, their compositions tend to converge (cf. Allan and Douglas, 1974).

(*a*) *Type I kerogens*

Type I kerogens give the highest S_2 yields and hydrogen indices at any level of maturity. Other bulk-flow pyrolysis techniques have also demonstrated that hydrogen-rich kerogens yield more pyrolysate than do kerogens that are depleted in hydrogen (Gransch and Eisma, 1970; Le Tran and van der Weide, 1969). The coal-petrological term "alginite" is synonymous with Type I, oil-prone kerogen (cf. Dormans *et al.*, 1957; Tissot *et al.*, 1974). Using programmed heating (P–FID), Larter *et al.* (1977), showed that hydrogen-rich exinite macerals, including alginite, yielded significantly more total and C_8+ compounds than did hydrogen-deficient macerals under uniform experimental conditions. It was considered that such differences in pyrolysate yield and composition reflected, though did not accurately represent, the molecular size distributions of volatilizable moieties in the original kerogen. It was also noted that the types and yield of volatile products released from kerogens during natural maturation appeared to reflect the composition and structure of the parent kerogen. Combining these ideas, it was concluded that the relative proportions of oil and gas generated from kerogen during its burial history can be predicted from high-temperature pyrolysis data.

In a multi-faceted analytical atudy, Allan *et al.* (1980) found that alginites gave high yields of solvent-soluble organic matter during static pyrolysis, and concluded that the structure of alginite contained very few thermally stable, condensed aromatic systems. Temperature-programmed and flash pyrolysis (P–FID, P–GC) of alginites (e.g. Larter *et al.*, 1977; Maters *et al.*, 1977; Bailey 1981) showed them to yield long-chain products, principally *n*-alkanes and *n*-alk-1-enes (Fig. 6). Further work using selected ion monitoring (P–MS) has shown that alkyl benzenes and alkyl naphthalenes are also generated (Solli *et al.*, 1980a, b).

Pyrolysis has not only confirmed that many kerogens of microbial origin are oil-prone and classified as Type I, but has also helped to demonstrate that this is not always the case. Using elemental analysis and pyrolysis–hydrogenation–gas chromatography, McKirdy *et al.* (1980 and references therein) showed that microbial kerogens could be classified as Types I, II, III and IV, depending on the composition of the precursor material. Type III kerogens, for example, were thought to be derived from the muco-polysaccharides in algal mucilage. Larter (1978) showed that torbanite and its peat-stage equivalent, coorongite, yielded alkanes and polyolefins when pyrolysed, consistent with their formation from long-chain alkadienes in the green active growth form of the alga *Botryococcus braunii* (Knights *et al.*, 1970; Cane and Albion, 1973). By contrast, so-called "Bb" rubber, an orange, rubbery deposit formed by oxidative polymerization of polyunsaturated hydrocarbons (botryococcenes) synthesized by the alga during the resting stage of its life cycle (Maxwell *et al.*, 1968; Douglas *et al.*, 1969) yielded mainly toluene and xylenes when pyrolysed (Larter and Douglas, 1980b). Torbanite is the ultimate geopolymeric product of the remains of the green alga, preserved under anoxic lacustrine conditions (Hutton *et al.*, 1980). The less prolific orange form might also be incorporated into geopolymers, as evidenced by the occurrence of botryococcenes in certain

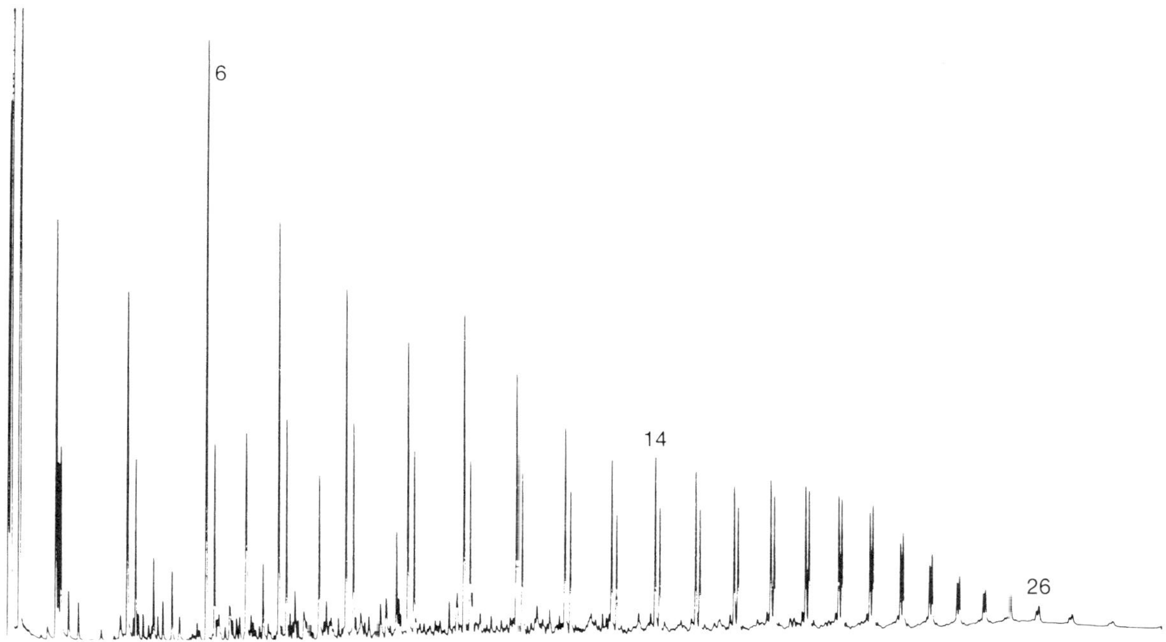

FIG. 6. Pyrolysis–gas chromatogram of the Type I kerogen, torbanite.

Indonesian crude oils (Seifert and Moldowan, 1981). Both green and orange forms of this alga are rich in hydrogen, yet, according to the pyrolytic evidence cited above, only the geopolymeric residues of the active green stage are truly oil-prone. Residues of the orange stage are probably too highly cross-linked to generate *significant* quantities of petroleum liquids, and appear to be gas-prone.

(b) Type II and Type III kerogens

Type II kerogens are compositionally equivalent to exinite coal macerals on the basis of atomic H/C and O/C ratios (cf. Dormans *et al.*, 1957; Tissot *et al.*, 1974). The so-called "marine" organic matter of microbial origin, and exinite coal macerals including sporinite (derived from spores), cutinite (derived from leaf cuticle) and resinite (derived from tree resins) fall into this category. Using the same frame of reference, Type III kerogens are equivalent to vitrinite or woody kerogen. Pyrolysis data indicate that these broad equivalences are indeed meaningful, particularly in relation to *hydrocarbon-generating potential* (cf. Espitalié *et al.*, 1977a). For instance, Given *et al.* (1960), Girling (1963) and Larter *et al.* (1977) observed that exinites gave higher yields of pyrolysate than did vitrinites when heated. Similarly, Bailey (1981) found that Type II sedimentary kerogens yielded more pyrolysate than did Type III, woody kerogens. Broad compositional differences between pyrolysates of Type II and Type III kerogens have also been found. For example, long-term and short-term heating experiments showed that a marine Type II kerogen (Rohrback and Kaplan, 1978) and sporinites from coal swamps (Horsfield, 1978; Larter, 1978; Horsfield and Douglas, 1980) yielded hydrocarbon gases containing relatively less methane than did certain Type III kerogens and coal macerals, under uniform experimental conditions. This finding is corroborated by field data (e.g. Powell, 1978). Bulk-flow pyrolysis products from exinites contain higher molecular weight compounds than do those from vitrinites (Larter *et al.*, 1977), a finding that is consistent with similar data obtained for Type II and Type III kerogens from fine-grained sediments (Horsfield *et al.*, 1983; Dembicki *et al.*, 1983). Amorphous or dispersed exinitic moieties have been detected in vitrinites and related materials such as "jet" (Traverse and Kolvoord, 1968; Hutton and Cook, 1980) using bulk-flow pyrolysis (Larter *et al.*, 1977).

Exinites and vitrinites differ from one another in aromaticity. Holden and Robb (1958) detected aromatic compounds when coals were heated in the ionization chamber of a mass spectrometer. Boyer *et al.* (1961) found that exinite pyrolysates were richer than vitrinite pyrolysates in aliphatic compounds. Bricteux (1966, 1967) also noted differences in the pyrograms of vitrinites and exinites. Romovacek and Kubat (1968) recognized that such differences were due to the more aromatic character of the vitrinite pyrolysates and proposed interpretive schemes for coal maceral pyrolysates.

Many detailed studies of *dispersed* Types II and III kerogens have been undertaken using deltaic and marine sediments (e.g. Vandenbroucke *et al.*, 1976) and results related to source quality. Thus Bordenave *et al.* (1970) and Giraud (1970) found that pyrolysis–gas chromatograms of terrestrial kerogens featured more aromatic compounds than did marine kerogens. It was concluded that terrestrial kerogens are gas-prone and marine kerogens are oil-prone. Using P–GC and P–GC–MS, van de Meent *et al.* (1980b) reported that autochthonous marine organic matter contains abundant long-chain hydrocarbons, whereas kerogens derived from land plants commonly contain phenolic and methoxyphenolic species, indicative of lignin precursors. Larter (1978) observed that the relative abundance of aromatic compounds, such as toluene and xylenes, in pyrolysates of terrigenous kerogens was higher than in those of aquatic kerogens.

2. *Kerogen classification using a combined analytical approach*

The classification of kerogen into four types, based on elemental composition, is a succinct way to show how kerogen composition evolves during maturation. It has thus provided a strong conceptual basis for the routine application of pyrolysis to source-rock evaluation (Espitalié *et al.*, 1977b). This simple scheme can, however, be taken to imply that all types of sedimentary organic matter evolve along one of four defined pathways during maturation, and that any given point on the H/C versus O/C plot describes a unique chemical composition. This is not so. Almost any composition on a van Krevelen diagram can be produced by mechanically mixing together two or more of the chemically distinct (as defined by Tissot *et al.* (1974) and Harwood (1977)) kerogen types. It is also incorrect to assume that all Type III kerogens are gas-prone. Hydrogen deficiency is synonymous only with low yield (see Espitalié *et al.*, 1977a); the parameter does not accurately define source quality.

Microscopic classifications of kerogen or coal components using both transmitted and reflected light are either too simplistic (Staplin, 1969) or too complex (Alpern *et al.*, 1972; International Committee for Coal Petrology, 1971) for accurate definition of source quality on a routine basis. For instance, herbaceous and woody components of kerogen are not necessarily derived solely from single plant components such as leaves and wood, as their names imply (Bujak *et al.*, 1977). In other words, morphologically defined classes of kerogen particles may be chemically quite diverse. At the other extreme, a bewildering array of macerals and submacerals have been identified in coals and kerogens, and described in terms of their habit

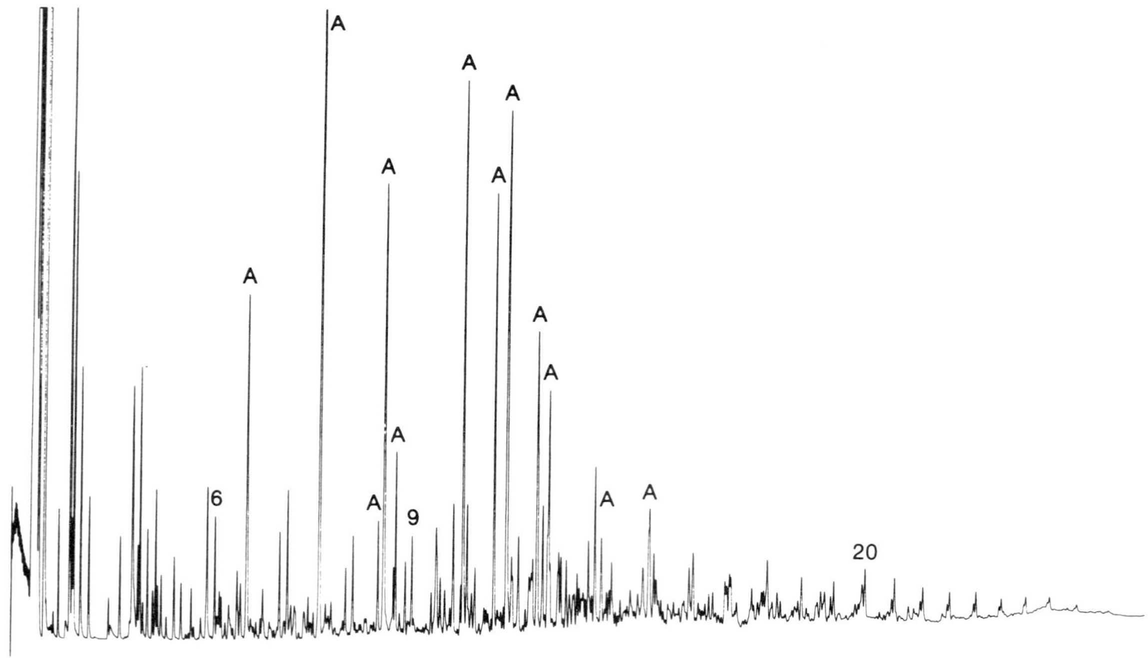

FIG. 7(*a*). Pyrolysis–gas chromatogram of vitrinite from a high-volatile bituminous coal.

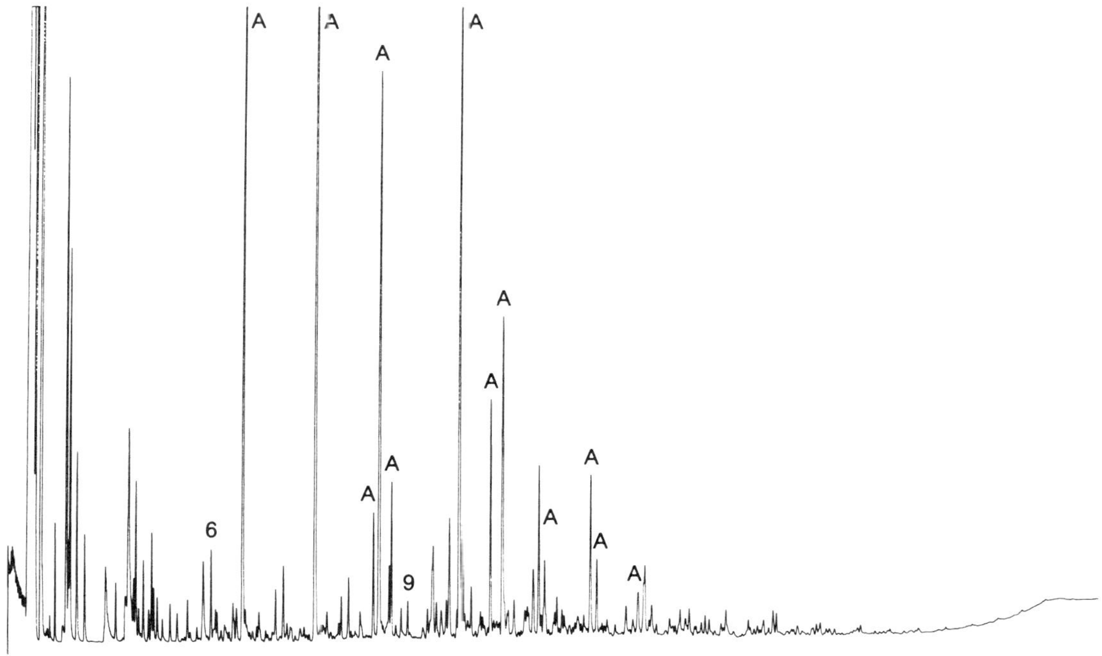

FIG. 7(*b*). Pyrolysis–gas chromatogram of vitrinite from another high-volatile bituminous coal.

and optical properties in transmitted and reflected light (Alpern *et al.*, 1972; International Committee for Coal Petrology, 1971). While sophisticated instrumentation like combined microscopy–laser pyrolysis–gas chromatography (Combaz, 1976) can chemically analyse individual kerogen particles, such schemes are too bulky and difficult to use in petroleum exploration.

A more definitive classification of kerogen types is possible, and, hence, also a more accurate prediction of source quality, if elemental and microscopy data are used in conjunction with pyrolysis data. Pyrolysis experiments, especially P–GC and sometimes also involving P–GC–MS or P–MS (see Maters *et al.*, 1977), have been used to clearly demonstrate that information other than bulk chemical data is required if the products of kerogen maturation are to be accurately predicted. This requirement applies to all kerogens, but particularly to those of terrigenous origin (Snowdon and Powell, 1982) which are deposited in transitional environments, such as deltas, swamps, lakes and lagoons.

Figure 7 shows pyrolysis–gas chromatograms of two vitrinites of high-volatile bituminous rank. The traces are very similar and consist almost entirely of gaseous and aromatic (including phenolic) compounds (see also Larter and Douglas, 1978). Minor differences in pyrolysate composition are evident, however, notably the relative abundance of long-chain, normal hydrocarbons. This observation may be related to the work of Allan (1975) and Allan *et al.* (1975) who, using extraction, saponification, oxidation and static pyrolysis, proposed that sedimentary lipids are mobile in coal swamp environments and may become incorporated in vitrinite kerogens. In support of this hypothesis, Allan and Douglas (1974) cited the fact that the alkanes in autoclave pyrolysates of vitrinites and sporinites had similar distributions of waxy $C_{23}+$ hydrocarbons. In the case of the vitrinite pyrolysates shown in Fig. 7, one appears to have incorporated more long alkyl moieties in its structure by homogenization effects than has the other.

Figure 8 shows pyrolysis–gas chromatograms of three kerogens isolated from fluviodeltaic sediments where this apparent homogenization effect is more pronounced. The kerogen from South Texas is hydrogen-poor (H/C = 0.71), yields predominantly gaseous hydrocarbons, aromatic hydrocarbons and phenolic compounds, and is thus typical of many woody Type III kerogens (van de Meent *et al.*, 1980b). In contrast, a hydrogen-deficient (H/C = 0.84) North African kerogen described as herbaceous contains long alkyl chains in its structure. The Indonesian kerogen is of the same maturity (immature) as the North African kerogen, is also described as herbaceous and has a similiar gas-chromatographic fingerprint, but is much more enriched in hydrogen (H/C = 1.04). Physico-chemical homogenization processes similar to those described for coal swamps (Allan, 1975) appear to

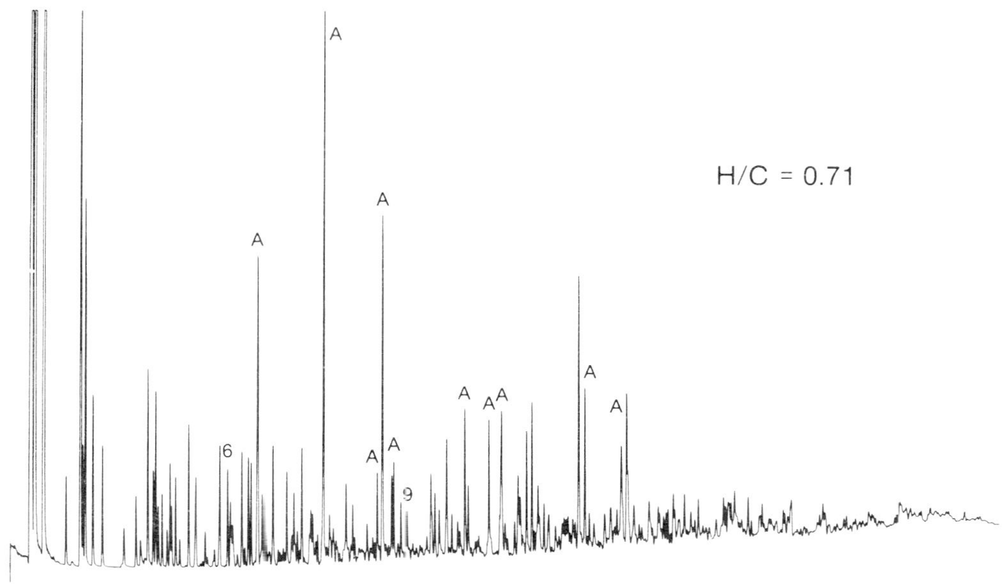

FIG. 8(*a*). Pyrolysis–gas chromatogram of terrigenous kerogen from South Texas.

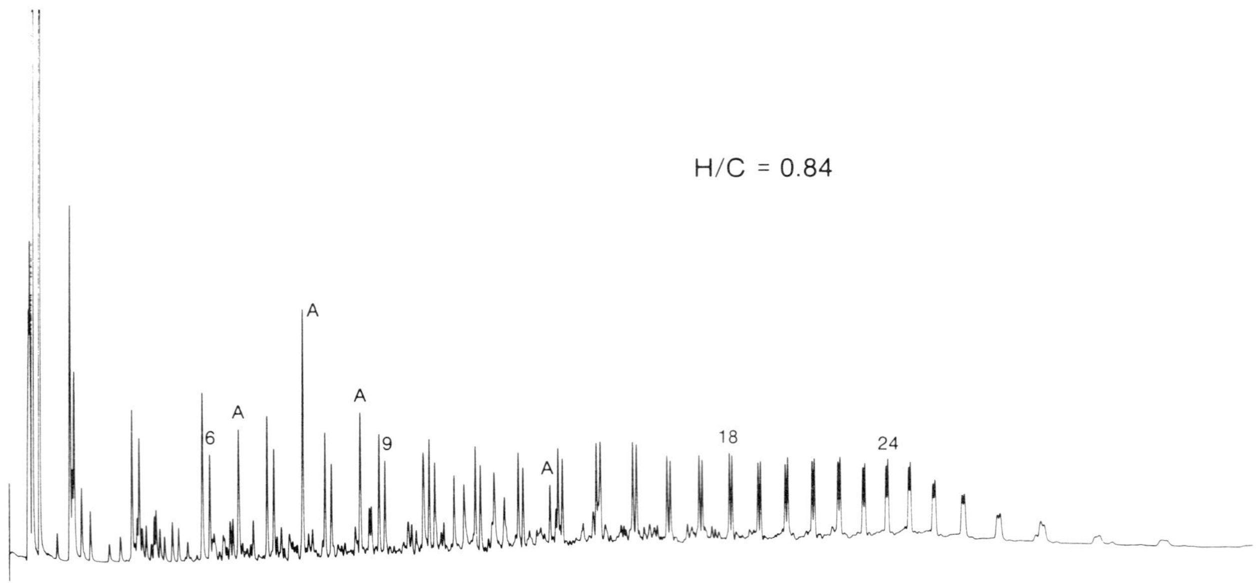

FIG. 8(*b*). Pyrolysis–gas chromatogram of terrigenous kerogen from North Africa.

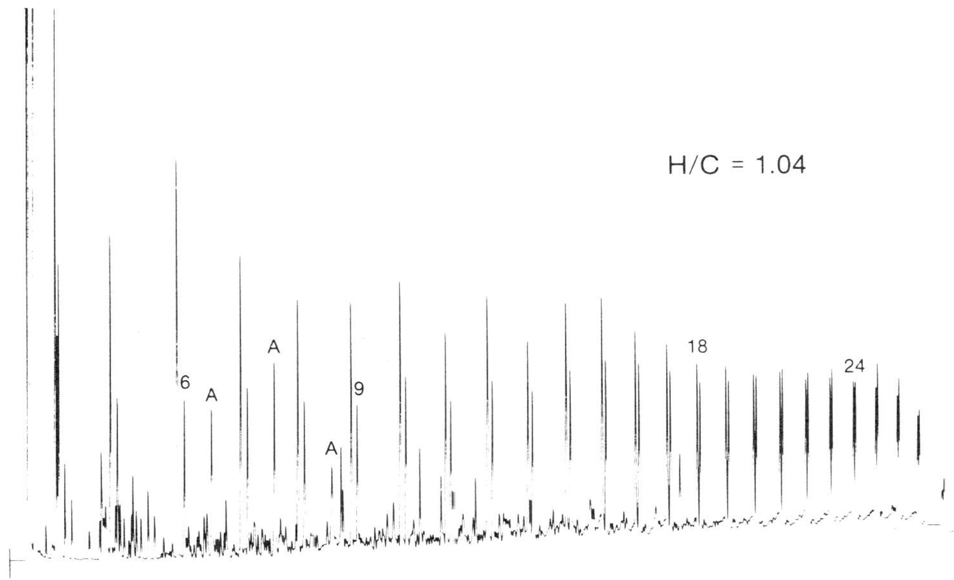

FIG. 8(c). Pyrolysis–gas chrcmatogram of terrigenous kerogen from Indonesia.

relate the compositions of these fluviodeltaic kerogens, whereby depolymerization of lignocellulosic tissues and their repolymerization as geopolymers (Welte, 1974; Stevenson 1974) provides the partly or totally restructured nucleus to which cuticular wax alcohols and esters can become attached. Larter and Douglas (1980a) found that alcohols and acids may be readily incorporated in the structures of synthetic melanoidins and impart an aliphatic character to their pyrolysis–gas chromatograms. The degree to which the periphery of the kerogen nucleus is substituted by alkyl moieties depends, in the present case (Fig. 8), on the relative abundances of woody and cuticular materials contributed to the original sedimentary environment. The kerogen from South Texas has a low H/C value and appears to roughly approximate the core or nucleus of the other two kerogens, to which waxy substituents have become attached. The Indonesian kerogen is inferred to be richest in cuticular lipids and has a high H/C ratio (H/C = 1.04, compared to values of ca. 2 in cutin and cuticular lipids). Such a kerogen is likely to be a potentially good source of waxy oil (cf. Hedberg, 1968). The North African kerogen has an aromatic core to which somewhat lesser amounts of cuticular lipids are grafted. Such a kerogen has some waxy oil potential, but is predominantly gas-prone.

The periodicity in *n*-alkene and *n*-alkane distributions (C_{20}–C_{30}) seen in the herbaceous kerogen pyrolysates (Fig. 8) may indicate that these hydrocarbons are derived from fatty acid and fatty alcohol moieties that were once attached to the kerogen nucleus by ester linkages (cf. van de Meent *et al.*, 1980b). Ester linkages are not now thought to attach the long alkyl chains to the kerogen, as the pyrolysate of the saponified kerogen (not shown) was identical to that shown in Fig. 8. The major mode of alkyl chain bonding appears to have changed from ester to stronger carbon–carbon or carbon–oxygen bonds via a mechanism that is closely analogous to that by which isoprenoid moieties are incorporated into kerogens (Larter *et al.*, 1979, 1983; van de Meent *et al.*, 1980a).

Theoretically, a wide range of H/C values is possible in terrigenous organic matter. The upper H/C limit of immature cuticular (herbaceous) kerogens may be as high as 1.85 (Neavel and Miller, 1960) or as low as 0.8 (Fig. 8), indicating that such organic matter may be classed as Type I, II or III; all varieties are, to varying degrees, waxy oil-prone. A modified version of the kerogen classification scheme of Tissot *et al.* (1974) using elemental data and pyrolysis–gas-chromatographic fingerprints of microscopically recognized precursors is proposed herein to more clearly define kerogen type. The scheme retains the Roman numerals I, II, III and IV, but includes a subscript to indicate the dominant kerogen precursor as defined by microscopy and/or pyrolysis (see Table II). Thus, terrigenous kerogen derived from cuticular materials is designated as I_C (e.g. Neavel

TABLE II. *A modified scheme for classifying kerogens based on elemental, microscopic and pyrolysis data.*

Kerogen Type		Observed in
I	I_A	Tissot et al. (1974)
	I_C	Neavel and Miller (1960)
	I_R	Murchison and Jones (1964)
II	II_A	Tissot et al. (1974)
	II_C	This Article
	II_R	Murchison and Jones (1964)
	II_S	Allan, 1975. Larter and Douglas (1978)
	II_W	Schuhmacher et al (1960)
III	III_A	McKirdy et al. (1980)
	III_C	This Article
	III_W	Tissot et al. (1974)
IV		Harwood (1977)

and Miller, 1960), II_C (e.g. Indonesian kerogen, Fig. 8) or III_C (e.g. North African kerogen, Fig. 8), depending on the H/C value; whereas those derived from woody precursors (e.g. South Texas kerogen, Fig. 8) are designated as III_W. Additional members may be added to the scheme as they are recognized and defined.

Woody precursors can, in theory, generate kerogens of variable composition. Using low-temperature, long-term pyrolysis experiments, Schuhmacher *et al.* (1960) and, later, Davis and Spackman (1964), found that pH and Eh appeared to control the elemental composition of products formed during the simulated diagenesis of wood, lignin and cellulose. These hydrous pyrolysis (cf. Lewan *et al.*, 1979; Winters *et al.*, 1983) or "carbonification" experiments showed that, under acidic conditions, a material compositionally resembling vitrinite was produced from cellulose. Under neutral conditions and then in calcareous water, the products were increasingly enriched in hydrogen, plotting on an H/C versus O/C diagram in the alginite/exinite field. Even lignin formed a product in calcareous water that was more enriched in hydrogen than vitrinite. There is, therefore, the possibility that in a marine environment detrital wood could be partially transformed to a material which resembles exinite in elemental composition and might be classified as Type II_W. However, the susceptibility of cellulose to microbiological degradation probably limits the amount of cellulose than can undergo such changes. Lignin appeared to form vitrinite under marine pH conditions.

The exinite maceral resinite has recently been proposed (Snowdon, 1980; Snowdon and Powell, 1982) as a precursor of so-called immature condensates that occur in the Tertiary sandstone–shale sequences of the MacKenzie Delta (Powell and Snowdon, 1980) and parts of Indonesia (Connan and Cassou, 1980). The atomic H/C and O/C ratios and pyrolytic yield of resinites in lignites are comparable with those of classical Type I or Type II kerogens (cf. Murchison and Jones, 1964; Tissot *et al.*, 1974; Allan, 1975). Clearly, the different maturation thresholds of, and products from, algal Type I (I_A) and resinite Type I (I_R) kerogen require that these two organic inputs to sediments be distinguished. Pyrolysis–gas chromatography readily fingerprints resinite (Snowdon, 1980; Horsfield *et al.*, 1983). For example, PGC traces of two resinites, one isolated from a bituminous coal and the other detected in an Indonesian sediment (Fig. 9) were found to include major "humps" of unresolved components in the sesquiterpenoid (cf. Douglas and Grantham, 1974) and triterpenoid hydrocarbon ranges.

The exinite maceral sporinite (Type II_S, Table II) is composed of the biopolymer sporopollenin which has been proposed to be an oxidative polymer of carotenoids and carotenoid esters (Brooks and Shaw, 1968). Indeed, the presence of ionene and substituted naphthalenes in sporopollenin pyrolysates has been used to support this hypothesis (Achari *et al.*, 1973; Day and Erdman, 1963). In terms of elemental composition sporinite is intermediate between vitrinite and classical Type II kerogen (Allan, 1975; Tissot *et al.*, 1974). Moreover, sporinite shows other similarities to vitrinite (Dryden, 1963), including the presence of methyl naphthalenes, dimethylnaphthalenes and other aromatic species in high-temperature pyrolysates (Larter *et al.*, 1978; Schenck *et al.*, 1981). The relative abundance of these light aromatics is, however, less than in vitrinites of equivalent rank (Larter and Douglas, 1980b), and low molecular weight *n*-alkyl moieties are abundant (see Fig. 10). The presence of predominantly $C_{15}+$ compounds in sporinite pyrolysates (Larter *et al.*, 1977), including petroleum-like hydrocarbons (Combaz, 1971; Allan and Douglas, 1974) is consistent with sporinite being considered as a potential oil-source kerogen, albeit not very prolific (cf. elemental composition).

3. *Amorphous kerogen*

It is now accepted that amorphous kerogen can form from a variety of precursors (Powell *et al.*, 1982) by, *inter alia*, mechanical disintegration of structured detritus and by random polymerization and condensation of biopolymers, their enzymatically degraded products and biolipids (Welte, 1974). Thus, not all amorphous kerogens are oil-prone, and pyrolysis–gas chromatography has been used to readily identify their precursors (Dembicki *et al.*, 1983).

The pyrolysis–gas chromatograms of amorphous kerogens isolated from two marine sediments are shown in Fig. 11. The Palaeozoic kerogen from Oklahoma is thought to be of autochthonous algal origin, and is oil-prone, based on its compound distribution (cf. Bailey, 1981; Horsfield *et al.*, 1983). By contrast, the pyrolysate of the Louisiana kerogen suggests a more gas-prone character (cf. van de Meent *et al.*, 1980b), and derivation from allochthonous humic (higher plant) debris.

V. Onset of Petroleum Generation for Different Kerogen Types

Differences in the threshold of hydrocarbon generation for various kerogen types (see Vassoyevich *et al.*, 1969; Snowdon and Powell, 1982; Monnier *et al.*, 1983) are a direct reflection of the diversity of kerogen type and structure discussed in the previous section. A knowledge of the type and amount of thermally labile species in kerogens is an important consideration when studying the relative onsets of maturity. In this regard, autoclave and other pyrolysis techniques have been used to show that the following types of moiety may be present in kerogens: fatty acid (e.g. Harrison, 1976; Connan, 1974), alcohol (e.g. Brooks and Smith, 1969; Douglas *et al.*, 1970, 1977; Ikan *et al.*, 1975), pigment (e.g. Oehler *et al.*, 1974), isoprenoid (e.g. van de Meent *et al.*, 1980a; Allan and Douglas, 1974; Scrima *et al.*, 1974; Ishiwatari *et al.*, 1977) and triterpenoid (van Dorsselaer *et al.* 1977; Gallegos, 1975), in addition to normal paraffin chains and aromatic nuclei. The facility with which some of these moieties are released from kerogen during maturation has been examined by programmed and stepwise heating experiments.

Tissot *et al.* (1974) and Vandenbroucke *et al.* (1976) heated kerogens from ambient to 500°C at 4°C per minute. Their resultant thermogravimetric data showed that major degradation for both Type II and Type III kerogens took place over the same temperature range (350° to 470°C), though product compositions were different. Thermogravimetric analysis was also used by Snowdon (1980) to study weight loss of an algal coal, a sapropel, a vitrain and a resinite concentrate during programmed heating (20°C per minute). Degradation took place over a similar temperature range for all kerogens except the resinite, which completely volatilized below 450°C

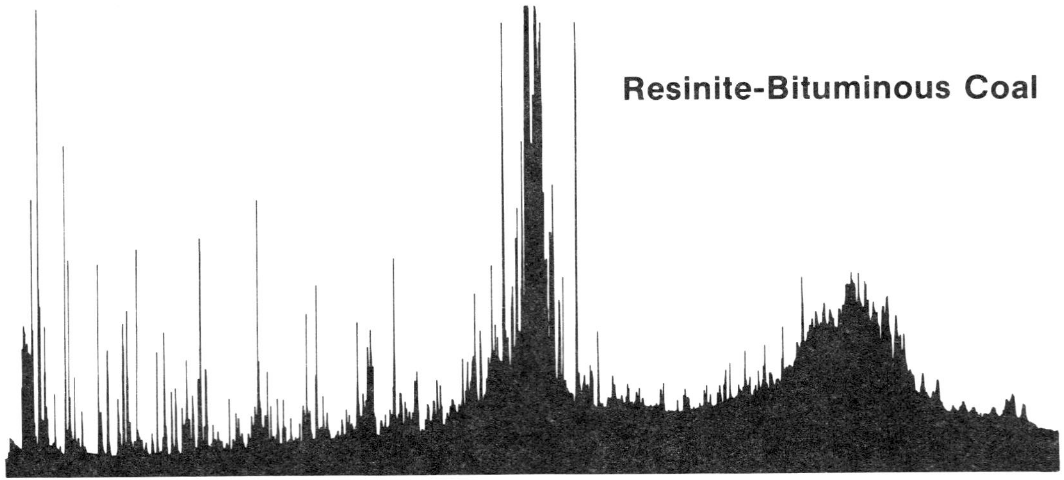

FIG. 9(*a*). Pyrolysis–gas chromatographic fingerprint of resinite in a bituminous coal.

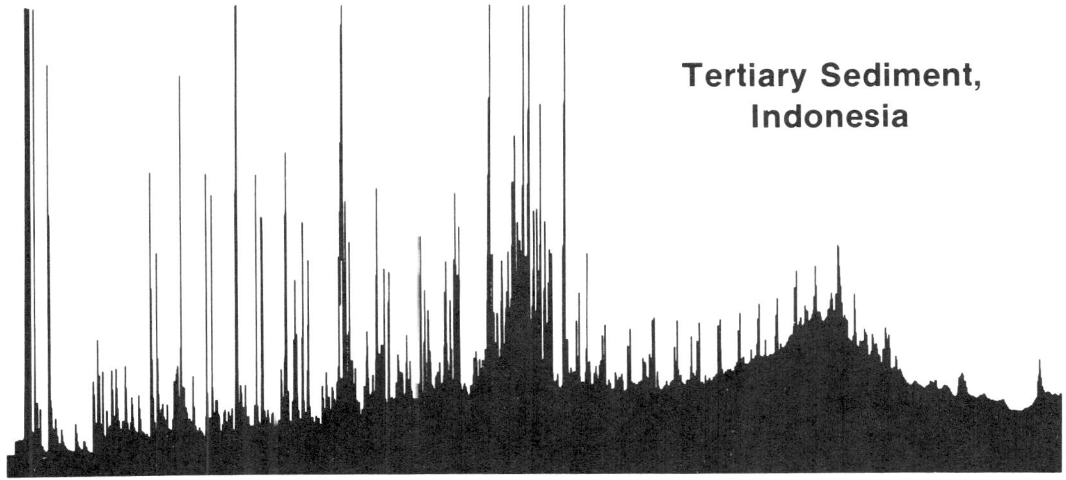

FIG. 9(*b*). Pyrolysis–gas chromatographic fingerprint of resinite in a Tertiary sediment.

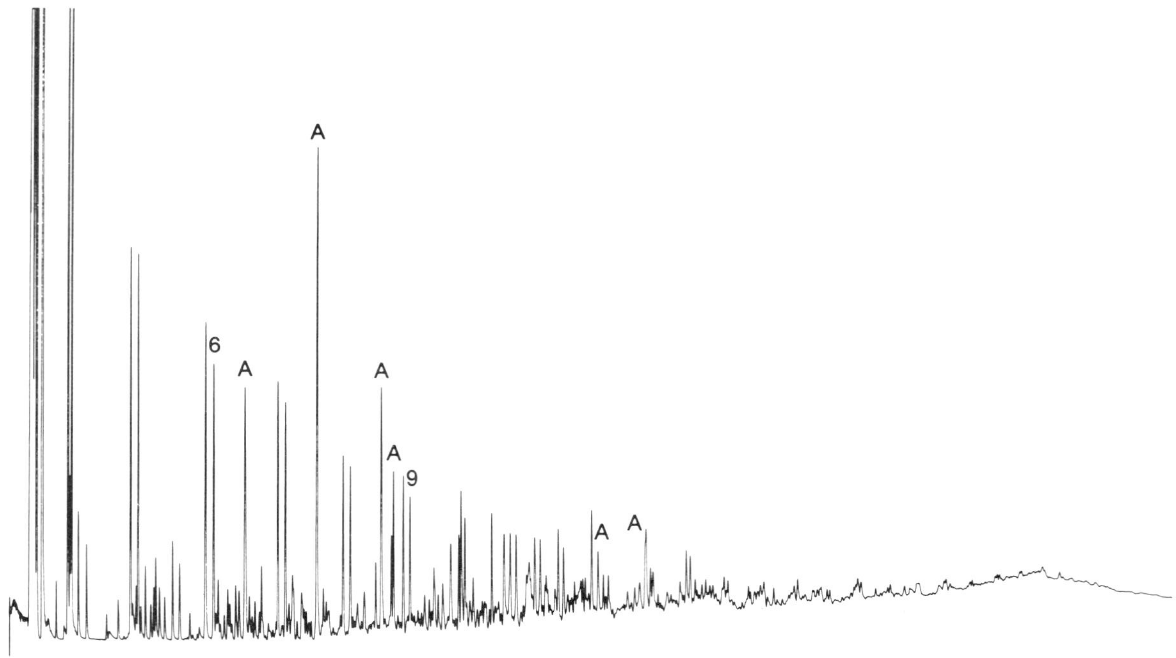

FIG. 10. Pyrolysis–gas chromatogram of sporinite from a high-volatile bituminous coal.

(see Fig. 12). This data was used by Snowdon to illustrate the thermal lability of resinite and to support other geochemical evidence of its role in petroleum generation.

More work is required to determine the onsets of hydrocarbon generation for different kerogen types. This might be achieved using stepwise heating experiments. For instance, Leventhal (1976) pyrolysed selected kerogens for 5 to 10 s at 450°C, 600°C, 750°C, 900°C, then 1000°C. Using this approach, deductions regarding structure could be made based on the varying distributions of pyrolysis products at each temperature. Implicit in this approach is the assumption that weaker bonds are broken at lower pyrolysis temperatures. Leventhal found that long-chain hydrocarbons were produced at lower temperatures than were short-chain hydrocarbons. Similarly, Larter (1978) pyrolysed a coorongite sample at 540°C, 600°C, 660°C, then 740°C, each time for 5 s. Alkanes and polyolefins, eluting as sextuplets, were generated at lower temperatures, whereas the higher-temperature pyrolysates contained successively more aromatic hydrocarbons. Stepwise pyrolysis generates a great deal of information from a single sample in a rapid way and is hence a potentially important tool for evaluating kerogen structure at different levels of maturity.

VI. Role of Catalysts in Petroleum Generation

There is considerable evidence that kerogen pyrolysates may be altered on mineral surfaces (see section IV.A.1 above). Minerals such as montmorillonite and kaolinite have been used as catalysts in many industrial processes, including the cracking of petroleum to gas, gasoline and coke (Thomas, 1970). This implies that hydrocarbon chain-length distributions in pyrolysates might differ, depending on whether the kerogen has been isolated or is still contained within its rock matrix. Of particular interest is the fact that mineral catalysts promote the dehydrogenation of paraffins to aromatic hydrocarbons (Fogelberg *et al.*, 1967; Thomas, 1970), suggesting that the relative proportions of aromatic and aliphatic constituents in kerogen pyrolysates may be influenced by the nature and concentration of the minerals present. The total pyrolysis yield may also be affected. For example, black polymeric residues form on mineral surfaces when fatty acids and alcohols are heated in the presence of clays (Eltantawy and Arnold, 1973; Shimoyama and Johns, 1971; Johns and Shimoyama, 1972), and catalyst fouling is a common problem in industrial isomerization processes (Thomas, 1970).

Most work on the role of catalysis in pyrolysis reactions has been directed towards the behaviour of individual compound classes (e.g. fatty acids and

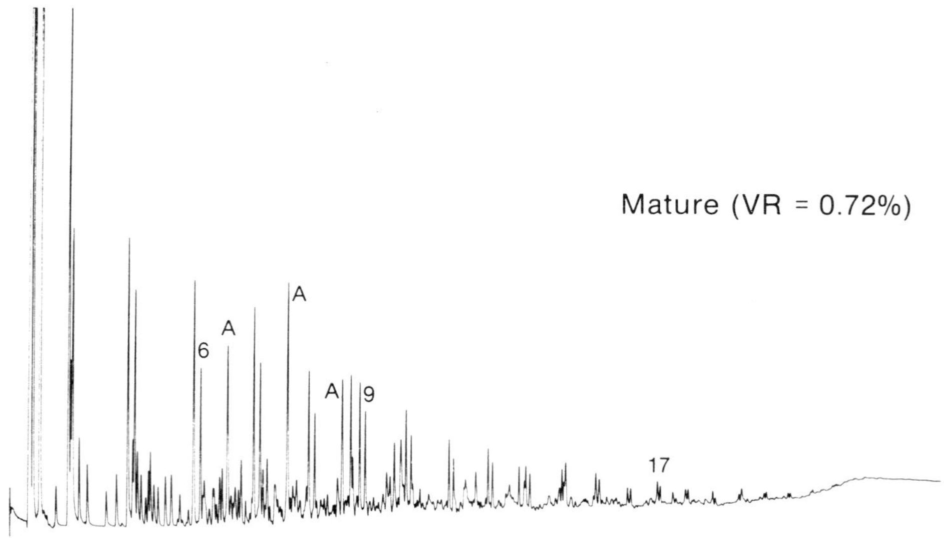

FIG. 11(*a*). Pyrolysis–gas chromatogram of an amorphous Palaeozoic kerogen from Oklahoma.

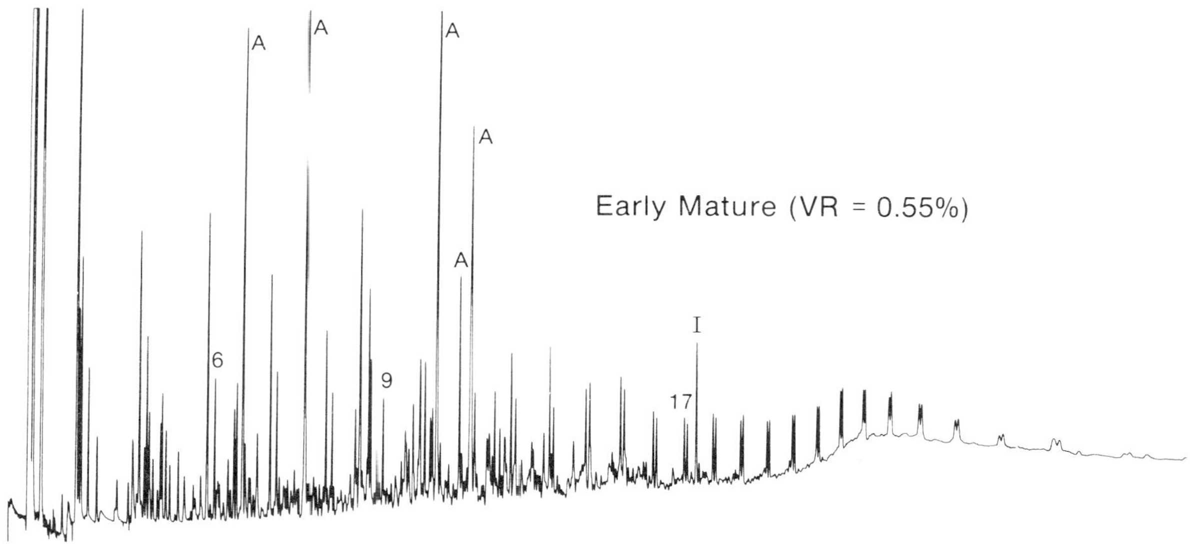

FIG. 11(*b*). Pyrolysis–gas chromatogram of amorphous Mesozoic kerogen from Louisiana.

alcohols) alone or mixed with minerals such as calcium carbonate, kaolinite and montmorillonite (e.g. Louis, 1966; Eisma and Jurg, 1969; Galway, 1969a, b; Shimoyama and Johns, 1971, 1972; Almon and Johns, 1977). However, neither Hoering and Abelson (1963a) nor Giraud (1970) found discernable differences between the pyrograms of whole rocks and their

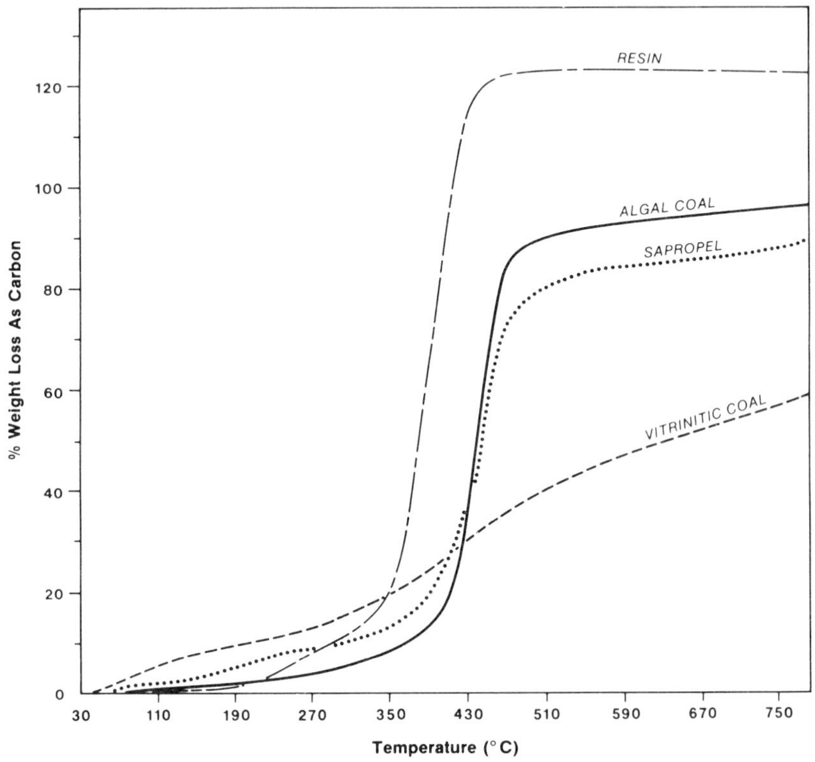

FIG. 12. Thermogravimetric analysis curves for pine resin, an algal coal, a sapropel-rich kerogen and a vitrinite-rich lignite (after Snowdon, 1980).

isolated kerogens. Souron *et al.* (1977) claimed that the rate of evolution of volatile products during kerogen pyrolysis was independent of mineralogy. In contrast to these results, Horsfield and Douglas (1980) and Espitalié *et al.* (1980) noted significant differences between whole rock and kerogen pyrolysates, with total product yield, relative yield of heavy (C_{15}+) hydrocarbons and relative aliphatic hydrocarbon contents being diminished in the presence of certain mineral matrices, notably smectite and illitic clays. If these retention and cracking processes also operate in geological systems, then the relative quantities of gas, gasoline and heavier petroleum fractions generated in source rocks ought to depend not only on the maturity and type of the organic matter but also upon the mineralogy of the source rock. The important question here is whether or not high-temperature pyrolysis products, notably unsaturated species, under anhydrous dynamic conditions, interact with the dewatered, dehydroxylated and, in some cases, decarbonated mineral species in such a way as to replicate subsurface thermal maturation processes involving clays and other minerals. Current opinion is that only low-temperature, preferably hydrous pyrolysis can satisfactorily approximate geological conditions (Lewan *et al.*, 1979; Winters *et al.*, 1983). Further work under these conditions is required to clarify the roles played by minerals in subsurface organic maturation.

VII. Petroleum Migration

While very little pyrolysis research has been aimed directly at petroleum migration, some interesting ideas regarding the possible roles of water and minerals in this process have evolved from kerogen and source-rock studies.

For example, Tissot *et al.* (1974) and Vandenbroucke *et al.* (1976) found that geologically significant volumes of water could be generated from kerogen, especially humic Type III kerogen. Ishiwatari *et al.* (1977) also observed this phenomenon and suggested that the water generated from kerogen during natural maturation might be a carrier for petroleum hydrocarbons and/or their functionalized precursors. Fatty acids which could be transported in such a manner are known to be generated from the mild heating of kerogen, and it has been suggested that these might subsequently decarboxylate in petroleum reservoirs to produce hydrocarbons (Harrison, 1976). Less polar compounds, such as light aromatic hydrocarbons, some of which may be generated from carotenoids (Day and Erdman, 1963) might also migrate in solution.

Rather than acting as a carrier for petroleum hydrocarbons (see McAuliffe, 1979), water appears to play a more active role in generating abnormal internal pressures that might bring about primary migration. Barker (1974a) studied this effect using a bulk-flow pyrolysis technique (P–FID) and concluded that petroleum hydrocarbons and water are contained in sealed microreservoirs within shales. During deep burial, aquathermally generated abnormal pressuring of these isolated fluids might bring about microfracturing of the source rock and expulsion of petroleum hydrocarbons. Ungerer *et al.* (1983) employed pyrolysis data to evaluate the theory that expansion of organic matter during maturation can also help to initiate primary migration. The type of organic matter, its level of maturity and the degree to which the system was closed were cited as important variables.

The effects of mineral matrices on the high-temperature pyrolysis of kerogens were extended to petroleum generation and migration by Espitalié *et al.* (1980). Clay-rich sediments were found to retain pyrolysis products, after which cracking to light hydrocarbons and pyrobitumen took place. Only when surface sites were de-activated was migration of heavy primary products thought to be possible. In contrast, retention by carbonates was low, and the expelled oils were heavy. As stated earlier, it is, however, uncertain whether high-temperature pyrolysis data can be used to study natural maturation and migration processes, particularly when analysing the roles played by minerals.

A novel and alternative approach to the study of petroleum migration is the use of fluid inclusion* data. McLimans (1981) found that coarse-grained calcite cements in rudist reef reservoirs of Fateh Field, Dubai, contained aqueous and oil-bearing inclusions. Geothermometry of the aqueous inclusions indicated a Miocene age for their formation, and because some oil was trapped during growth of the calcite, petroleum migration was also dated as Miocene. These oil-bearing inclusions were decrepitated ($\leq 300°C$) and analysed using a pyrolysis–bulk-flow/–gas-chromatographic instrument (McLimans and Horsfield, 1982; cf. Evans *et al.*, 1964), in preference to crushing and extraction (Kranz, 1969; Kvenvolden and Roedder, 1971 and references therein). Gas chromatograms of the oils revealed a complex history of migration and subsequent alteration (see Fig. 13; cf. Bailey *et al.*, 1973). Interestingly, a homologous series of alkenes (n-C_6 to n-C_{15}) was detected in the oil inclusions of some carbonate cements, particularly those that appeared to contain altered petroleum. The alkenes are not thought to be artefacts of the thermal decrepitation procedure, as a similar n-alkene distribution has been documented in a crude oil from the Etosha Basin (McKirdy *et al.*, 1983).

VIII. Conclusions

Over the last several decades, pyrolysis has played an active and important role in petroleum geochemistry, particularly as regards source-rock characterization. There is every indication that pyrolysis systems and analytical methods will continue to be developed so that pyrolysis can be used in a more diverse range of applications. For example, static pyrolysis, particularly under hydrous conditions, now provides the means by which immature

* Fluid inclusions are small volumes (10^{-10} cc) of fluid that are trapped during the initial growth of a mineral crystal or during its later recrystallization.

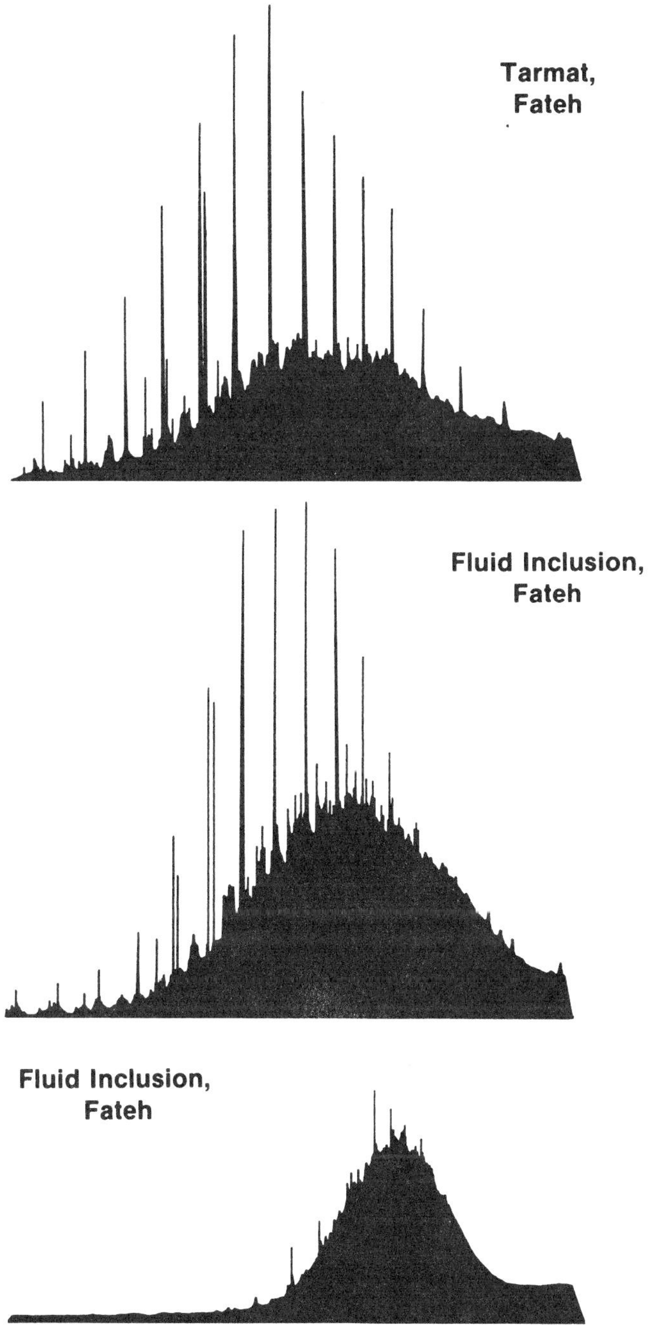

FIG. 13. Thermal decrepitation products from fluid inclusions compared with thermal evaporation products from a tarmat.

source rocks can be artificially matured for oil-to-source correlations (Winters *et al.*, 1983); further documentation of this application is likely to be prolific. In the light of its close approximation to natural maturation, hydrous pyrolysis might also be used to study maturation and migration, particularly as regards the roles played by minerals in these processes.

High-temperature pyrolysis systems seem likely to continue making important contributions in source-rock geochemistry by identifying kerogen types and elucidating their behaviour in different geothermal regimes. However, a new and important application of pyrolysis is likely to be in studying reservoir rocks. Just as in the case of static pyrolysis, high-temperature dynamic pyrolysis systems will probably be used to address the poorly understood processes of migration and alteration *in situ* from a new perspective.

Acknowledgments

I would like to acknowledge the advice and encouragement given to me over the past several years by friends and colleagues, many of whom are, or were, at Newcastle Upon Tyne University. Special thanks are extended to David McKirdy who painstakingly edited and reviewed early drafts of the manuscript and to Brian Cooper, whose suggested revisions were included in the final draft. I am grateful to Conoco Inc. for permission to publish this review.

References

Abelson, P. H. (1964). *Proc. Sixth World Petr. Cong. 1963, Frankfurt/Main*, **1**, 397–407.

Achari, R. G., Shaw, G. and Holleyhead, R. (1973). *Chem. Geol.* **12,** 229–234.

Albrecht, P. and Ourisson, G. (1969). *Geochim. Cosmochim. Acta* **33,** 138–142.

Allan, J. (1975). PhD Thesis, Univ. of Newcastle Upon Tyne.

Allan, J. and Douglas, A. G. (1974). *In* "Advances in Organic Geochemistry 1973" (Tissot, B. and Bienner, F., Eds.), 203–206. Editions Technip, Paris.

Allan, J., Murchison, D. G., Scott, E. and Watson, S. (1975). *Fuel* **54,** 283–287.

Allan, J., Bjorøy, M. and Douglas, A. G. (1980). *In* "Advances in Organic Geochemistry 1979" (Douglas, A. G. and Maxwell, J. R., Eds.), 599–618. Pergamon Press, Oxford.

Almon, W. R. and Johns, W. D. (1977). *In* "Advances in Organic Geochemistry 1975" (Campos, R. and Goni, J., Eds.), 157–177. ENADIMSA, Madrid.

Alpern, B., Durand, B., Espitalié, J. and Tissot, B. (1972). *In* "Advances in Organic Geochemistry 1971" (Gaertner, H. R. V. and Wehner, H., Eds.), 1–28. Pergamon Press, Oxford.

Anders, D. F., Claypool, G. E., Lubeck, C. M. and Patterson, J. M. (1978). *In* "Initial Reports of the Deep Sea Drilling Project" (Ross, D. A., Neprochnovy, P. *et al.*, Eds.), Vol. XLII, Pt. 2, 755–763. U.S. Government Printing Office, Washington.

Austen, D. E. G., Ingram, D. J. E. and Tapley, J. G. (1958). *Trans. Faraday Soc.* **54,** 400–408.

Bailey, N. J. L. (1981). *In* "Organic Maturation Studies and Fossil Fuel Exploration" (Brooks, J., Ed.), 284–302. Academic Press, London, Orlando and New York.

Bailey, N. J. L., Krouse, H. R., Evans, C. R. and Rogers, M. A. (1973). *AAPG Bull.* **57,** 1276.

Bajor, M., Roquebert, M. H. and van der Weide, B. M. (1969). *Bull. Centre Rech. Pau SNPA* **3** (1), 113–124.

Baker, E. W. and Louda, J. W. (1980). *In* "Advances in Organic Geochemistry 1979" (Douglas, A. G. and Maxwell, J. R., Eds.), 295–319. Pergamon Press, Oxford.

Barker, C. (1974a). *AAPG Bull.* **58,** 2349–2361.

Barker, C. (1974b). *Fuel* **53,** 176–177.

Barker, C. (1978). *Earth Sciences Miscell.*, Publ. 78–1, Univ. of Tulsa.

Bordenave, M., Combaz, A. and Giraud, A. (1970). *In* "Advances in Organic Geochemistry 1966" (Hobson, G. D. and Speers, G. C., Eds.), 389–405. Pergamon Press, Oxford.

Boulmier, J. L., Oberlin, A. and Durand, B. (1977). *In* "Advances in Organic Geochemistry 1975" (Campos, R. and Goni, J., Eds.), 781–796. ENADIMSA, Madrid.

Boyer, A. F., Ferrand, R. and Payen, P. (1961). *Chimie et Industrie* **86** (5), 523–530.

Bray, E. E. and Evans, E. D. (1961). *Geochim. Cosmochim. Acta* **22,** 2–15.

Bricteux, J. (1966). *Annales. Mines. Belg.* **12,** 1543.

Bricteux, J. (1967). *Annales. Mines. Belg.* **13,** 761.

Brooks, J. and Shaw, G. (1968). *Nature* **219,** 532.

Brooks, J. D. and Smith, J. W. (1967). *Geochim. Cosmochim. Acta* **31,** 2389–2397.

Brooks, J. D. and Smith, J. W. (1969). *Geochim. Cosmochim. Acta* **33,** 1189–1194.

Bujak, J. P., Barss, M. S. and Williams, G. L. (1977). *Oil and Gas J.* **75** (14), 198–202.

Burlingame, A. L., Haug, P. A., Schnoes, H. K. and Simoneit, B. R. (1969). *In* "Advances in Organic Geochemistry 1968" (Schenck, P. A. and Havenaar, I., Eds.), 85–129. Pergamon Press, Oxford.

Cane, R. F. and Albion, P. R. (1973). *Geochim. Cosmochim. Acta* **37,** 1543.

Cardwell, A. L. (1977). *Quarterly of Colorado School of Mines* **72,** 3.

Claypool, G. E. and Baysinger, J. P. (1978). *In* "Initial Reports of the Deep Sea Drilling Project", (eds Benson, W. E. and Sheridan, R. E., Eds.), Vol. XLIV, 635–638. U.S. Government Printing Office, Washington.

Claypool, G. E. and Baysinger, J. P. (1980). *In* "Initial Reports of the Deep Sea Drilling Project" (Laucelot, Y. and Winterer, E. L., Eds.), Vol. L, 605–608. U.S. Government Printing Office, Washington.

Claypool, G. E. and Reed, P. R. (1976). *AAPG Bull.* **60,** 608–626.

Claypool, G. E., Lubeck, C. M., Patterson, J. M. and Baysinger, J. P. (1976). *U.S. Department of the Interior Geological Survey, Open File Report* 76–232, 45–56. U.S. Government Printing Office, Washington.

Claypool, G. E., Lubeck, C. M., Baysinger, J. P. and Ging, T. G. (1977). *In*

"Geological Studies on the Cost No. B-2 Well, U.S. Mid-Atlantic Outer Continental Shelf Area" (Scholle, P. A., Ed.), Geol. Surv. Circular, 750.
Clementz, D. M. (1978). *AAPG Bull.* **63,** 2227–2232.
Clementz, D. M., Demaison, G. J. and Daly, A. R. (1979a). *Proc. Eleventh Offshore Tech. Conf.* **1,** 465–470, OTC 3410.
Clementz, D. M., Demaison, G. J. and Daly, A. R. (1979b). *Oil and Gas J.* **77,** 142–146.
Combaz, A. (1971). *In* "Sporopollenin" (Brooks, J., Grant, P. R., Muir, M. D., van Gijzel, P. and Shaw, G., Eds.), 621. Academic Press, London, Orlando and New York.
Combaz, A. (1976). U.S. Patent 3,941,567.
Connan, J. (1972). *Bull. Centre. Rech. Pau SNPA* **6** (1), 195–214.
Connan, J. (1974. *In* "Advances in Organic Geochemistry 1973" (Tissot, B. and Bienner, F., Eds.), 73–75. Editions Technip, Paris.
Connan, J. and Cassou, A. M. (1980). *Geochim. Cosmochim. Acta* **44,** 1–23.
Connan, J. and van der Weide, B. M. (1974). *Can. Soc. Petr. Geol., Oil Sands Fuel of the Future*, 134–147.
Connan, J., Le Tran, K. and van der Weide, B. M. (1975). *Proc. Ninth World Petr. Cong.* **2,** 171–178. Applied Science Publishers, London.
Davis, A. and Spackman, W. (1974). *Fuel* **43,** 215–224.
Davis, J. B. and Stanley, J. P. (1982). *AAPG Annual Convention, Calgary, (abstract).*
Day, W. C. and Erdman, J. E. (1963). *Science* **141,** 982–984.
Dembicki, H. and Woods, R. A. (1982). U.S. Patent 4,325,907.
Dembicki, H., Horsfield, B. and Ho, T. T. Y. (1983). *AAPG Bull.* **67** (7), 1094–1103.
Dormans, H. N. M., Huntjens, F. J. and van Krevelen, D. W. (1957). *Fuel* **36,** 321.
Douglas, A. G. and Grantham, P. J. (1974). *In* "Advances in Organic Geochemistry 1973" (Tissot, B. and Bienner, F., Eds.), 261–276. Editions Technip, Paris.
Douglas, A. G., Eglinton, G. and Maxwell, J. R. (1969). *Geochim. Cosmochim. Acta* **33,** 569–577.
Douglas, A. G., Eglinton, G. and Henderson, W. (1970). *In* "Advances in Organic Geochemistry 1966" (Hobson, G. D. and Speers, G. D., Eds.), 369–388. Pergamon Press, Oxford.
Douglas, A. G., Coates, R. C., Bowler, B. F. J. and Hall, K. (1977). *In* "Advances in Organic Geochemistry 1975" (Campos, R. and Goni, J., Eds.), 357–374. ENADIMSA, Madrid.
Dryden, I. G. C. (1963). *In* "Chemistry of Coal Utilisation" (Lowry, H. H., Ed.), Supplementary Volume, 232. John Wiley, New York.
Dungworth, G. and Schwartz, A. W. (1972). *In* "Advances in Organic Geochemistry 1971" (Gaertner, H. R. V. and Wehner, H., Eds.), 699–706. Pergamon Press, Oxford.
Durand, B. (1980). "Kerogen—Insoluble Organic Matter from Sedimentary Rocks". Editions Technip, Paris.
Durand, B., Barsony, I., Monin, J. C., Nicaise, G., Robin, P. and Souron, C. (1976). *I. F. P. Division Geologic Ref. 23*, 825.
Durand, B., Nicaise, G., Roucache, J., Vandenbroucke, M. and Hagemann, H. W. (1977). *In* "Advances in Organic Geochemistry 1975" (Campos, R. and Goni, G., Eds.), 601–631. ENADIMSA, Madrid.

Eglinton, G. (1972). *In* "Advances in Organic Geochemistry 1971" (Gaertner, H. R. V. and Wehner, H., Eds.), 29–48. Pergamon Press, Oxford.
Eglinton, G. and Hamilton, R. J. (1963). *In* "Chemical Plant Taxonomy" (Swain, T., Ed.), 187–217. Academic Press, London, Orlando and New York.
Eisma, E. and Jurg, J. W. (1969). *In* "Organic Geochemistry" (Eglinton, G. and Murphy, M. T. J., Eds.), 676, 698. Springer Verlag, Berlin.
Eltantaway, I. M. and Arnold, P. W. (1973). *Nature* **244,** 144.
Espitalié, J., Laporte, J. L., Madec, M., Marquis, F., Le Plat, P., Paulet, J. and Boutefeu, A. (1977a). *Rev. Inst. Fran. Petr.* **32,** 23–42.
Espitalié, J., Madec, M. and Tissot, B. (1977b). *Proc. Ninth Offshore Tech. Conf.*, 439–444, OTC 2935.
Espitalié, J., Madec, M. and Tissot, B. (1980). *AAPG Bull.* **64,** 59–66.
Evans, W. D., Cooper, B. S. and Gunn, R. K. (1964). *In* "Advances in Organic Geochemistry 1962" (Columbo, U. and Hobson, G. D., Eds.), 417–421. Pergamon Press, Oxford.
Fisher, M. A. (1980). *AAPG Bull.* **64,** 1140–1157.
Fisher, M. A. (1982). *AAPG Bull.* **66** (3), 286–301.
Fogelberg, L. G., Gore, R. and Ranby, B. (1967). *Acta Chem. Scand.* **21,** 2041–2049.
Gallegos, E. J. (1975). *Anal. Chem.* **47** (9), 1524–1528.
Galway, A. K. (1969a). *Nature* **223,** 1257, 1259.
Galway, A. K. (1969b). *J. Chem. Soc. D* **11,** 577–578.
George, A. E., Banerjee, R. C., Smiley, G. T. and Sawatzky, H. (1977). *Bull. Can. Petr. Geol.* **25,** 1085–1096.
Giraud, A. (1970). *AAPG Bull.* **54,** 439–451.
Girling, G. W. (1963). *J. Appl. Chem.* **13,** 77–91.
Given, P. H., Peover, M. E. and Nyss, W. R. (1960). *Fuel* **39,** 323–340.
Gransch, J. A. and Eisma, E. (1970). *In* "Advances in Organic Geochemistry 1966" (Hobson, G. D. and Speers, G. C., Eds.), 407–426. Pergamon Press, Oxford.
Harrison, W. E. (1976). *AAPG Bull.* **60** (3), 452–457.
Harwood, R. J. (1977). *AAPG Bull.* **61,** 2082–2102.
Heacock, R. L. and Hood, A. (1970). U.S. Patent 3,508,877.
Hedberg, H. D. (1968). *AAPG Bull.* **52,** 736–750.
Henderson, W., Eglinton, G., Simmonds, P. and Lovelock, J. E. (1968). *Nature* **219,** 1012–1016.
Hirsch, P. B. (1954). *Proc. Roy. Soc., London*, A226, 143–169.
Ho, T. T. Y. (1978). *United Nations ESCAP, CCOP Tech. Publ.*, No. 6, 54–80.
Hoering, T. C. and Abelson, P. H. (1963a). *Carnegie Inst. Wash. Yrbk.* **66,** 510–515.
Hoering, T. C. and Abelson, P. H. (1963b). *Carnegie Inst. Wash. Yrbk.* **66,** 256–258.
Holden, H. W. and Robb, J. C. (1958). *Nature* **182,** 340.
Horsfield, B. (1978). PhD Thesis, Univ. of Newcastle Upon Tyne.
Horsfield, B. and Douglas, A. G. (1980). *Geochim. Cosmochim. Acta* **44,** 1119–1131.
Horsfield, B. and Larter, S. R. (1981). "An Interlaboratory Pyrolysis Study: Collation of Results", unpublished.
Horsfield, B., Dembicki, H. and Ho, T. T. Y. (1983). *J. Geol. Soc., London*, **140** (3), 431–443.

Huc, A. Y. and Hunt, J. M. (1980). *Geochim. Cosmochim. Acta* **44** (8), 1081–1090.
Huc, A. Y., Hunt, J. M. and Whelan, J. K. (1981). *J. Geochem. Expl.* **15,** 671–681.
Hummell, D. O. (1977). *In* "Analytical Pyrolysis" (Jones, C. E. R. and Cramers, C. A., Eds.), 117–138. Elsevier, Amsterdam.
Hunt, J. M. (1972). *AAPG Bull.* **56,** 2273–2277.
Hunt, J. M. (1979a). "Petroleum Geochemistry and Geology". Freeman, San Francisco.
Hunt, J. M. (1979b). *Woods Hole Oceanographic Inst. Contrib.* No. 4492.
Hutton, A. C. and Cook, A. C. (1980). *Fuel* **59,** 711–714.
Hutton, A. C., Kantsler, A. J., Cook, A. C. and McKirdy, D. M. (1980). *J. Aust. Petr. Expl. Assoc.* **20,** 68–86.
Ikan, R., Baedecker, M. J. and Kaplan, I. R. (1975). *Geochim. Cosmochim. Acta* **39,** 195–203.
International Committee for Coal Petrology (1971). "International Handbook of Coal Petrograph", Supplement to the Second Edition. Centre National de la Recherche Scientifique, France.
Ishiwatari, R., Ishiwatari, M., Kaplan, I. R. and Rohrback, B. G. (1976). *Nature* **264** (5584), 347–349.
Ishiwatari, R., Ishiwatari, M., Rohrback, B. G. and Kaplan, I. R. (1977). *Geochim. Cosmochim. Acta* **41,** 815–828.
Johns, W. D. and Shimoyama, A. (1972). *AAPG Bull.* **56,** 2160, 2167.
Jonathan, D., L'Hote, G. and Du Rouchet, J. (1975). *Rev. Inst. Fran. Petr.* **30,** 65–88.
Jones, C. E. R. and Cramers, C. A. (1977). "Analytical Pyrolysis". Elsevier, Amsterdam.
Juntgen, H. and van Heek, K. H. (1968). *Fuel* **47,** 103–117.
Katz, B. J. (1981). *AAPG Bull.* **65** (5), 944.
Kimber, R. W. L. and Searle, P. L. (1970a). *Geoderma* **4** (1), 47–55.
Kimber, R. W. L. and Searle, P. L. (1970b). *Geoderma* **4** (1), 57–71.
Knights, B. A., Brown, A. C., Conway, E. and Middleditch, B. S. (1970). *Phytochem.* **9,** 1317–1324.
Kranz, R. (1969). *In* "Organic Geochemistry" (Eglinton, G. and Murphy, M. T. J., Eds.), 521–533. Springer Verlag, Berlin.
Kvenvolden, K. A. and Roedder, E. (1971). *Geochim. Cosmochim. Acta* **35,** 1209–1220.
Landes, K. K. (1967). *AAPG Bull.* **51,** 828–841.
Larter, S. R. (1978). PhD Thesis, Univ. of Newcastle Upon Tyne.
Larter, S. R. (1981). Unpublished data.
Larter, S. R. and Douglas, A. G. (1978). *In* "Environmental Biogeochemistry and Geomicrobiology" (Krumbein, W. E., Ed.), 373–386. Ann Arbor, Science, Woburn, Massachusetts.
Larter, S. R. and Douglas, A. G. (1980a). *Geochim. Cosmochim. Acta* **44,** 2087–2095.
Larter, S. R. and Douglas, A. G. (1980b). *In* "Advances in Organic Geochemistry 1979" (Douglas, A. G. and Maxwell, J. R., Eds.), 579–584. Pergamon Press, Oxford.
Larter, S. R., Horsfield, B. and Douglas, A. G. (1977). *In* "Analytical Pyrolysis" (Jones, C. E. R. and Cramers, C. A., Eds.), 189–202. Elsevier, Amsterdam.

Larter, S. R., Solli, H. and Douglas, A. G. (1978). *J. Chromatog.* **167,** 423–431.
Larter, S. R., Solli, H., Douglas, A. G., Delange, F. and de Leeuw, J. W. (1979). *Nature* **279,** 405–408.
Larter, S. R., Solli, H. and Douglas, A. G. (1983). *In* "Advances in Organic Geochemistry 1981" (Bjorøy, M. *et al.*, Eds.), 513–523. John Wiley, Chichester.
Le Plat, P. (1967). *J. Gas Chromatog.* **5,** 128–135.
Le Tran, K. and van der Weide, B. M. (1969). *Bull. Centre Rech. Pau SNPA* **3** (2), 449–457.
Le Tran, K., Connan, J. and van der Weide, B. (1974). *Bull. Centre Rech. Pau SNPA* **8,** 111–137.
Leventhal, J. S. (1976). *Chem. Geol.* **18,** 5–20.
Leventhal, J., Suess, S. E. and Cloud, P. (1975). *Proc. Nat. Acad. Sci. U.S.A.* **72,** 4706–4710.
Levy, R. L., Wolf, C. J. and Cro, J. (1970). *J. Chromatog. Sci.* **8,** 524–526.
Lewan, M. D., Winters, J. C. and McDonald, J. H. (1979). *Science* **203,** 897–899.
Louis, M. (1966). *In* "Advances in Organic Geochemistry 1964" (Hobson, G. D. and Louis, M., Eds.), 261–278. Pergamon Press, Oxford.
Louis, M. C. and Tissot, B. P. (1967). *Proc. Seventh World Petr. Cong., Mexico,* **2,** 47–60.
McAuliffe, C. D. (1979). *AAPG Bull.* **63** (5), 761–781.
McHugh, D. J., Saxby, J. D. and Tardiff, J. W. (1976). *Chem. Geol.* **17,** 243–259.
McHugh, D. J., Saxby, J. D. and Tardiff, J. W. (1978). *Chem. Geol.* **21,** 1–14.
McIver, R. D. (1967). *Proc. Seventh World Petr. Cong.* **2,** 25–36.
Mackenzie, A. S., Lewis, C. A. and Maxwell, J. R. (1981). *Geochim. Cosmochim. Acta* **45** (12), 2369–2376.
McKirdy, D. M., McHugh, D. J. and Tardif, J. W. (1980). *In* "Biogeochemistry of Ancient and Modern Environments" (Trudinger, P. A., Walter, M. R. and Ralph, B. J., Eds.), 187–200. Australian Academy of Science and Springer Verlag, Berlin.
McKirdy, D. M., Aldridge, A. K. and Ypma, P. J. M. (1983). *In* "Advances in Organic Geochemistry 1981" (Bjorøy, M. *et al.*, Eds.), 99–107. John Wiley, Chichester.
McLimans, R. K. (1981). *AAPG Bull.* **65** (5), 957.
McLimans, R. K. and Horsfield, B. (1982). *183rd A. C. S. National Meeting, Las Vegas*, (abstract).
Magoon, L. B. and Claypool, G. E. (1981). *AAPG Bull.* **65,** 1043–1061.
Martin, S. J. (1977). *In* "Advances in Organic Geochemistry 1975" (Campos, R. and Goni, J., Eds.), 677–692. ENADIMSA, Madrid.
Maters, W. L., van de Meent, D., Schuyl, P. J. W., DeLeeuw, J. W. and Schenk, P. A. (1977). *In* "Analytical Pyrolysis" (Jones, C. E. R and Cramers, C. A., Eds.), 203. Elsevier, Amsterdam.
Maxwell, J. R., Douglas, A. G., Eglinton, G. and McCormick, A. (1968). *Phytochem.* **7,** 2157.
Merewether, E. A. and Claypool, G. E. (1980). *AAPG Bull.* **64,** 488–500.
Mitterer, R. M. and Hoering, T. C. (1968). *Carnegie Inst. Wash. Yrbk.* **66,** 510–515.
Monin, J. C., Durand, B., Vandenbroucke, M. and Huc, A. Y. (1980). *In* "Advances in Organic Geochemistry 1979" (Douglas, A. G. and Maxwell, J. R., Eds.), 517–530. Pergamon Press, Oxford.

Monnier, F., Powell, T. G. and Snowdon, L. R. (1982). *In* "Advances in Organic Geochemistry 1981" (Bjorøy, M. *et al.*, Eds.), 487–495. John Wiley, Chichester.
Moschopedis, S. E., Parkash, S. and Speight, J. G. (1978). *Fuel* **57,** 431–434.
Murchison, D. G. and Jones, J. M. (1964). *In* "Advances in Organic Geochemistry 1962" (Columbo, U. and Hobson, G. D., Eds.), 1–21. Pergamon Press, Oxford.
Murphy, M. T. J. (1969). *In* "Organic Geochemistry"" (Eglinton, G. and Murphy, M. T. J., Eds.), 74–86. Springer Verlag, Berlin.
Neavel, R. C. and Miller, L. V. (1960). *Fuel* **39,** 217–222.
Oberlin, A., Boulmier, J. L. and Durand, B. (1974a). *In* "Advances in Organic Geochemistry 1973" (eds Tissot, B. and Bienner, F., Eds.), 15–27. Editions Technip, Paris.
Oberlin, A., Boulmier, J. L. and Durand, B. (1974b). *Geochem. Cosmochim. Acta* **38,** 647–650.
Oehler, J., Aizenschtat, Z. and Schopf, J. W. (1974). *AAPG Bull.* **58** (1), 124–132.
Orr, W. L. (1983). *In* "Advances in Organic Geochemistry 1981" (Bjorøy, M. *et al.*, Eds.), 775–787. John Wiley, Chichester.
Palmer, S. E. and Zumberge, J. E. (1981). *In* "Organic Maturation Studies and Fossil Fuel Exploration" (Brooks, J., Ed.), 393–426. Academic Press, London, Orlando and New York.
Peters, K. E., Simoneit, B. R. T., Brenner, S. and Kaplan, I. E. (1978). "Symposium in Geochemistry: Low Temperature Metamorphism of Kerogen and Clay Minerals" (Oltz, D. F., Ed.), 53–58. SEPM Pacific Section, Los Angeles.
Peters, K. E., Rohrback, B. G. and Kaplan, I. R. (1980). *In* "Advances in Organic Geochemistry 1979" (Douglas, A. G. and Maxwell, J. R., Eds.), 547–558. Pergamon Press, Oxford.
Peters, K. E., Rohrback, B. G. and Kaplan, I. R. (1981). *AAPG Bull.* **65** (4), 688–705.
Philippi, G. T. (1965). *Geochim. Cosmochim. Acta* **29,** 1021–1049.
Philp, R. P. and Saxby, J. D. (1980). *In* "Advances in Organic Geochemistry 1979" (Douglas, A. G. and Maxwell, J. R., Eds.), 639–651. Pergamon Press, Oxford.
Philp, R. P., Gilbert, I. D. and Russell, N. J. (1982). *Fuel* **61,** 221–226.
Powell, T. G. (1978). *Geol. Surv. Can. Paper* 78–12.
Powell, T. G. and Snowdon, L. R. (1980). *Can. Soc. Petr. Geol. Mem.* **6,** 421–446.
Powell, T. G., Creaney, S. and Snowdon, L. R. (1982). *AAPG Bull.* **66,** 430–435.
Poxon, D. W. and Wright, R. G. (1971). *J. Chromatog.* **61,** 142–144.
Price, L. C., Clayton, J. L. and Rumen, L. L. (1981). *Org. Geochem.* **3** (3), 59–78.
Pusey, W. C. (1973). *Petr. Times*, January 1973.
Retcofsky, A. L., Stark, J. M. and Friedal, R. A. (1968). *Anal. Chem.* **40,** 1699–1704.
Rhead, M. M., Eglinton, G. and Draffan, G. H. (1971). *Chem. Geol.* **8,** 277–297.
Rohrback, B. G. (1978). "Symposium in Geochemistry: Low Temperature Metamorphism of Kerogen and Clay Minerals" (Oltz, D. F., Ed.), 47–52. SEPM Pacific Section, Los Angeles.
Rohrback, B. G. and Kaplan, I. R. (1978). "Symposium in Geochemistry: Low Temperature Metamorphism of Kerogen and Clay Minerals (Oltz, D. F., Ed.), 13–17. SEPM Pacific Section, Los Angeles.
Romovacek, J. and Kubat, J. (1968). *Anal. Chem.* **40,** 1119.
Rouxhet, P. G. and Robin, P. L. (1978). *Fuel* **57,** 533–540.

Rubenstein, I., Spyckerelle, C. and Strausz, O. P. (1979). *Geochim. Cosmochim. Acta* **43,** 1–6.

Saint-Paul, C., Monin, J. C. and Durand, B. (1980). *Rev. Inst. Fran. Petr.* **35** (6), 1065–1078.

Samer, S. F. (1972). "A bibliography on solids pyrolysis with selected references to vapor-phase pyrolysis." Chemical Data Systems, Oxford, Pennsylvania.

Schenck, P. A., de Leeuw, J. W., van Graas, G., Haverkamp, J. and Bouman, M. (1981). *In* "Organic Maturation and Fossil Fuel Explanation" (Brooks, J., Ed.), 225–237. Academic Press, London, Orlando and New York.

Schuhmacher, J. P., Huntjens, F. J. and van Krevelen, D. W. (1960). *Fuel* **39,** 223–234.

Scott, W. M., Modzeleski, V. E. and Nagy, B. (1970). *Nature* **225** (5238), 1129–1130.

Scrima, D. A., Yen, T. F. and Warren, P. L. (1974). *Energy Sources* **1** (3).

Seifert, W. K. and Moldowan, J. M. (1980). *In* "Advances in Organic Geochemistry 1979" (Douglas, A. G. and Maxwell, J. R., Eds.), 229–238. Pergamon Press, Oxford.

Seifert, W. K. and Moldowan, J. M. (1981). *Geochim. Cosmochim. Acta* **45,** 783–794.

Shimoyama, A. and Johns, W. D. (1971). *Nature* **232,** 140–144.

Shimoyama, A. and Johns, W. D. (1972). *Geochim. Cosmochim. Acta* **36,** 87–91.

Snowdon, L. R. (1979). *AAPG Bull.* **63** (7), 1128–1134.

Snowdon, L. R. (1980). *Can Soc. Petr. Geol. Mem.* **6,** 509–521.

Snowdon, L. R. and Powell, T. G. (1982). *AAPG Bull.* **66,** 775–788.

Solli, H., Larter, S. R. and Douglas, A. G. (1980a). *J. Anal. Appl. Pyrol.* **1,** 231–241.

Solli, H., Larter, S. R. and Douglas, A. G. (1980b). *In* "Advances in Organic Geochemistry 1979" (Douglas, A. G. and Maxwell, J. R., Eds.), 591–598. Pergamon Press, Oxford.

Souron, C., Boulet, R. and Espitalié, J. (1977). *In* "Advances in Organic Geochemistry 1975" (Campos, R. and Goni, J., Eds.), 797–820. ENADIMSA, Madrid.

Stahl, E. (1977). *In* "Analytical Pyrolysis" (Jones, C. E. R. and Cramers, C. A., Eds.), 29–38. Elsevier, Amsterdam.

Staplin, F. L. (1969). *Bull. Can. Geol.* **17,** 47–66.

Stevenson, F. J. (1974). *In* "Advances in Organic Geochemistry 1973" (Tissot, B. and Bienner, F., Eds.), 701–714. Editions Technip, Paris.

Summerhayes, C. P. (1981). *AAPG Bull.* **65,** 2364–2380.

Tarafa, M. E., Hunt, J. M. and Ericsson, I. (1983). *J. Geochem. Expl.* **18,** 75–85.

Thomas, C. L. (1970). "Catalytic Processes and Proven Catalysts". Academic Press, London, Orlando and New York.

Tissot, B. (1969). *Rev. Inst. Fran. Petr.* **24,** 470.

Tissot, B. P. and Welte, D. (1978). "Petroleum Formation and Occurrence." Springer Verlag, Berlin.

Tissot, B., Califet-Debyser, Y., Deroo, G. and Oudin, J. L. (1971). *AAPG Bull.* **55,** 2177–2193.

Tissot, B., Durand, B., Espitalié, J. and Combaz, A. (1974). *AAPG Bull.* **58,** 499–506.

Tissot, B., Deroo, G. and Hood, A. (1978). *Geochim. Cosmochim. Acta* **42,** 1469–1485.

Tissot, B., Demaison, G., Masson, P., Delteil, J. R. and Combaz, A. (1980). *AAPG Bull.* **64** (12), 2051–2063.
Traverse, A. and Kolvoord, R. W. (1968). *Science* **159,** 302–305.
Treibs, A. (1934). *Ann. Chem.* **510,** 42–62.
Ungerer, P., Behar, E. and Discamps, D. (1983). *In* "Advances in Organic Geochemistry 1981" (Bjorøy, M. *et al.*, Eds.), 129–135. John Wiley, Chichester.
Vandenbroucke, M., Albrecht, P. and Durand, B. (1976). *Geochim. Cosmochim. Acta* **40,** 1241–1249.
van de Meent, D., de Leeuw, J. W. and Schenck, P. A. (1980a). *In* "Advances in Organic Geochemistry 1979" (Douglas, A. G. and Maxwell, J. R., Eds.), 469–474. Pergamon Press, Oxford.
van de Meent, D., Brown, S. C., Philp, R. P. and Simoneit, B. R. T. (1980b). *Geochim. Cosmochim. Acta* **44,** 999–1014.
van Dorsselaer, A., Albrecht, P. and Connan, J. (1977). *In* "Advances in Organic Geochemistry 1975" (Campos, R. and Goni, J., Eds.), 53–59. ENADIMSA, Madrid.
van Graas, G., de Leeuw, J. W. and Schenck, P. A. (1980a). *J. Anal. Appl. Pyrol.* **2,** 265–276.
van Graas, G., de Leeuw, J. W. and Schenck, P. A. (1980b). *In* "Advances in Organic Geochemistry 1979" (Douglas, A. G. and Maxwell, J. R., Eds.), 485–494. Pergamon Press, Oxford.
van Graas, G., de Leeuw, J. W., Schenck, P. A. and Haverkamp, J. (1981). *Geochim. Cosmochim. Acta* **45,** 2465–2474.
van Krevelen, D. W., van Heerden, C. and Huntjens, F. J. (1951). *Fuel* **30,** 253–259.
Vassoyevich, N. B., Korchagina, Yu. I., Lopatin, N. V. and Chernyshev, V. V. (1969). *Internat. Geol. Rev.* **12** (11).
Walker, J. Q. (1972). *Chromatographia* **5,** 547.
Welte, D. H. (1974). "Advances in Organic Geochemistry 1973" (Tissot, B. and Bienner, F., Eds.), 3–13. Editions Technip, Paris.
Whelan, J. K., Hunt, J. M. and Huc, A. Y. (1980). *J. Anal. Appl. Pyrol.* **2,** 79–96.
White, D. (1915). *Wash. Acad. Sci. J.* **5**(6), 189–212.
Winters, J. C., Williams, J. A. and Lewan, M. D. (1983). *In* "Advances in Organic Geochemistry 1981" (Bjorøy, M. *et al.*, Eds.), 524–533. John Wiley, Chichester.
Yellow, P. C. (1965). *B.C.U.R.A. Mon. Bull.* **29,** 285.

Biodegradation of Crude Oils in Reservoirs

Jacques Connan
Société Nationale Elf-Aquitaine (Production), Pau, France

I. Introduction

The history of petroleum does not end when petroleum is pooled in reservoirs. Several post-accumulation processes may drastically modify the original properties of the crude oil. The main processes which affect crude oil in reservoirs are summarized in Fig. 1. Compositional changes in petroleum may be due to temperature and pressure (in-reservoir matur ation), moving waters (water washing and biodegradation) or natural gas (deasphalting by precipitation of solid bitumen within the reservoir). Physical and chemical transformations caused by the above-mentioned phenomena have been reviewed by several authors, namely Milner *et al.* (1977), Tissot and Welte (1978), Hunt (1979) and Price (1980).

Among the possible alteration processes, water washing and biodegradation are widespread phenomena, observed in shallow reservoirs on the rim of many petroliferous basins. The importance of microbial transformation of crude oil in the subsurface must be stressed.

Demaison (1977) pointed out that the seven largest supergiant accumulations of tar sands, i.e. biodegraded crudes, contain as much oil as the 264 biggest conventional oil fields in the world. Alberta's oil sands contain

ADVANCES IN PETROLEUM GEOCHEMISTRY Vol. 1.
ISBN 0-12-032001-0

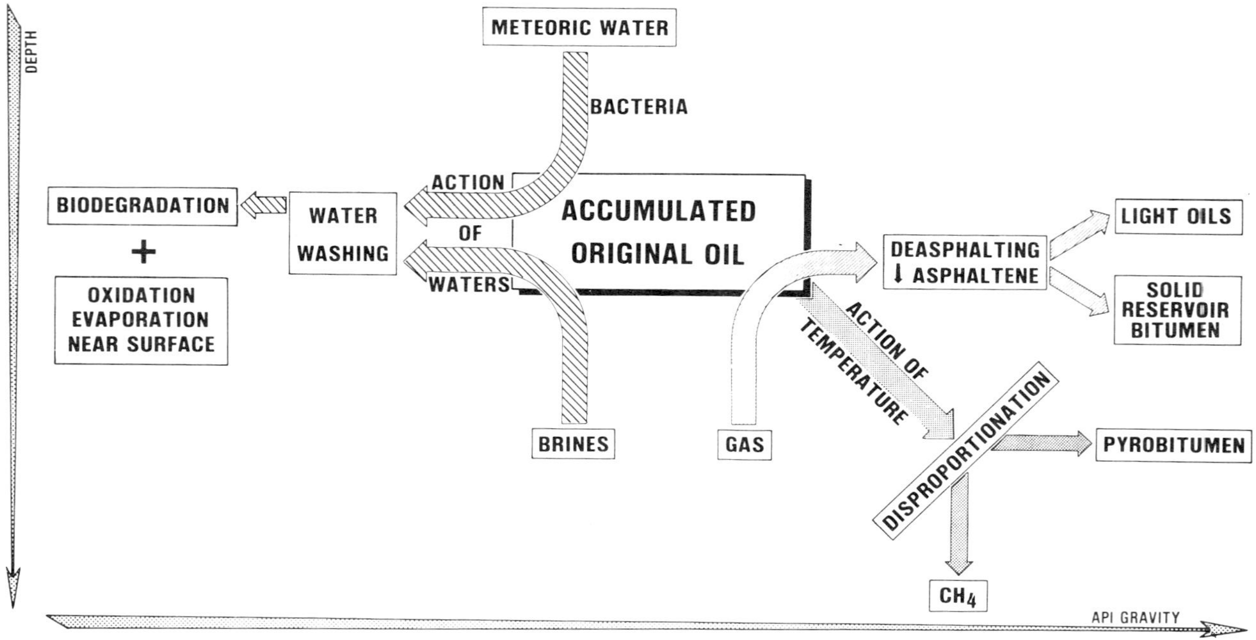

FIG. 1. Alteration of Crude oils in reservoirs (adapted from Milner *et al.*, 1977).

approximately 10^{12} barrels of petroleum *in situ*. The Orinoco oil belt, considered to be one of the world's most important deposits of extra-heavy oil (10° API) is thought to contain between 700×10^9 and 1000×10^9 barrels *in situ* (Zamora and Zambrano, 1982). Thousands of millions of tons of heavy (10°–20° API) to extra-heavy oils (less than 10° API) in Western Canada (Athabasca tar sands and subordinate deposits) and eastern Venezuela originate from microbial transformation of medium- to high-gravity crudes.

Since the first outstanding review paper by Milner *et al.* (1977), no updated synthesis has been published. This paper reviews and critically evaluates the recent literature pertaining to biodegradation in reservoirs, by referring to the author's experience in that field. It includes a discussion of unpublished case histories. The aim is to carry out a survey of several examples which can provide a sequence illustrating the step-by-step biodegradation of crude oil. Degrees of alteration ranging from incipient to drastic changes of both alkanes and aromatics will be presented.

II. Conditions for Aerobic Microbial Degradation of Crude Oil

Transformation of crude oil in reservoirs by aerobic bacteria requires five major conditions (Philippi, 1977; Barker, 1980), which may be summarized as follows:

(i) *Moving waters either by hydrodynamism or by compaction.* Drastic biodegradations are generally observed in shallow reservoirs which are flooded by meteoric waters (low salinity and high sulphate). Evans *et al.* (1971) have shown in Mississippian reservoirs of Saskatchewan (Canada) that the degree of biodegradation of crude oils increases when formation waters become less saline and more sulphate-rich. Fresh, sodium-sulphate invasion waters, i.e. meteoric waters, are found in association with the most biologically degraded crudes.

(ii) *Oil–water contact.* Bacteria live in the aqueous phase and do not thrive in oil. Bacterial degradation takes place at the oil–water interface.

(iii) *Sufficient supply of nutrients* (nitrate, phosphate) and of dissolved oxygen in moving waters.

(iv) *Presence of microbes.*

(v) *Subsurface temperature allowing activity of bacteria.* 100°C appears to mark the upper limit of biodegradation when all bacteria should be killed. Figure 2 summarizes subsurface maximum temperatures of biodegraded oils in several basins. These data are unfortunately not fully comparable because they came both from the literature and the author's files. Maximum temperature values, dependent upon methods of recording (tests, logging instruments, etc.) span from 54°C to 88°C. This result, however, does not mean that bacteria are active within the temperature range but only that typical features related to bacterial degradation may still be recognized

MAXIMUM SUBSURFACE TEMPERATURES OF BIODEGRADED OILS IN SEVERAL BASINS

77° C	LOS ANGELES BASIN (U.S.A.)
62° C	SAN JOAQUIN BASIN (U.S.A.)
82° C	GULF COAST (U.S.A.)
54° C	SOUTH SUMATRA- SOUTH BORNEO
88° C	DAMPIER BASIN (AUSTRALIA)
80° C	AQUITAINE BASIN (FRANCE)
85° C	MACKENZIE DELTA (CANADA)
71° C	BEAUFORT BASIN (CANADA)

FIG. 2. Maximum subsurface temperatures of biodegraded oils in several basins.

within those subsurface temperatures in crude oils. Upper temperature limits of different taxonomic groups have been reviewed by Brock (1978). If eucaryotic micro-organisms such as protozoa, algae and fungi do not survive at temperatures higher than 62°C, procaryotic micro-organisms (cyanobacteria, photosynthetic, chemolithotrophic and heterotrophic bacteria) live and reproduce in the 70–90°C range.

It must be pointed out, however, that the effects of bacterial degradation tend to be less and less important when temperature increases. This statement is documented herein by results obtained in the Aquitaine Basin (south-west France) where severe biodegradation of both alkanes and aromatics has been recorded between 20°C and 60°C (exceptionally 75°C) whereas a slight alteration of alkanes is generally attached to the 61–77°C range (see Table I). In that particular basin, 80°C seems to be the upper limit for detection of any trace of biodegradation processes.

To summarize, extensive biodegradation of crude oils is encountered in reservoirs with a temperature range from 20°C to 60–75°C.

Bacterial degradation of limited extent is the most common phenomenon at higher temperature, i.e. within the 60–88°C range. The occurrence of slightly biodegraded crudes in the 60–80°C range may be due to either less

TABLE I. *Depth and temperature range for biodegraded and unaltered oils in the Aquitaine Basin (S.W. France).*

DEPTH (m)	LOGGING TEMPERATURE (°C)	ALKANES	AROMATICS
SURFACE TO 2700	SURFACE TO 75	DRASTICALLY BIODEGRADED	DRASTICALLY BIODEGRADED
1900 TO 2700	61 TO 77	SLIGHTLY BIODEGRADED	SLIGHTLY OR DRASTICALLY BIODEGRADED
>2800	82	NO BIODEGRADATION	NO BIODEGRADATION

bacterial activity (Barker, 1980) or neogenesis of alkanes by catagenesis which obscured the original pattern of the biodegraded crude (Connan, 1972).

III. Changes in Crude Oil Properties through Biodegradation

A. Biodegradation effects on gross properties

Action of bacteria entails changes in the properties of crude oils, which are listed in Fig. 3.

1. *Compositional changes*

Preferential removal of gases (C_1–C_6) and gasoline range compounds (C_6–C_{15}) by both water washing and biodegradation evolves residual crude oils with lower API gravities and higher viscosities. Oils in which API gravities are lower than 20° at 60°F (15.6°C) under atmospheric pressure and viscosities are higher than 100 centipoises (0.1 Pas) at reservoir temperatures should be termed heavy to extra-heavy oils (Gibson, 1982). Extra-heavy oils are often encountered in outcropping reservoirs, in which evaporation coupled with photo-oxidation and oxidation by atmospheric oxygen contribute significantly to inspissation.

When biodegradation proceeds, further compositional changes are then recorded in the $C_{15}+$ fractions. Alkanes and aromatics are degraded to various extents, whereas NSO compounds and asphaltenes appear to be

1 - GASES (C_1 - C_6) ↘
2 - GOR (GAS/OIL RATIO) ↘
3 - GASOLINE RANGE (C_6 - C_{15}) ↘
4 - API GRAVITY ↘
5 - VISCOSITY ↗
6 - CHANGES IN GROSS COMPOSITION OF C_{15}^+ COMPOUNDS
alkanes ↘
aromatics ↘
NSO's compounds ↗
asphaltenes ↗
7 - SULPHUR CONTENT ↗
8 - NITROGEN CONTENT ↗
9 - V AND Ni ↗
10 - OPTICAL ACTIVITY ↗
alkanes ↗
11 - POUR POINT ↘
12 - $\delta^{13}C$
whole oil ↗
alkanes ↗
aromatics ≃ or ↘
asphaltenes ↘
13 - CHANGES IN OIL TYPES
paraffinic oil → naphthenic oils
paraffinic or paraffinic-naphthenic oils → aromatic-naphthenic oils
paraffinic condensates → naphthenic condensates
condensates → light oils
aromatic - intermediate oils → aromatic-asphaltic oils

FIG. 3. Biodegradation effects on gross properties.

fairly resistant to bacterial attack. The hydrocarbon impoverishment in bacterially altered crudes (natural asphalt) is clearly shown in Fig. 4.

$^{13}C/^{12}C$ fractionations during biodegradation of crude oil have been reported by Stahl (1977, 1980), Schoell (1978), and Hahn-Weinheimer and Hirner (1980). In Stahl's *in vitro* experiments (1980), performed on Ekofisk crude inoculated with North Sea water, $\delta^{13}C$ was enriched in alkanes, depleted in asphaltenes, and remain unchanged in aromatics.

These results, considered as typical trends by Stahl (1977), are probably not valid in all cases. Hahn-Weinheimer and Hirner (1980), in their study of crude oils from the Molasse basin of Southern Germany, showed that biodegradation shifted $\delta^{13}C$ ratios of whole oil, total alkanes, branched-cyclic alkanes and aromatics towards the heavier isotope. This result suggests that aromatic fractions of biodegraded crudes from the Molasse basin are probably severely altered. Hartman and Hammond (1981) concluded

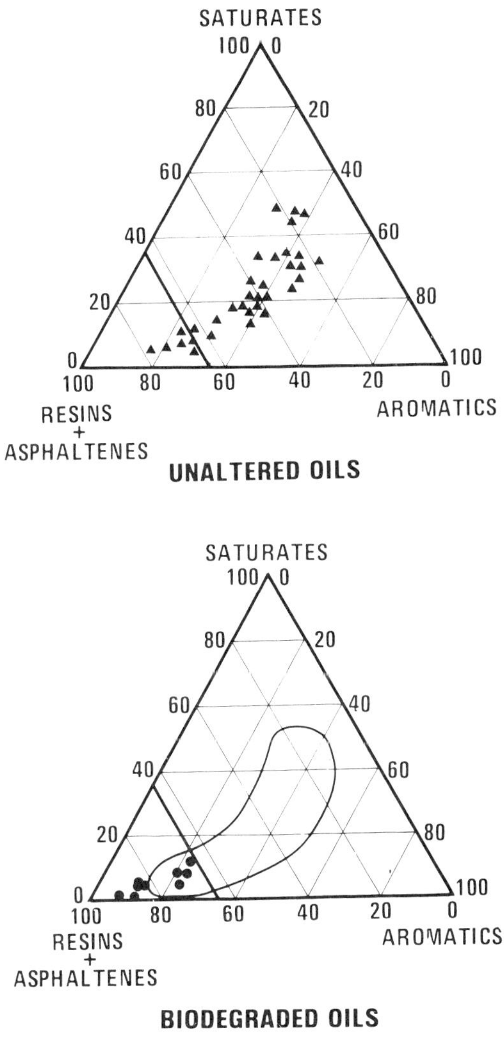

FIG. 4. Gross composition of unaltered and biodegraded oils from the Aquitaine Basin (S.W. France).

from their study of beach tars from the Southern California borderland that $\delta^{13}C$ and $\delta^{34}S$ of asphaltenes are preserved after a three-week period of weathering.

2. *Changes in oil type*

According to Tissot and Welte (1978), conventional non-biodegraded oils should be classified into paraffinic, paraffinic–naphthenic and aromatic–intermediate oils. The changes in crude oil composition due to biodegradation lead to changes in oil type.

Selective removal of alkanes, especially normal alkanes, transforms paraffinic oils into heavy- to medium-gravity naphthenic oils (Philippi, 1977; Snowdon and Powell, 1979). Condensates, also paraffinic in character (Connan and Cassou, 1980), become very light naphthenic oils (Snowdon and Powell, 1979). An example of such a type of oil is presented in Fig. 5. This light oil (38.8° api, 55% volatile fraction) is very rich in alkanes, which represent 87% of the topped crude. Their alkanes, however, display a typical pattern of biodegraded oil.

In the gasoline range *n*-alkanes are completely lacking (Fig. 5), whereas low-boiling naphthenes, i.e. cycloalkanes (cyclopentane, methylcyclopentanes, cyclohexane, methylcyclohexanes, ethylcyclohexanes, etc.) predominate. Paraffinic–naphthenic oils, when severely degraded, turn into aromatic–naphthenic oils (Claret *et al.*, 1977), Aromatic–asphaltic oils, rich in polar compounds (NSO's and asphaltenes) and in sulphur, are generally the altered counterpart of aromatic–intermediate oils (Tissot and Welte, 1978). A representative example of such a type is shown in Fig. 4. The natural asphalts from the Aquitaine tar belt are viscous or quasi-solid oils which result from sulphur-rich aromatic–intermediate oils.

The degradability of NSO's and asphaltenes still remains a subject of debate among geochemists and microbiologists. Kallio (1982) summarized the problem by writing "There is little evidence of bacteria contributing asphaltenes to crude oils or degrading asphaltenes". Bailey *et al.* (1973b) claimed that bacteria can produce asphaltenes. The results of Rubinstein *et al.* (1977) led to an opposite conclusion: that polar compounds are intact after biodegradation. Asphaltene, however, indeed seems degradable. Zajic *et al.* (1977) established that *Pseudomonas* species, i.e. typical hydrocarbon oxidizing bacteria, are able to metabolize asphaltene. Finnerty *et al.* (1982) confirmed that point of view by revealing bacteria which can grow at the expense of asphaltene substrate.

Nevertheless, even if asphaltenes are degradable, biodegradation of crude oil obviously starts with both alkane and aromatic fractions. This preferential hydrocarbon uptake ends with a correlative non-hydrocarbon enrichment in the residual oil. The enrichment in NSO's and asphaltenes

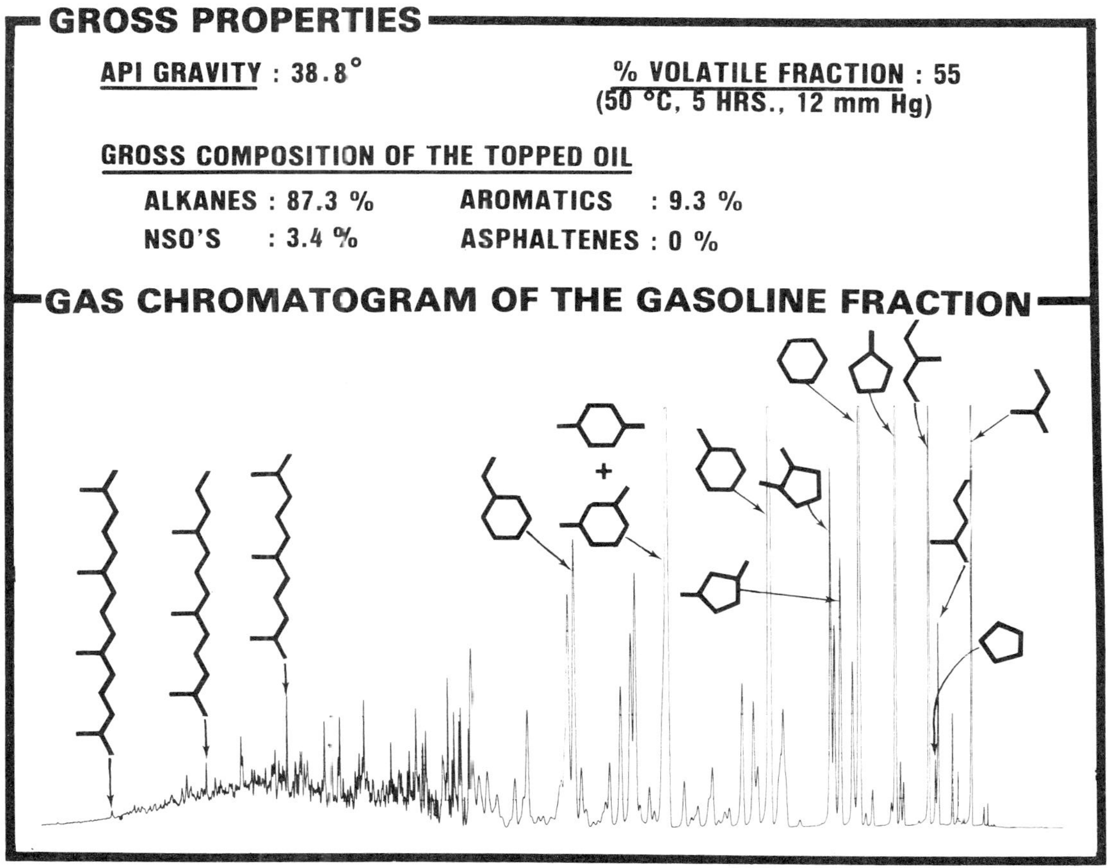

FIG. 5. An example of a biodegraded condensate.

is responsible for the increase of nickel, vanadium, sulphur and nitrogen in the biodegraded residue.

3. *Increase in optical activity*
Bacterially degraded crudes, and their associated alkane fraction, show an increase in optical activity. Winters and Williams (1969) suggest that this may be due to the addition of optically active compounds produced by microbes. Seifert and Moldowan (1979) explained high optical activities in their biodegraded crudes as a result of one or more of the following three possibilities:

(i) removal of non-active structures, i.e. *n*-alkanes;

(ii) immaturity of original crudes which were rich in optically active polycyclic alkanes;

(iii) transformation of steranes and terpanes into other optically active structures.

4. *Lowering of the pour point*
The above-mentioned uptake of *n*-alkanes, especially the waxy ones (C_{22} to C_{30}), lowers the pour point of biodegraded crudes. This feature may be highly beneficial in very cold areas (Alaska, Arctic islands, etc.), in which high-wax oils are produced. Burns *et al.* (1975) emphasized this point in a study of Beaufort basin oils. They showed that the lowering of the pour point by biodegradation from +25°F (4°C) to −35°F (−37°C) reduced flow problems in production and transportation.

B. Biodegradation effects at a molecular level

Some of the initial metabolic products of some alkanes and aromatics are listed in Fig. 6. Alkanes and aromatics are initially degraded into oxygen bearing derivatives (fatty acids, alcohols, ketones, phenols, etc.) The degradation of hydrocarbons by micro-organisms is a bio-oxidation process (see Fig. 6).

1. *Biodegradation of alkanes* (Fig. 7)
Biodegradation of alkanes covers a wide range of molecular weights from C_1 (methane) to C_{27}–C_{35}, including steranes and triterpanes.

(*a*) *Gaseous alkanes*—C_1–C_5
Bacterial methane oxidation is generally considered as a true aerobic phenomenon; however, some anaerobic pathways are also possible. Methane-oxidizing bacteria change the isotopic composition of the residual

methane. Both carbon and hydrogen in the methane remaining after bacterial oxidation are enriched in heavier isotopes (Stahl, 1977; Coleman *et al.*, 1981).

Methane-utilizers also co-oxidize C_2–C_4 alkanes. Other varieties of micro-organisms (ethane-, propane- and butane-utilizing bacteria, moulds

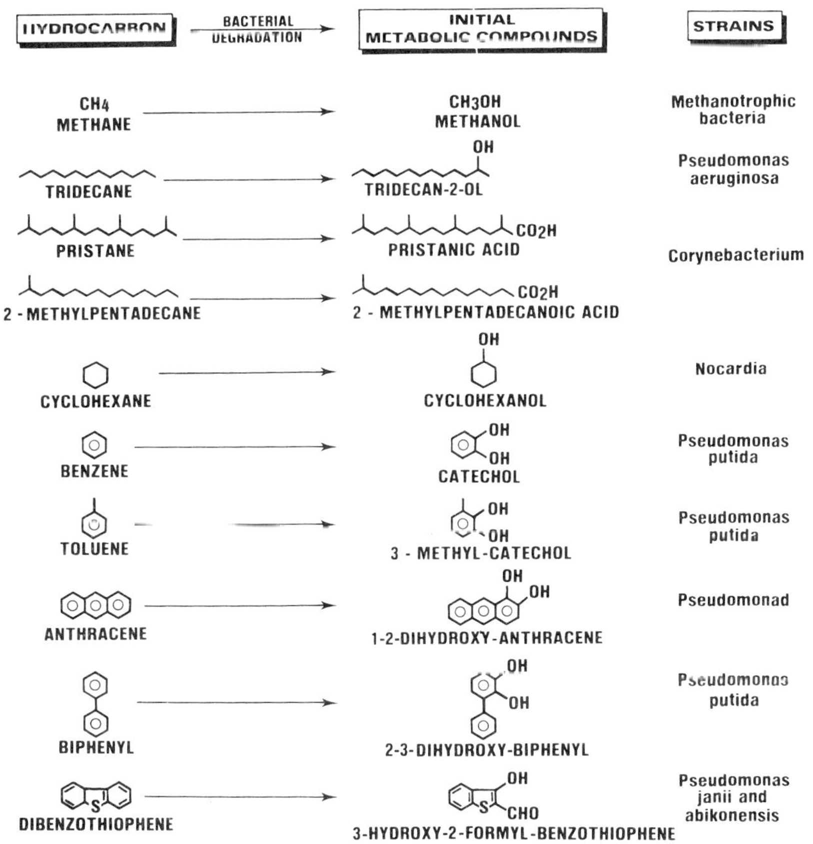

FIG. 6. Initial metabolic fate of some alkanes and aromatics (after Higgins and Gilbert, 1978; Hopper, 1978; Radledge, 1978; Cripps and Watkinson, 1978; Kodama *et al.*, 1970).

and yeasts) also metabolize these compounds to grow (Higgins and Gilbert, 1978). Iso-butane generally degrades at a slower rate than normal-butane (Bopp *et al.*, 1981). Increasing *i*-C_4 to *n*-C_4 ratios from 1.8 to 14.0 in sediments at 42–58°C were ascribed by Alexander *et al.* (1981) to a selective removal of *n*-butane by bacteria. A similar explanation is also proposed by Bayliss and Smith (1980) in their source-rock evaluation reference manual. It may also be extended to i-C_5 to n-C_5 ratios, which increase with biodegradation (Stahl, 1980).

(b) Gasoline fraction—C_6–C_{15}

Regarding the gasoline fraction (C_6–C_{15}), Higgins and Gilbert (1978) wrote "The C_{10}–C_{18} compounds are biodegraded most readily but most hydro-

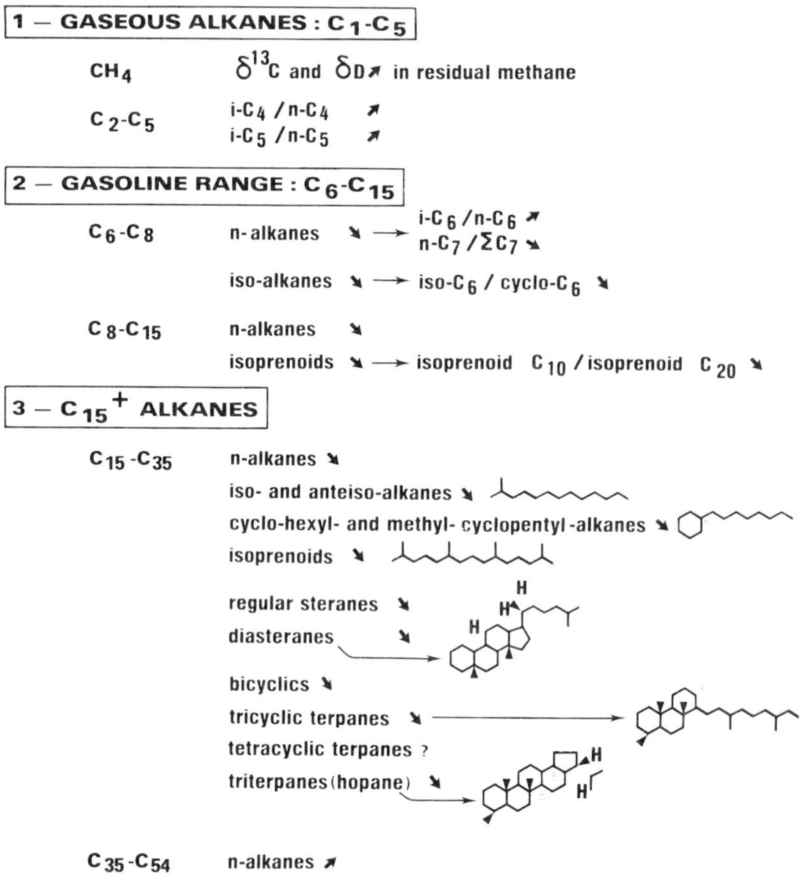

FIG. 7. Biodegradation effects on alkanes.

carbon-utilizing yeasts seem unable to metabolize the shorter-chain liquid hydrocarbons (C_5–C_9) although some bacteria and fungi grow on these compounds". Milner *et al.* (1977) also favour this idea by quoting data from

Perry and Cerniglia in which C_{10}–C_{19} *n*-alkanes are biodegraded first.

Schaefer and Leythaeuser's results (1980) suggest an opposite sequence of degradability among alkanes by showing a biodegraded crude in which *n*-alkanes are severely depleted in the C_6–C_8 range but not affected in the C_{15} + fraction. The preferential uptake of C_6–C_8 *n*-alkanes, observed also by Philippi (1977) and Magoon and Claypool (1981), who used *n*-$C_7/\Sigma C_7$ as an index of biodegradation, is not restricted to this narrow molecular range. In fact the *n*-alkane removal seems to affect the whole gasoline range (C_6–C_{15}) prior to the degradation of branched and cyclic alkanes. Figure 5 confirms this statement by presenting a light oil in which *n*-alkanes are lacking within the C_6–C_{18} range, whereas cycloalkanes with subordinate branched and isoprenoid alkanes may be clearly identified. According to the results in Fig. 5, confirmed by Snowdon and Powell (1979), the removal sequence is light *n*-alkanes followed by heavy *n*-alkanes. Biodegradation of isoprenoids apparently followed a similar trend for Illich *et al.* (1977), who used the C_{10}-isoprenoid to C_{20}-isoprenoid ratio as a measure of the degree of biodegradation in crude oils from the Marañon Basin (Peru).

(*c*) C_{15}+ *alkanes*

The preferential removal of C_{15}–C_{35} *n*-alkanes relative to branched and cyclic alkanes, well documented for the first time by Winters and William (1969) in their study of the Bell Creek field in the Powder Basin (U.S.A.), is now universally recognized as a standard feature in assessing biodegradation of crude oils. Numerous examples are listed in the review literature by Tissot and Welte (1978), Hunt (1979) and Milner *et al.* (1977), so this well known work will not be dwelt on at length here. An example of a step-by-step removal of *n*-alkanes is depicted in Fig. 8. In that well from offshore Canada, oil and condensate shows have been produced between 2290 m and 3005 m. DST 3 (2952–2959 m) condensate as well as DST 2 oil (3000–3005 m) display gas chromatograms of C_{15}+ alkanes in which *n*-alkanes are very abundant. Both fluids are obviously non-biodegraded (DST 3) or very mildly biodegraded (DST 2). In FIT 1 oil, accumulated in a shallower reservoir (2462 m), the biological uptake of *n*-alkanes has begun. They have been completely removed in DST 4 oil. This oil, pooled in the shallowest reservoir at a depth of 2290.50–2293.50 m, is the most biodegraded oil among the four fluids analysed herein. The gradual removal of *n*-alkanes by bacteria is often followed by measuring ratios of pristane to *n*-C_{17} and phytane to *n*-C_{18}, which increase with increasing biodegradation (Bailey *et al.*, 1973a; Jobson *et al.*, 1972). An example of such an evolution is depicted in Fig. 8. Within this set of samples, the classification of crude with regard to their increasing degree of biodegradation is as follows: DST 3, DST 2, FIT 1 and DST 4. The corresponding ratios of pristane to *n*-C_{17}

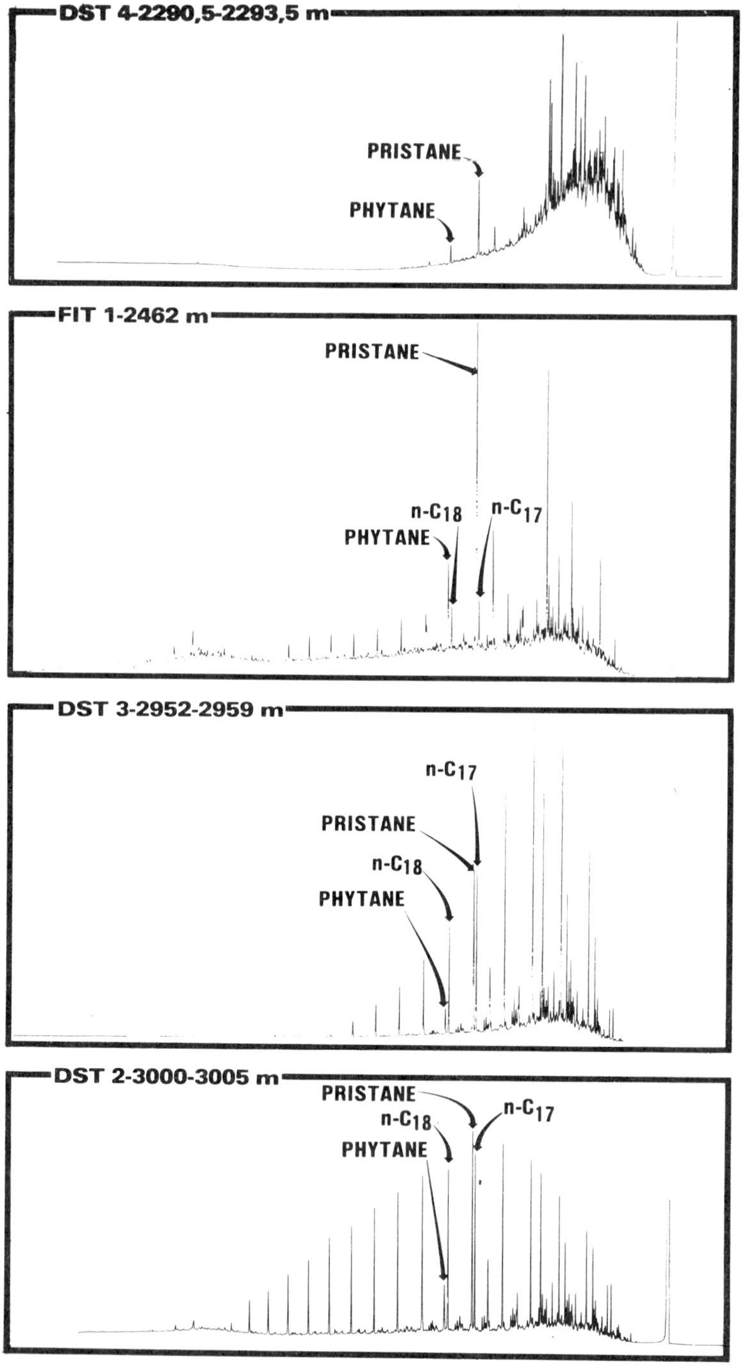

FIG. 8. Compositional changes in the total alkanes of oils by biodegradation of limited extent.

Regarding both tricyclic (21/3 to 26/3) and hopane (29, 30, etc.) spectra, no significant changes may be noticed in Les Epoisats asphalt; nevertheless the relative amount of an unknown C_{30} triterpane is enhanced. This feature seems to indicate an incipient biodegradation of hopane compounds. Removal of the hopane family is completed in Val de Travers asphalt. In this severely biodegraded crude the C_{30} triterpane is concentrated and the tricyclic family still survives. This bacterially resistant family (Reed, 1977; Seifert and Moldowan, 1979; Connan *et al.*, 1979), found by Goodwin *et al.* (1981) as unaffected in samples in which regular steranes and hopane are missing, are indeed degradable as shown in the St Aubin asphalt (Fig. 11).

Biodegradation of steranes and triterpanes, originally discovered in naturally biodegraded samples (Reed, 1977; Seifert and Moldowan, 1979) failed to be reproduced in the laboratory by Rubinstein *et al.* (1977) and Connan *et al.* (1979), Goodwin *et al.* (1981) made a definite contribution by demonstrating that both steranes and triterpanes (hopane) can be degraded under laboratory conditions. The conclusions of their *in vitro* experiments are as follows:

(i) Regular steranes are degraded at a faster rate than rearranged steranes, as already claimed by Seifert and Moldowan (1979).

(ii) Among triterpanes, hopanes are eliminated more rapidly than moretanes and the 22R hopane is preferentially degraded.

(iii) Hopanes are not converted into ring A/B demethylated hopanes.

Neogenesis of ring A/B demethylated hopanes (Seifert and Moldowan, 1979) was not observed in their naturally degraded samples from Switzerland, either. They do not occur in our natural asphalts from Switzerland, namely Val de Travers and St Aubin (Fig. 11). What happens to hopanes as a result of biodegradation? Demethylated hopanes may be one possibility, as postulated by Seifert and Moldowan (1979); however, some other routes via skeleton cleavage may be also invoked.

To summarize information available from different sources, including the author's records, it has been attempted to set up a removal sequence in alkanes. Figure 13 lists the different classes of molecular structures, beginning with the most readily degradable compounds (C_6–C_{15} *n*-alkanes) and ending with the most bacterially resistant (tricyclic terpanes). It should be stressed that this classification is not valid in every situation, as emphasized by Goodwin *et al.* (1981); the relative rate of biodegradation of steranes and triterpanes for instance depends upon specific environmental conditions. In one of their samples C_{27}–C_{29} diasteranes were found to disappear before hopanes, whilst in another sample the hopane degradation took place before diasterane consumption.

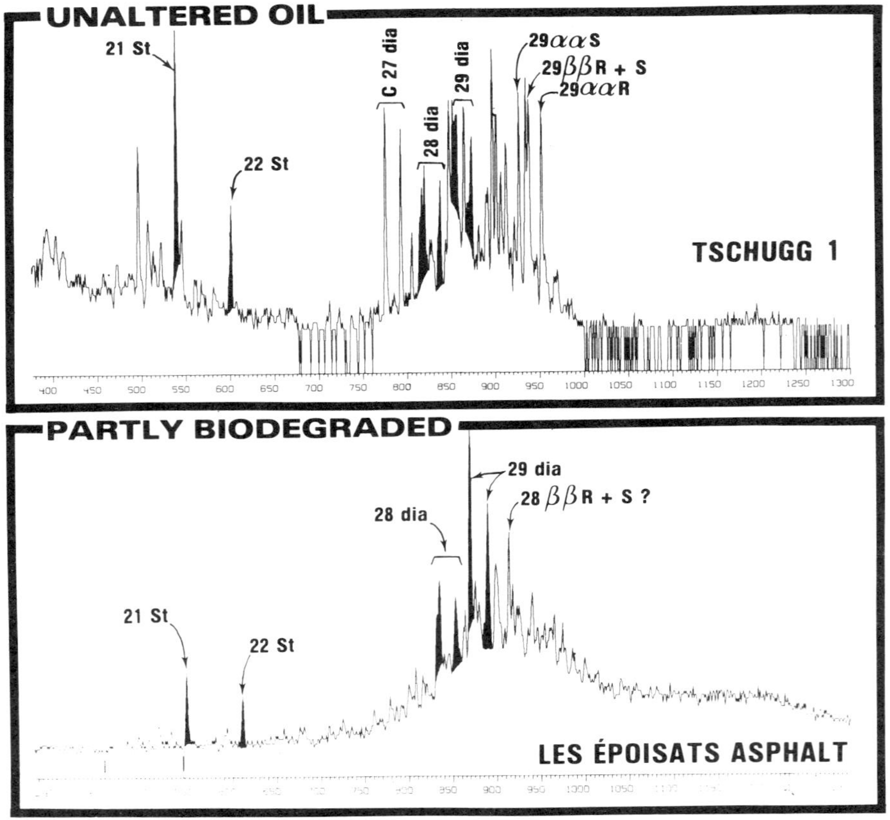
UNALTERED OIL
21 St
22 St
C 27 dia
28 dia
29 dia
29αα S
29ββ R + S
29αα R
TSCHUGG 1
400 450 500 550 600 650 700 750 800 850 900 950 1000 1050 1100 1150 1200 1250 1300
PARTLY BIODEGRADED
21 St
22 St
28 dia
29 dia
28 ββ R + S ?
LES ÉPOISATS ASPHALT

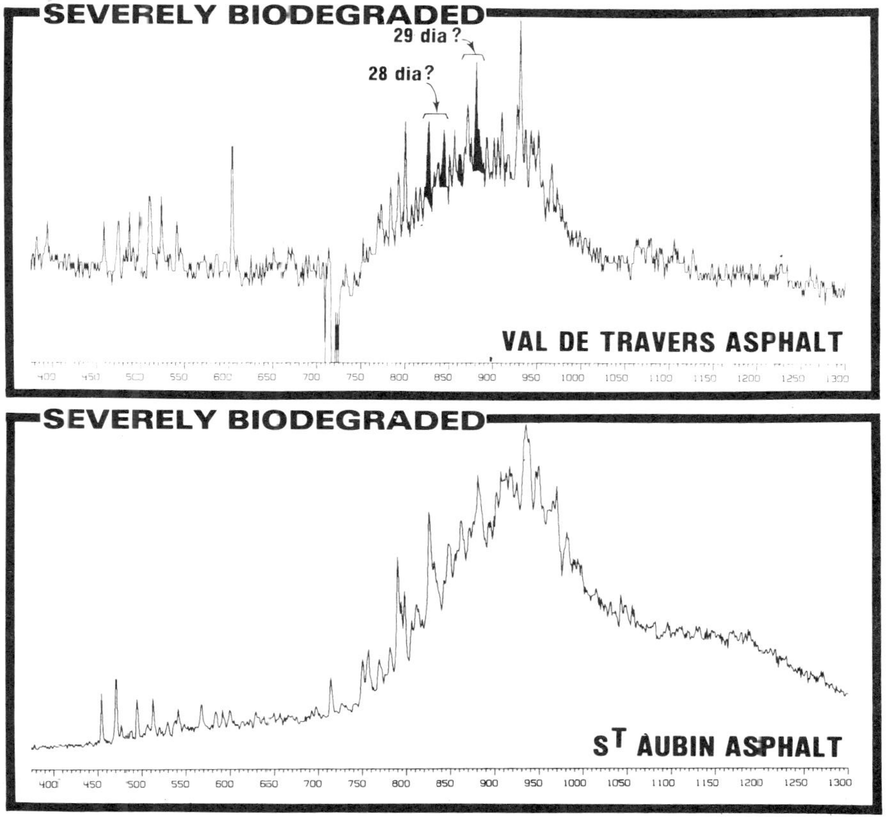

FIG. 10. Sterane distribution as shown by the plot of mass 217. An example of biodegradation of steranes in severely altered oils from Switzerland.

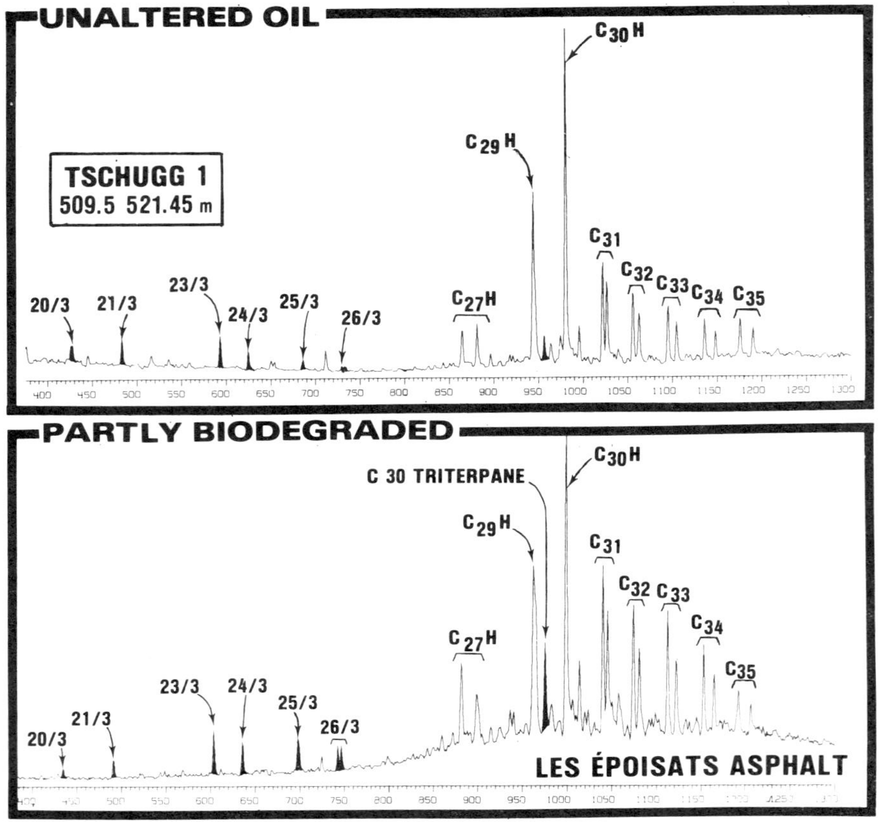
UNALTERED OIL
TSCHUGG 1
509.5 521.45 m
20/3
21/3
23/3
24/3
25/3
26/3
C27H
C29H
C30H
C31
C32
C33
C34
C35
PARTLY BIODEGRADED
C 30 TRITERPANE
LES ÉPOISATS ASPHALT

and phytane to n-C_{18} for the first three samples are respectively 0.98, 1.1, 6.4 and 0.23, 0.29, 2.0.

If C_{15}–C_{35} *n*-alkanes are rapidly eliminated from crude oils under natural conditions (ten days with *Pseudomonas oleovorans* in the laboratory (Connan *et al.*, 1979), one year after the Amoco Cadiz oil spill in the sandy sediments of the Roscoff beach (Oudot *et al.*, 1981) very-long-chain alkanes (n-C_{35}–n-C_{55}) appear in high concentration in the residual oil (Winters, 1978, personal communication; Oudot *et al.*, 1981; Kallio, 1982). The main problem that arises from the occurrence of these very-long-chain alkanes in crude oil lies in deciding whether these compounds pre-existed as undetectable amounts in the original crude or whether they are *de novo* synthesized structures resulting from bacterial activity. The second hypothesis is considered as very likely by Winters, Oudot *et al.* and Kallio. Even predominance of *n*-alkanes, exclusively found in the C_{38}–C_{54} range according to Winters (1978, personal communication) and Kallio (1982), is also considered by Oudot *et al.* (1981) as a proof of the bacterial production of very-long-chain alkanes. Production of even *n*-alkanes by some bacterial strains is already reported in literature. Following the observations he made in the marine environment, Oudot *et al.* (1981) recommended the use of very-long-chain alkanes as long-term indications in oil-polluted areas. Their occurrence in significant amounts is detectable later when weathering has reached a very advanced stage. To the author's knowledge, however, nobody except Winters has reported results relating to oils biodegraded in reservoirs.

Among other homologous series of molecular structures which are readily attacked by bacteria, one must mention iso-, anteiso-, cyclohexyl-/methylcyclopentylalkanes (Aldridge, 1977; Philp and Gilbert, 1980). According to Radledge (1978), single-branched chain alkanes, i.e. iso- and anteisoalkanes, are attacked at the end of the molecule which does not contain the methyl substituent. The fatty acid generated is then incorporated into the cell. Screening of samples for the presence of iso-, anteiso-, alkylcyclohexyl-, methylcyclopentylalkanes should be done by recording a single ion plot for mass 85 (iso- and anteisoalkanes) and mass 83 (alkylcyclohexyl- and/or methylcyclopentylalkanes).

Figure 9 depicts a typical example of incipient biodegradation of $C_{15}+$ alkanes. Both samples have come from the Vic Bilh oil field (Aquitaine Basin). VBH.9 oil may be considered as a biodegraded counterpart of VBH.2 oil. The gas chromatogram of total alkanes clearly indicates that the *n*-alkane uptake has begun in VBH.9; nevertheless, one must rely on a computerized GC–MS analysis to state whether biodegradation of other homologous series of long-chain alkanes has started. The ion plot of mass 85 and 83 (Fig. 9) reveals that the four homologous series other than

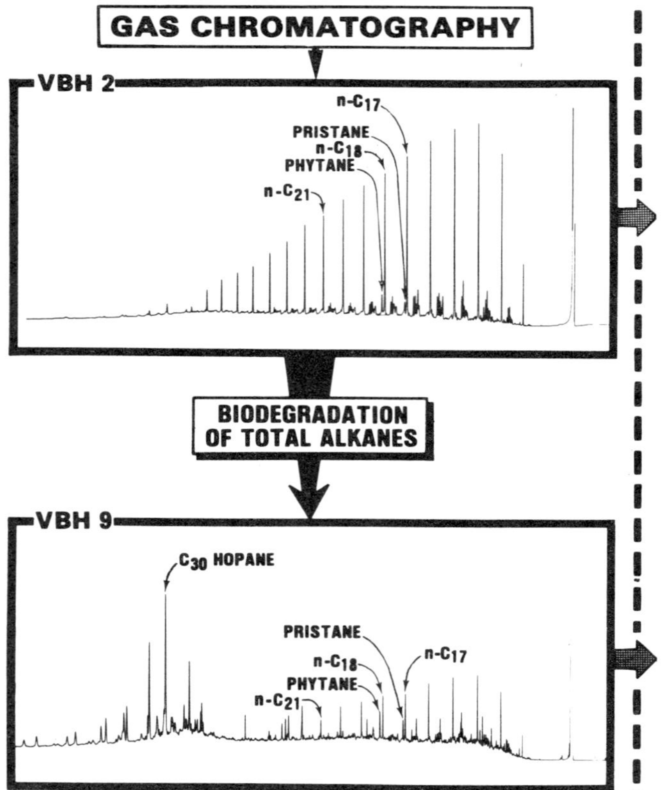

GAS CHROMATOGRAM OF TOTAL ALKANES

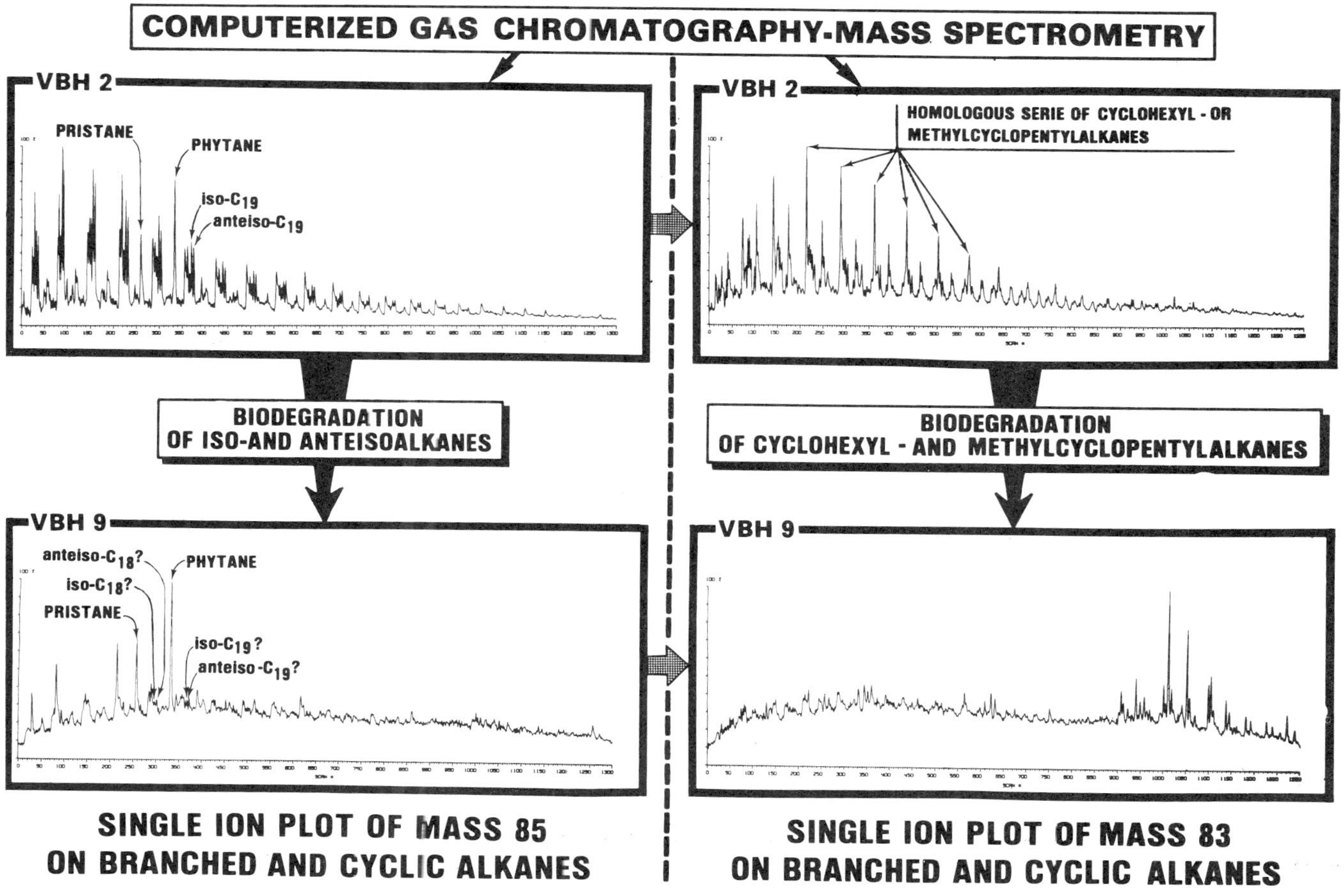

FIG. 9. Removal of iso-, anteiso-, cyclohexyl- and methylcyclopentylalkanes by biodegradation of limited extent.

n-alkanes have undergone significant biodegradation. One should emphasize that a careful screening of the distribution of the four above-mentioned series of iso- and cycloalkanes is very useful in assessing biodegradation of very limited extent.

Multiple branched alkanes such as isoprenoids (farnesane, C_{16}, C_{18}, pristane, phytane), less readily degradable than single-branched or *n*-alkanes, are oxidized by bacteria when biodegradation of crude oil proceeds one step further (Bailey *et al.*, 1973a, b; Jobson *et al.*, 1972). Isoprenoid uptake is a well known phenomenon, widely documented in literature, especially in textbooks by Tissot and Welte (1978) and Hunt (1979), and has been clearly demonstrated in the laboratory (Cox *et al.*, 1974, 1976). Removal of isoprenoids seems to gradually progress from C_{15} to C_{20} (Illich *et al.*, 1977; Blanc *et al.*, 1982) and there appears to be no effect on their stereochemistry as indicated by studies of pristane (Mackenzie *et al.*, 1981).

Tetra- and pentacyclic alkanes, namely steranes and triterpanes, more resistant to bacterial attack (Rubinstein *et al.*, 1977; Connan *et al.*, 1979) are indeed degraded under natural conditions in the subsurface (Reed, 1977; Seifert and Moldowan, 1979; Goodwin *et al.*, 1981; Mackenzie *et al.*, 1981). Figures 10 and 11 depict a typical case history from Switzerland (Fig. 12) in which both steranes and terpanes are gradually affected. The Tschugg oil (core extract, 509.50 m–521.45 m, Urgonian) possesses the usual molecular pattern of crude oils from both the German (Hufnagel *et al.*, 1980) and the Swiss (Goodwin *et al.*, 1981) Molasse basin. The sterane distribution monitored by the plot of mass 217 (Fig. 10) shows the occurrence of C_{27}–C_{28}–C_{29} diasteranes, $C_{29}\alpha\alpha$R and S, and $C_{29}\beta\beta$R and S regular steranes. This mass fragmentogram matches perfectly with those published by Goodwin *et al.* (1981) for sample 6, i.e. a medium-degraded oil from the Geneva area. C_{21} and C_{22} low molecular weight steranes, previously identified in the Aquitaine basin (Connan *et al.*, 1979), are also present. Terpane analysis achieved by monitoring mass 191 (Fig. 11) shows the hopane and the tricylic terpane families as the dominant ones.

In Les Epoisats asphalt (tectonic breccia in Bathonian sediments, outcrop sample), the sterane alteration has taken place (Fig. 10). C_{27} diasteranes are obviously missing, whereas $C_{28}\alpha\alpha$ and $\beta\beta$ regular steranes as well as C_{28} and C_{29} diasteranes have been seriously reduced. This result is consistent with the data of Seifert and Moldowan (1979), who observed that C_{29} diasteranes survive better than C_{27} diasteranes. In Val de Travers and St Aubin asphalt (Urgonian, mine and outcrop sample), biodegradation has proceeded one step further by eliminating C_{21}–C_{22} steranes as well as C_{28}–C_{29} diasteranes. Some peaks still occur; however, they could not be definitely ascribed to steranes.

Figure 11 refers to the evolution of related terpane distribution patterns.

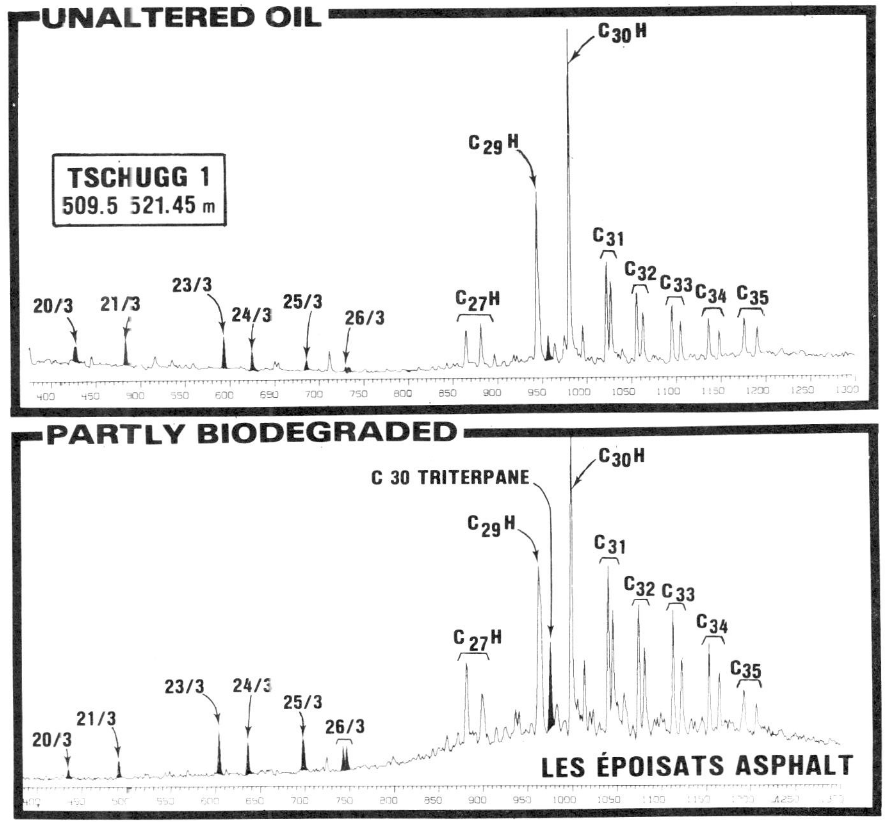
UNALTERED OIL
TSCHUGG 1
509.5 521.45 m
20/3
21/3
23/3
24/3
25/3
26/3
C27H
C29H
C30H
C31
C32
C33
C34
C35
400
450
500
550
600
650
700
750
800
850
900
950
1000
1050
1100
1150
1200
1250
1300
PARTLY BIODEGRADED
C 30 TRITERPANE
LES ÉPOISATS ASPHALT

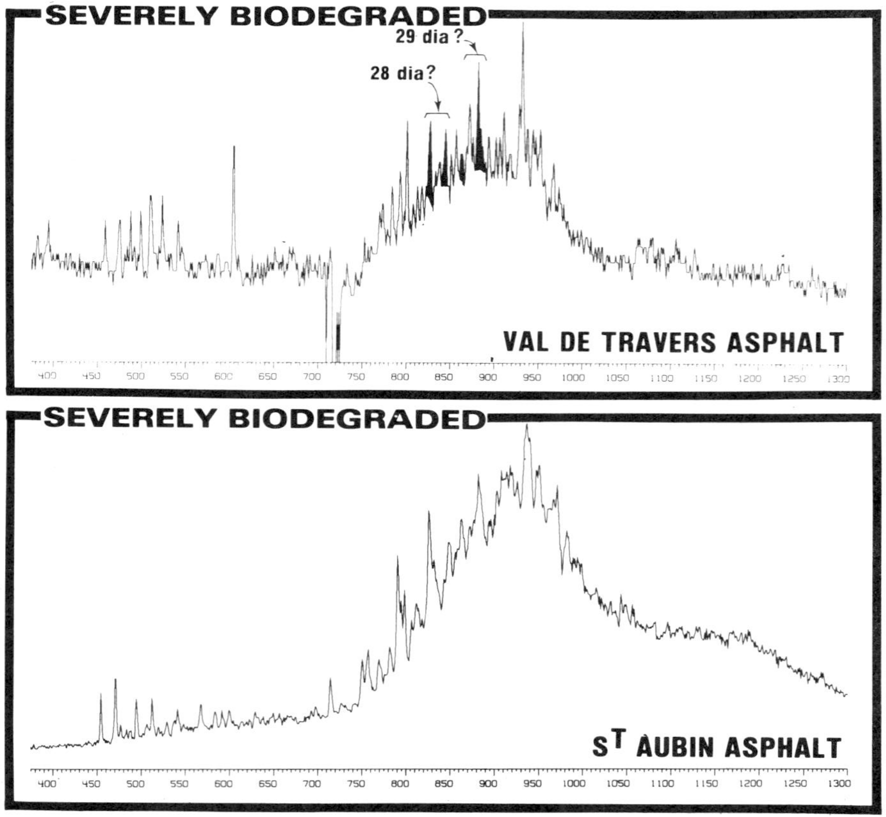

FIG. 10. Sterane distribution as shown by the plot of mass 217. An example of biodegradation of steranes in severely altered oils from Switzerland.

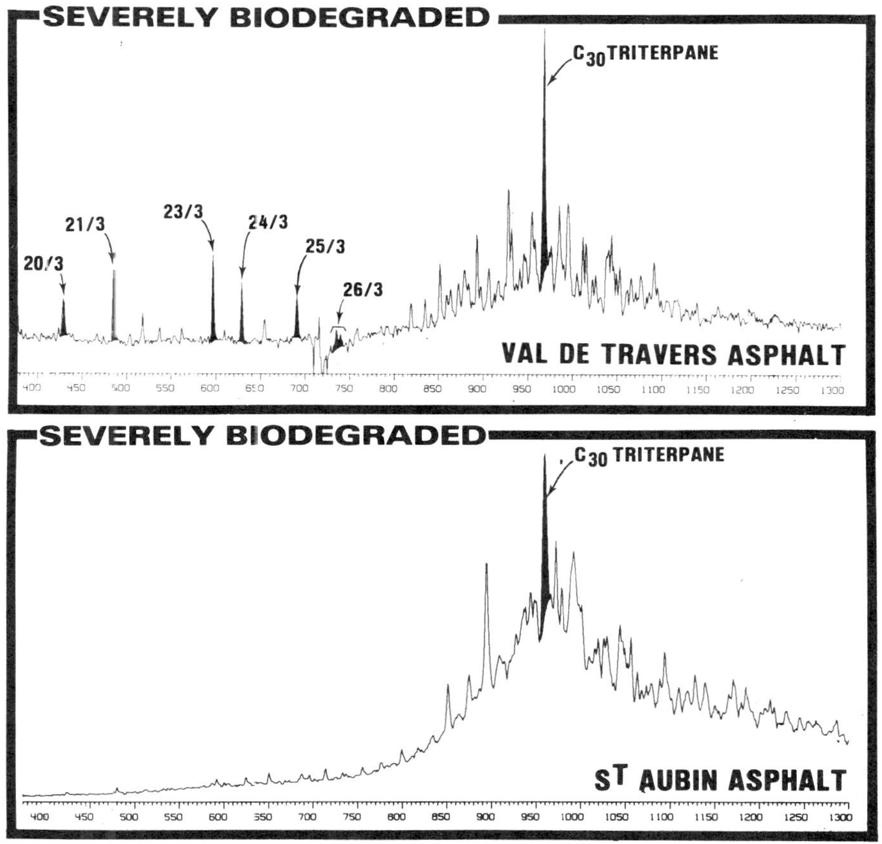

FIG. 11. Terpane distribution as shown by the plot of mass 191. An example of biodegradation of terpanes in severely altered oils from Switzerland.

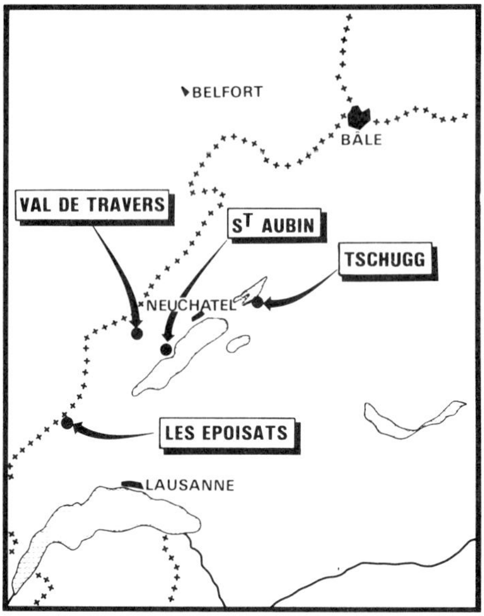

FIG. 12. Location of unaltered and biodegraded oils from Switzerland.

C_6-C_{15} N-ALKANES

C_6-C_{15} ISOALKANES

C_6-C_{15} CYCLOALKANES
ISOPRENOIDS

C_{15}-C_{35} N-ALKANES
ISOALKANES
ANTEISOALKANES
CYCLOHEXYL-AND METHYLCYCLOPENTYLALKANES

C_{15}-C_{21} ISOPRENOIDS

C_{27}-C_{29} REGULAR STERANES ($C_{27} > C_{28} > C_{29}$)
C_{30}-C_{35} HOPANE (C_{35}-C_{34}-$C_{33} > C_{32}$-C_{31}-C_{30} and 22 R >22 S)
C_{27} DIASTERANES (20 S > 20 R)

C_{28}-C_{29} DIASTERANES ($C_{28} > C_{29}$*)
C_{27}-C_{29} HOPANE
C_{21}-C_{22} STERANES ($C_{21} > C_{22}$)

TRICYCLIC TERPANES

*$C_{28} > C_{29}$:C_{28} is biodegraded at a faster rate than C_{29}

FIG. 13. Step-by-step biodegradation of alkanes: a proposal of removal sequence in alkanes.

2. *Biodegradation of aromatics* (Fig. 14)
If, during the last decade, special attention has been paid to the bacterial degradation of alkanes, much less effort has been devoted to elucidate the molecular changes that take place within the aromatic fractions.

Most of the basic papers on bacterial degradation of crude oils in reservoirs, i.e. those by Winters and Williams (1969), Bailey *et al.* (1973), Jobson *et al.* (1972, 1979), Milner *et al.* (1977), Sassen (1980), etc., refer only to progressive compositional changes among alkanes. Few papers (Claret *et al.*, 1977; Deroo *et al.*, 1977; Rubinstein *et al.*, 1977) provide detailed information on the microbial transformation of aromatics.

Biodegradation of aromatic hydrocarbons is extensively described by Higgins and Gilbert (1978), Cripps and Watkinson (1978), Cain (1980) and Callely (1978). Many organisms can grow at the expense of aromatics and oxidize them to CO_2 and H_2O. Metabolic pathways of benzene, toluene, xylenes, naphthalene and substituted naphthalenes, phenanthrene and anthracene, biphenyl, thiophene and dibenzothiophene are described in the above-mentioned books.

Mechanisms involved in the biodegradation of hydrocarbons containing more than three fused rings are not known; however, according to Cripps and Watkinson (1978), benzopyrene (five rings) and benzoanthracene (five rings) have been found to be degraded in soils.

(*a*) *Low-boiling aromatics*
Depletion of benzene, toluene and, to a lesser extent, of xylenes in crude oils is generally ascribed to water washing rather than biodegradation (Barker, 1980; Schaefer and Leythaeuser, 1980). Benzene and toluene, i.e. the most soluble low-boiling hydrocarbons according to Price (1973, 1976) and McAuliffe (1966), are indeed the most susceptible to water washing. In order to circumvent the effect of either water washing or biodegradation, which often act simultaneously, the gasoline range of the slightly altered oils from Fig. 8 has been carefully analysed.

The gas chromatograms recorded by the so-called "thermovaporization" technique are reproduced in Fig. 15. Specific ratios related to compounds identified in Fig. 15 are listed in Table II. The gradual changes in ratios when comparing unaltered (DST 3) to mildly altered (DST 2 and FIT 1) oils do not match the evolution predicted on the basis of solubility data. Removal of benzene, which is the most soluble hydrocarbon, may be seen as mainly due to water washing (refer to the ratios of benzene to n-C_6 and benzene to cyclohexane, Table II). Evolution of toluene and xylenes in comparison to that of their parent n- and cycloalkanes cannot be explained by differential solubility processes. Toluene and xylenes, impoverished by reference to cycloalkanes, are concentrated with respect to n-alkanes.

The last result, the reverse of that predicted according to solubility data, means that the preferential biodegradation of n-C_7 and n-C_8 is more efficient than the water washing of toluene and xylenes. In addition, ratios of *o*-xylene to *p*-xylene and *p*-xylene to *m*-xylene (Table II), unchanged in

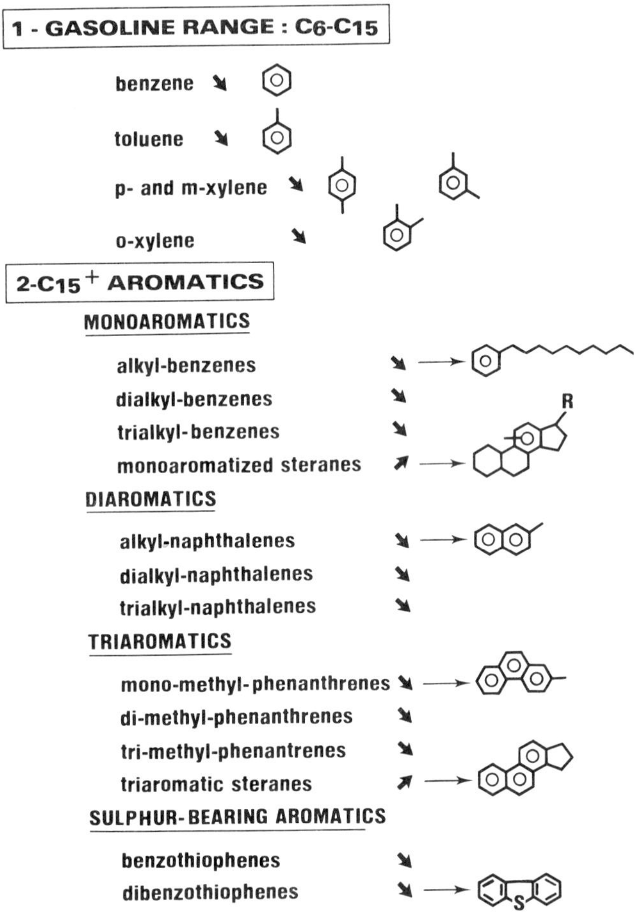

FIG. 14. Biodegradation effects on aromatics.

FIT 1, seem to indicate that xylenes did not undergo biodegradation. *p*- and *m*-xylenes have been used as substrate for pure cultures, whereas growing organisms on *o*-xylene was unsuccessful (Hopper, 1978). The occurrence of two adjacent methyl substituents in *ortho*-xylene decreases the biodegradation rate. Consequently, ratios of *o*- to *p*- or *m*-xylene should be a useful parameter to assess early stages of biodegradation.

(b) C_{15}+ *aromatics*

Among C_{15}+ aromatics, homologous series of alkylbenzenes, dialkylbenzenes and triakylbenzenes are highly susceptible to bacterial attack, either in recent sediments contaminated by oil spills (Oudot *et al.*, 1981) or in oils pooled in reservoirs (Aldridge, 1977; Snowdon and Powell, 1979; Philp and Gilbert, 1980; Connan, 1981a, b). Long-chain *n*-alkylbenzene degradation mimics *n*-alkane metabolism. 1-phenylnonane, for instance, is initially converted to the corresponding phenylalkanoic acid by oxidation of the terminal methyl group (Higgins and Gilbert, 1978). The occurrence of long-chain alkyl substituents in the three families of alkylbenzenes explains why these compounds are readily degraded by bacteria adapted to *n*-alkane consumption. These families are generally missing in mildly biodegraded oil (Aldridge, 1977; Connan, 1981a). Other classes of monoaromatic structures such as the monoaromatized steranes display an opposite behaviour with regard to biodegradation. These tetracyclic molecules are presumably bacterially resistant (Rubinstein *et al.*, 1977; Philp and Gilbert, 1980; Connan, 1981a).

Di- and triaromatic fractions containing complex mixtures of alkylated naphthalenes, anthracenes and phenanthrenes are less susceptible to biodegradation than mono-, di- and trialkylbenzene series (Rubinstein *et al.*, 1977; Aldridge, 1977; Connan, 1981a). Methylated derivatives of napthalene (two rings), are however, fairly rapidly degraded, either in weathered, oil-stained seashore sediments (Oudot *et al.*, 1981) or in oils biodegraded *in vitro* or in reservoirs (Rubinstein *et al.*, 1977). As far as methylated naphthalenes are concerned, one may point out that bacteria preferentially remove monomethylated structures (Rubinstein *et al.*, 1977; Connan, 1981a).

Methylated phenanthrenes, i.e. three aromatic rings, persist for a longer period of time in oils biodegraded at sea or in reservoirs. Triaromatic steranes seem to be unchanged in heavily biodegraded oils (Rubinstein *et al.*, 1977; Connan, 1981a); however, much more accurate investigations should be made before this statement can be confirmed.

Alkylchrysene series (four rings) suffer compositional changes in oils biodegraded under laboratory conditions (Rubinstein *et al.*, 1977).

Sulphur-bearing aromatics, i.e. alkylated thiophene, benzothiophene, dibenzothiophene and naphthobenzothiophene series, have been reported to be partially attacked by bacteria (Deroo *et al.*, 1977; Claret *et al.*, 1977). As generally observed in the subsurface, benzothiophenes are eliminated prior to dibenzothiophenes and naphthobenzothiophenes (Deroo *et al.*,

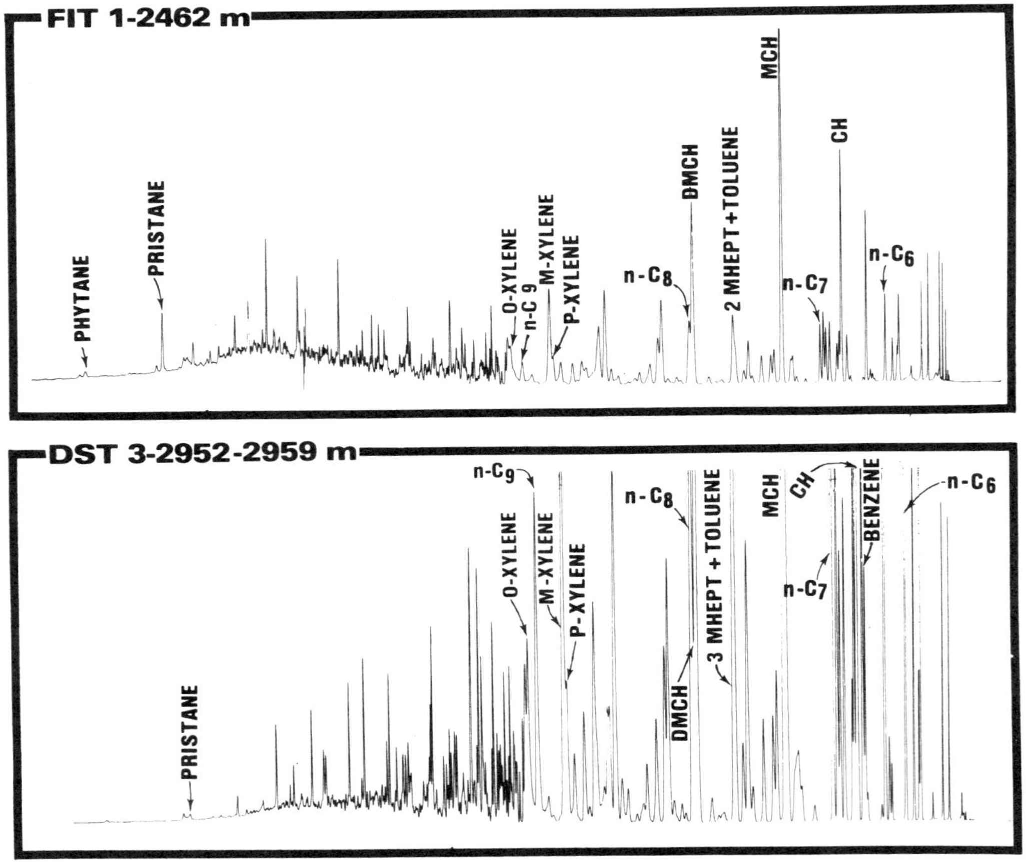

FIT 1-2462 m
PHYTANE
PRISTANE
O-XYLENE
n-C 9
M-XYLENE
P-XYLENE
n-C8
DMCH
2 MHEPT + TOLUENE
MCH
n-C7
CH
n-C6
DST 3-2952-2959 m
PRISTANE
O-XYLENE
n-C9
M-XYLENE
P-XYLENE
n-C8
DMCH
3 MHEPT + TOLUENE
MCH
n-C7
CH
BENZENE
n-C6

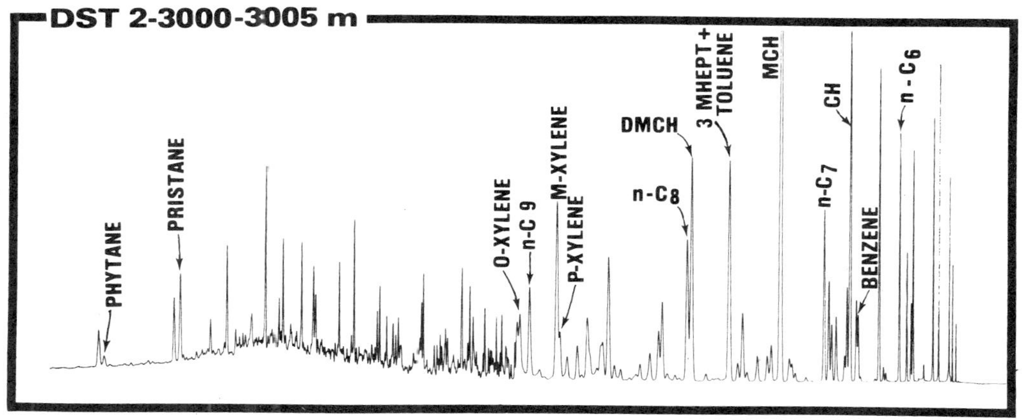

FIG. 15. Compositional changes in the gasoline fraction of oils by biodegradation of limited extent.

Note: 3-methylheptane and toluene coelute on the column used for this analysis. Other chromatographic conditions have been applied in order to separate toluene from 3-methylheptane.

TABLE II. *Changes in ratios of aromatics to n- and cycloalkane by water washing and biodegradation.*

RATIOS / SAMPLES	AROMATICS / N-ALKANES				AROMATICS / CYCLOALKANES				XYLENE RATIOS	
	BZ / $n\text{-}C_6$	TOL / $n\text{-}C_7$	P+M-XYL / $n\text{-}C_8$	O-XYL / $n\text{-}C_8$	BZ / CH	TOL / MCH	P+M-XYL / DMCH	O-XYL / DMCH	P-XYL / M-XYL	O-XYL / P-XYL
FIT 1 2462.6 m	0	1.73	2.32	0.72	0	0.12	0.64	0.19	0.21	0.28
DST 3 2952 m	0.81	0.97	1.15	0.35	0.35	0.24	0.87	0.26	0.23	0.28
DST 2 3000 m	0.85	1.50	1.53	0.39	0.33	0.26	0.97	0.24	0.22	0.27
EXPECTED CHANGES BY SOLUBILITY*	↘↘↘	↘↘↘	↘↘↘	↘↘↘	↘↘	↘↘	↘↘	↘↘	≃	≃
RESULTS	↘	↗	↗	↗	↘	↘	↘	↘	≃	≃

* in TISSOT & WELTE (1978), p. 277

1977; Blanc *et al.*, 1982). Benzothiophene and dibenzothiophene compounds are not degraded in Aquitaine crude oils, incubated for three months by *Pseudomonas oleovorans* under aerobic conditions (Connan, 1981a). Both families of sulphur-bearing aromatics are indeed degradable by bacteria (Callely, 1978; Connan, 1981a).

Micro-organisms belonging to the genus *Pseudomonas* were proved to be capable of converting dibenzothiophene into sulphur-containing oxygenated compounds (mainly 3-hydroxy-2-formyl-benzothiophene, dibenzothiophene-S-oxide, Fig. 6) under aerobic conditions (Yamada *et al.*, 1968; Kodama *et al.*, 1970). Anaerobic conversion of thiophene and of other sulphur-bearing compounds of petroleum (asphaltene, polysulphides, heavy residue) with production of H_2S has also been demonstrated (Kurita *et al.*, 1971).

Examples of severely biodegraded aromatics including sulphur compounds are presented in Fig. 16. They refer to natural asphalts from the Aquitaine Basin, France (Fig. 4, biodegraded oils). In both samples the complete removal of benzothiophenes and dibenzothiophenes series has been confirmed by computerized GC–MS analysis (Connan, 1981a). It should be noticed that the severe biodegradation of aromatics is not always coupled with a drastic alteration of alkanes. Case histories in which both aromatics and alkanes are intensively altered are generally encountered at the surface (Bastennes Gaujacq asphalt, Fig. 16) or in shallow depth reservoirs (at 835 m in Cl.6, for instance). In deeper sediments, highly degraded aromatics may be related to mildly altered alkanes. Such a situation is shown in BRS.1 asphalt, collected from a calcitic geode at a depth of 2013 m (Fig. 16). The preferential biodegradation of aromatics including sulphur-bearing compounds is not clearly understood. It has been ascribed to sulphur-decomposing bacteria and possibly some species of sulphate-reducing bacteria. Sulphate-reducers unable to metabolize *n*-alkanes can grow on the residues of crude oils previously altered under aerobic conditions (Jobson *et al.*, 1979). The depth range (1900–2700 m, Table II) in which the preferential biotransformations of sulphur-bearing aromatics was observed suggests that degradation processes may have taken place under anaerobic conditions. Decomposition of sulphur-bearing aromatics are also encountered in outcrop samples (Bastennes Gaujacq, Fig. 16). Therefore an aerobic metabolic pathway cannot be definitely ruled out.

IV. Conclusions

Biodegradation of crude oils in shallow reservoirs is a common phenomenon in many basins. It is generally considered as very active under aerobic

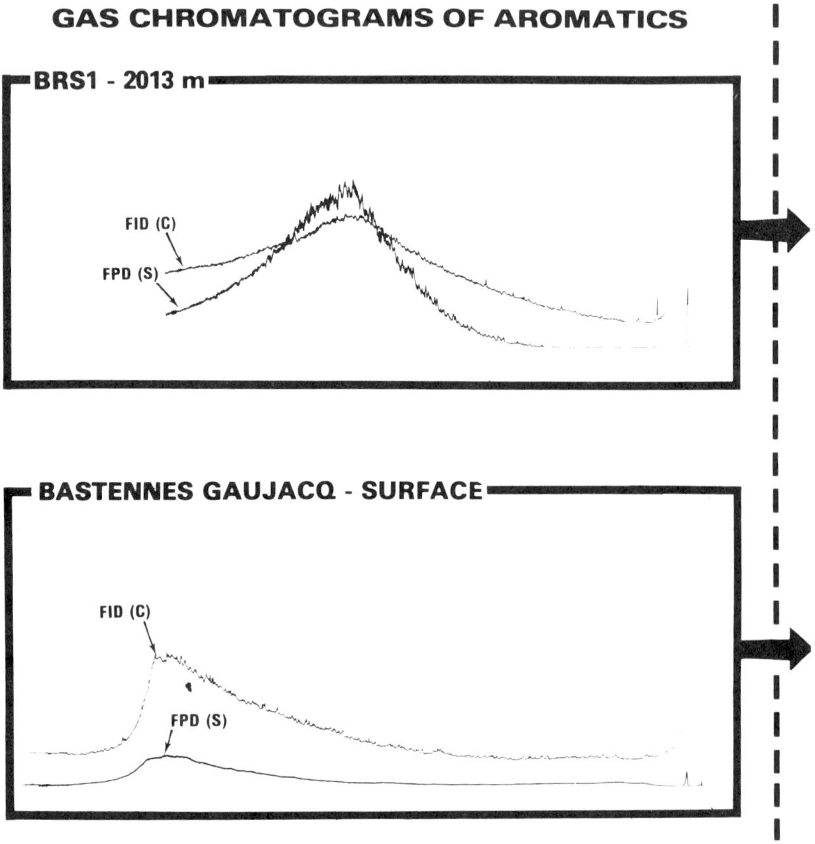

FIG. 16. Biodegradation of aromatics and alkanes in natural asphalts from the Aquitaine Basin, France.

GAS CHROMATOGRAMS OF TOTAL ALKANES

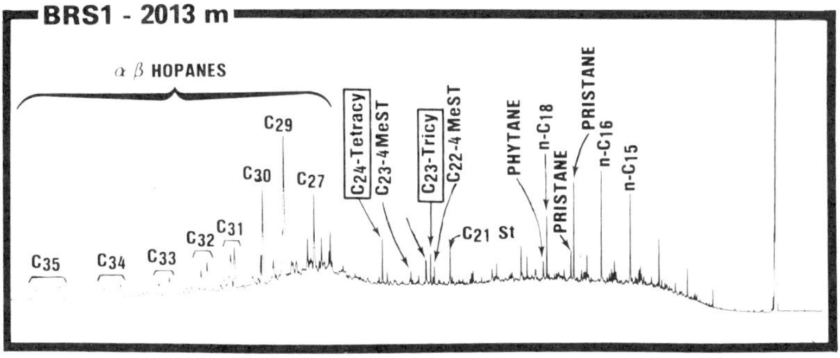

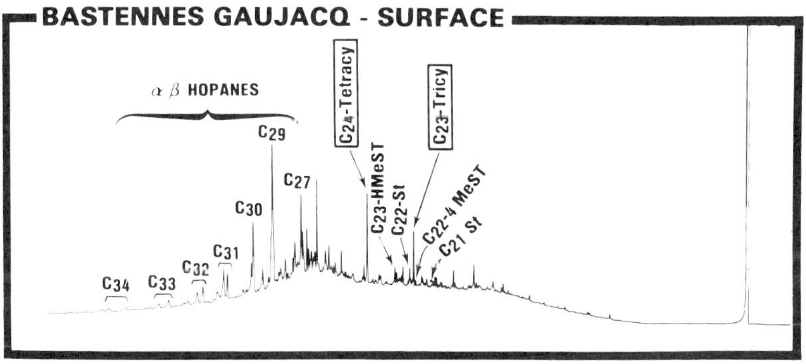

conditions. Alkanes and aromatics are initially transformed into oxygen-bearing derivatives (fatty acids, alcohols, ketones, phenols, etc.) which undergo subsequent changes (oxidation, ring cleavage, etc.) according to various metabolic pathways. The biodegradation of hydrocarbons is a bio-oxidation on a chemical standpoint. Compositional changes in crude oils pooled in reservoirs result both from the effect of water washing and biodegradation in relation to the influx of meteoric waters.

Depletion of low-boiling aromatics, namely benzene, toluene, and, to a lesser extent, xylenes, is generally ascribed to water washing rather than to biodegradation. In order to circumvent the contribution of water washing, a careful study has been undertaken on a set of slightly biodegraded crude oils derived from terrestrial organic matter. Solubility data on individual low molecular weight hydrocarbons (C_6–C_8), have been computed. Results, interpreted by referring to these data, show that the benzene evolution may be partly explained by water washing, whereas toluene and xylenes impoverishment is not consistent with the predicted evolution based on solubilities. In fact, the biodegradation of low-boiling alkanes and especially n-C_7 and n-C_8 is much more important than the removal of toluene and xylenes by solubility. In that particular example, changes within the gasoline range were much more influenced by biodegradation than by water washing.

Biodegradation effects can be recorded either on gross properties or at a molecular level.

Gas and gasoline-range compounds are preferentially removed. The residual oils become generally more viscous, richer in sulphur and nitrogen, in NOS's and asphaltenes. The $\delta^{13}C$ is shifted towards heavier isotopes. The API gravity drops and optical activity increases. In extensively biodegraded crudes, biodegradation proceeeds further by degrading $C_{15}+$ alkanes as well as $C_{15}+$ aromatics. Biodegradation of compounds in both fractions has been reviewed by referring to literature and case histories. A tentative sequence of biodegradability of alkanes has been proposed. n-alkanes are the most readily degraded structures, whereas tricyclic terpanes appear as the most bacterially resistant. This class of molecules has also been recognized as degradable in severely biodegraded oils.

Biodegradation of $C_{15}+$ aromatics including sulphur-bearing structures has also been documented. Series of alkyl-, dialkyl- and trialkylbenzenes are easily consumed by bacteria, whereas naphthalene, phenanthrene and chrysene are more difficult to metabolize. Sulphur-bearing aromatics, i.e. benzothiophenes and dibenzothiophenes are also completely eliminated; however, the conditions of degradation are not fully understood. More *in vitro* experiments should be performed with organic sulphur-decomposing bacteria, to better assess the subsurface conditions which prevail when sulphur-aromatic structures are degraded (either by aerobic or anaerobic processes).

Acknowledgements

I am indebted to J. Maxwell of the University of Bristol and to S. Jardine and J. Claret of Elf-Aquitaine, for their review and constructive criticisms of the manuscript.

I am grateful to J. Brooks and D. Welte, the editors of Advances in Petroleum Geochemistry, Vol. 1, who invited me to write the chapter.

The management of Societé Nationale Elf-Aquitaine (Production) kindly gave permission for this paper to be published.

References

Aldridge, A. K. (1977). "Analysis of hydrocarbons and petroleum", Information leaflet No. 6, Masspec Analytical, Unpublished.

Alexander, R., Kagi, R. I. and Woodhouse, G. W. (1981). *Tenth Int. Meeting on Organic Geochemistry, Bergen, 14–18 September 1981.*

Bailey, N.J.L., Krouse, H. R., Evans, C. R. and Rogers, M. A. (1973a). *AAPG Bull.* **57,** 1276–1290.

Bailey, N. J. L., Jobson, A. M. and Rogers, M. A. (1973b). *Chem. Geol.* **11,** 203–221.

Barker, C. (1980). *Application of Geochemistry in Exploration, Maidenhead, 28 July 1980.*

Bayliss, G. S. and Smith, M. R. (1980). "Source Rock Evaluation Reference Manual". Geochem. Laboratories Inc., Houston.

Blanc, R., Coustau, H., Connan, J., Ebanks, W. J. and Roux, C. (1982). *Second Int. Conf. on Heavy Crude and Tar Sands, Caracas, 7–17 February 1982,* in press.

Bopp, R. F., Santschi, P. H., Li, Y. H. and Deck, B. L. (1981). *Org. Geochem.* **3,** 9–14.

Brock, T. B. (1978). "Thermophilic Microorganisms and Life at High Temperatures". Springer Verlag, Berlin.

Burns, B. J., Hogarth, J. T. C. and Milner, C. W. D. (1975). *Bull. Can. Petr. Geol.* **23** (2), 295–303.

Cain, R. B. (1980). "Hydrocarbons in Biotechnology". Heyden and Son Ltd., London.

Callely, A. G. (1978). "Progress in Industrial Microbiology", Vol. 14. Elsevier, Amsterdam.

Claret, J. Tchikaya, J. B., Tissot, B., Deroo G. and van Dorsselaer, A. (1977). *In* "Advances in Organic Geochemistry 1975" (Campos, R. and Gõni, J., Eds.), 509–522. *ENADIMSA*, Madrid.

Coleman, D. D., Risatti, J. B. and Schoell, M. (1981). *Geochim. Cosmochim. Acta* **45** (7), 1033–1037.

Connan, J. (1972). *Bull. Centre Rech. Pau SNPA* **6**(1), 195–214.

Connan, J. (1981a). *Bull. Centre Rech. Explor.-Prod. Elf-Aquitaine* **5** (1), 151–171.
Connan, J. (1981b). *In* "Petroleum Geology in China" (Mason, J. F., Ed.), 48–70, 257–258. Pennwell Books, Tulsa.
Connan, J. and Cassou, A. M. (1980). *Geochim. Cosmochim. Acta* **44** (1), 1–23.
Connan, J., Restlé, A. and Albrecht, P. (1979). *In* "Advances in Organic Geochemistry 1979" (Douglas, A. G. and Maxwell, J. R., Eds.), 1–17. Pergamon Press, Oxford.
Cox, R. E., Maxwell, J. R., Ackman, R. G. and Hooper, S. N. (1974). *Biochim. Biophys. Acta.* **V** (360), 166–173.
Cox, R. E., Maxwell, J. R. and Myers, R. N. (1976). *Lipids* **11,** 72–76.
Cripps, R. E. and Watkinson, R. J. (1978). "Developments in Biodegradation of Hydrocarbons", Vol. 1. Applied Science Publishers, London.
Demaison, G. J. (1977). *AAPG Bull.* **61** (11), 1950–1961.
Deroo, G., Powell, T. G., Tissot, B. and McGrossan, R. G. (1977). *Geol. Surv. of Can. Bull.* **262.**
Evans, C. R., Rogers, M. A. and Bailey, N. J. L. (1971). *Chem. Geol.* **8,** 147–170.
Finnerty, W. R., Singer, M. E. and King, A. D. (1982). *Second Int. Conf. on Heavy Crude and Tar Sands, Caracas, 7–17 February 1982,* in press.
Gibson, B. J. (1982). *Second Int. Conf. on Heavy Crude and Tar Sands, Caracas, 7–17 February 1982,* in press.
Goodwin, N. S., Park, P. J. D. and Rawlinson, A. P. (1981). *Tenth Int. Meeting on Organic Geochemistry, Bergen, 14–18 September 1981.*
Hahn-Weinheimer, P. and Hirner, A. (1980). *Geochim. Cosmochim. Acta* **2,** 45–53.
Hartman, B. and Hammond, D. E. (1981). *Geochim. Cosmochim. Acta* **45,** 309–319.
Higgins, I. J. and Gilbert, P. D. (1978). "The Oil Industry and Microbial Ecosystems". Heyden and Son, London.
Hopper, D. J. (1978). "Developments in Biodegradation of Hydrocarbons", Vol. 1. Applied Science Publishers, London.
Hufnagel, H., Teschner, M. and Wehner, H. (1980). *In* "Advances in Organic Geochemistry 1979" (Douglas, A. G. and Maxwell, J. R. Eds.), 51–66. Pergamon Press, Oxford.
Hunt, J. (1979). "Petroleum Geochemistry and Geology". W. H. Freeman and Company, San Francisco.
Illich, H. A., Haney, F. R. and Jackson, T. J. (1977). *AAPG Bull.* **61** (12), 2103–2114.
Jobson, A., Cook, F. D. and Westlake, D. W. S. (1972). *Applied Microbiology* **23** (6), 1082–1089.
Jobson, A. M., Cook, F. D. and Westlake, D. W. S. (1979). *Chem. Geol.* **24,** 355–365.
Kallio, R. E. (1982). *Second Int. Conf. on Heavy Crude and Tar Sands, Caracas, 7–17 February 1982,* in press.
Kodama, K., Nakatani, S., Umehara, K., Shimizu, K., Minoda, Y. and Yamada, K. (1970). *Agr. Biol. Chem.* **34** (9), 1320–1324.
Kurita, S., Endo, T., Nakamura, M., Yagi, T. and Tamiya, N. (1971). *J. Gen. Appl. Microbiol.* **17,** 185–198.
McAuliffe, C. (1966). *J. Phys. Chem.* **70,** 1267–1275.

Mackenzie, A. S., Wolff, G. A. and Maxwell, J. R. (1981). *Tenth Meeting on Organic Geochemistry, Bergen, 14–18 September 1981.*
Magoon, L. B. and Claypool, G. E. (1981). *AAPG Bull.* **65** (4), 644–652.
Milner, C. W. D., Rogers, M. A. and Evans, C. R. (1977). *J. Geochem. Explor.* **7,** 101–153.
Oudot, J., Fusey, P., van Praet, M., Feral, J. P. and Gaill, F. (1981). *Environ. Pollut. Ser. A.* **26,** 93–110.
Philippi, G. T. (1977). *Geochim. Cosmochim. Acta* **41,** 33–52.
Philp, R. P. and Gilbert, T. D. (1980). *APEA J.* **20,** 221–228.
Price, L. C. (1973). Ph.D. Thesis Geol., Univ. California, Riverside.
Price, L. C. (1976). *AAPG Bull.* **60,** 213–244.
Price, L. C. (1980). *Chem. Geol.* **28,** 1–30.
Radledge, C. (1978). "Developments in Biodegradation of Hydrocarbons", Vol. 1, Applied Science Publishers, London.
Reed, W. E. (1977). *Geochim. Cosmochim. Acta* **41,** 237–247.
Rubinstein, I., Strausz, O. P., Spyckerelle, C., Crawford, R. J. and Westlake, D. W. S. (1977). *Geochim. Cosmochim. Acta* **41,** 1341–1353.
Sassen, R. (1980). *Org. Geochem.* **2,** 153–166.
Schaefer, R. G. and Leythaeuser, D. (1980). *In* "Advances in Organic Geochemistry 1979" (Douglas, A. G. and Maxwell, J. R., Eds.), 149–156. Pergamon Press, Oxford.
Schoell, M. (1978). *Erdöl Erdgas Zeitschr.* **94,** 119–125.
Seifert, W. K. and Moldowan, J. M. (1979). *Geochim. Cosmochim. Acta* **43,** 111–126.
Snowdon, L. R. and Powell, T. G. (1979). *Bull. Can. Petr. Geol.* **27** (2), 139–162.
Stahl, W. J. (1977). *Chem. Geol.* **20,** 121–149.
Stahl, W. J. (1980). *Geochim. Cosmochim. Acta* **44,** 1903–1907.
Tissot, B. P. and Welte, D. M. (1978). "Petroleum Formation and Occurrence". Springer Verlag, Berlin.
Winters, J. C. and Williams, J. A. (1969). *Am. Chem. Soc. Div. Petr. Chem., New York City Meeting Preprints,* E22–E31.
Yamada, K., Minoda, Y., Kodama, K., Nakatani, S. and Akasaki, T. (1968). *Agr. Biol. Chem.* **33**(7), 840–845.
Zajic, J. E., Gerson, D. F. and Camp, S. E. (1977). *Can. Fed. Soc.* **20,** 33.
Zamora, L. and Zambrano, G. (1982). *Second Int. Conf. on Heavy Crude and Tar Sands, Caracas, 7–17 February 1982,* in press.

Subject Index

V

W